Performance of Grounding Grids at Faulty and Lightning Strokes Conditions

Osama El-Sayed Gouda
Cairo University, Egypt

A volume in the Advances in Chemical and Materials Engineering (ACME) Book Series

IGI Global
Publishing Tomorrow's Research Today

Published in the United States of America by
IGI Global
Engineering Science Reference (an imprint of IGI Global)
701 E. Chocolate Avenue
Hershey PA, USA 17033
Tel: 717-533-8845
Fax: 717-533-8661
E-mail: cust@igi-global.com
Web site: http://www.igi-global.com

Library of Congress Cataloging-in-Publication Data

CIP DATA PENDING

ISBN13: 9798369329696
EISBN13: 9798369329702

British Cataloguing in Publication Data
A Cataloguing in Publication record for this book is available from the British Library.

All work contributed to this book is new, previously-unpublished material.
The views expressed in this book are those of the authors, but not necessarily of the publisher.

For electronic access to this publication, please contact: eresources@igi-global.com.

Table of Contents

Preface

Grounding system is one of the most important items of the power plants and substation systems design. The main purpose of power system substation grounding grids is to maintain reliable operation and provide protection for personnel and equipment during fault conditions. In any substation planning, economical and effective grounding system plays an essential role since absence of effective grounding grid can result in mal-operation of protective devices and hazardous for operators and substation equipment. Therefore, in this research, evolutional technique for unequally spaced grounding grid design is suggested. The design is based on hybridization of optimum compression ratio (OCR) and particle swarm optimization (PSO) techniques. This approach is purposed to control the grounding copper volume of the grounding system with the aid of controlling of unequally spaced grounding conductor lengths with respect to the safety measures. The proposed technique offers almost 20% of copper savings comparing with the optimum compression ratio method. It is found that each grid has its own OCR depending on the meshes number.

The power substations are usually exposed to many faults, whether internal in operation or external such as transient lightning stroke. The function of the protection system is to protect the equipment in the power substation against large fault currents. One of the important protection systems is the grounding system. The behavior of grounding systems at power frequencies is well understood, and detailed procedures for their design are widely accepted. Simple arrangements of vertical and horizontal grounding electrodes are often used for the lightning protection system earth termination.

This book presents the characteristics of vertical and horizontal grounding electrodes in uniform and two-layer soils when subjected to lightning including the impacts of soil ionization with frequency, soil resistivity, and permittivity variations. The critical breakdown field strength of two-layer soils with different values of the reflection factors was presented. The transient probabilistic grounding potential rise and the transient impedances of the horizontal and vertical electrodes were investigated in uniform and two-layer soils including the effect of soil ionization with frequency. The influences of resistivity and permittivity variations, which have substantial impacts on the electric field at the occurrence of the soil ionization, were considered. Moreover, transmission line approach (TL) with the aid of ATP has been used to compute the transient grounding voltages. The results indicated that the reflection factor has a significant impact on the equivalent radius of the grounding electrode. It is also observed that the peak values of the lightning induced voltages decrease sharply when the soil ionization phenomenon considering the variations in soil resistivity and permittivity with the stroke frequency changes.

This book presents the effect of high resistivity of non-homogeneous soils, considering the phenomenon of soil ionization. Among results presented to understand performances of the vertical and horizontal grounding electrode in this case are impulse impedance, earth surface potential, grounding potential rise and effective length of grounding electrode. Many factors affecting grounding electrode performance are presented in this book, such as: length of electrode, type of lightning current stroke, front time of lightning stroke, peak current of lightning stroke and type of grounding electrode.

As it is known, lightning strikes are considered one of the riskiest natural phenomena that can lead to human harm and the surrounding soil layers. To address this issue, this thesis investigates the impact of direct lightning characteristics on human body safety. Specifically, such investigation is carried out on the effect of two-layer soils resistivity on human safety when lightning stroke hits the towers of the high voltage transmission lines (HVTLs). Further, the study focuses on the current passing through the human heart when step and touch (contact) voltages are generated by grounding potential rise, caused by direct lightning strikes transmission tower, and the produced potential rise that a person could be exposed. Moreover, studying the effects of peak current and time of lightning strikes are investigated. Additionally, this thesis presents the effect of different reflection factors on human safety. For that purpose, the ATP program is used in the simulation of the grounding system as well as the human body model and gauss–Sedial method that is used to estimate earth surface potential (ESP) in non-homogenous soil. Numerous simulations were accomplished to examine the behavior of the current passing through the human heart.

Based on the simulation results, it is concluded that the soil characteristics have superior influences on the contact and step potentials and, accordingly, the survival threshold. Finally, the influences of the soil resistivity of homogeneous and heterogeneous soils on the performance of the grounding electrodes affected by lightning strokes considering soil ionization phenomenon are investigated. The investigations are done on the performance of the grounding electrode when the soil resistivity is taken between 500 and 2000 Ω.m in homogeneous soil and for positive and negative reflection factors in case of heterogeneous soils.

The function of the protection system is to protect the equipment in the power substation against large fault currents. One of the important protection systems is the grounding system. This thesis presents a comparison between two designs of the grounding grid in homogenous soil resistivity and non-homogenous soil resistivity. The first one is a square type grounding grid and the second one is a circular type grounding grid. Firstly, the two types are assumed to be had the same total area enclosed by the ground grid, and secondary, the two types are assumed to be had the same length of the total grounding grid conductor. The parameters under study are the grounding resistance, step voltage, touch voltage and earth surface potential. The charge simulation method (CSM) and the image method are used within this study, and the effects of the parameters such as soil resistivity, depth of grounding grid, the number of meshes and presence of the vertical rods, on the performance of grounding grids are taken into account. Finally, the validation of the simulation method is done by comparing the simulation results with the experimental results obtained in this study. When lightning strikes the overhead shield wire of a transmission line or tower, the lightning current flows into the ground from the grounding system through the tower. The lightning current waveform has pioneer influence on dynamic performance of grounding system. Also, the dynamic performance of grounding system is affected by configuration of grounding system, electrical permeates of soil, and frequency content.

Osama El-Sayed Gouda
Cairo University, Egypt

Introduction

1.1 THE BOOK OBJECTIVES

The Objectives of this book are:

i) Studying the parameters that are affecting effective electrode radius including soil ionization effect.
ii) Studying the dynamic performance of grounding electrodes under lightning current in two-layer soil when soil ionization is considered.
iii) Studying step and touch (contact) voltages that are generated by transient grounding potential rise when soil ionization is considered.
iv) Studying the parameters that are affecting the current passing through the human heart that is caused by direct lightning strikes transmission towers in two-layer soil.

1.2 THE BOOK CONTENTS

The book contains seven chapters their contents can be summarized as the following:

Chapter (1): This chapter gives a general introduction to the book, as well as the objective of this study, as well as some definitions related to grounding principles and the chapters' contents.

Chapter (2): Chapter two gives a general overview of the optimum design of grounding grids and studying the earth surface potentials and ground potential rise (GPR) of substation grounding in case of uniform and non-uniform soils.

Chapter (3): In this chapter, one of the most important types of faults that are exposed to the electrical network is presented, which is the phenomenon of lightning. This chapter presents an explanation of the phenomenon of lightning, the types of lightning, the types of lightning's effect on the electrical network, difference between direct and Indirect Lightning Strokes as well as the problems associated with the phenomenon of lightning and how to solve these problems. Also, this chapter presents a study of the effect of transient lightning stroke on the grounding electrodes.

Chapter (4): This chapter discusses the following: TT system (earthed neutral), automatic disconnection for TT system, TN system (exposed conductive parts connected to the neutral), TN-C system, TN-S system- TN-C-S system, IT system (isolated or impedance-earthed neutral), IT system (impedance-earthed neutral), grounding transformer connection and design, grounding of industrial and commercial generators, Zigzag grounding transformer earthed systems (solid grounding, resistance grounding, reactance grounding, resonant grounding (ground-fault neutralizer), location of system grounding points selection and grounding locations specified by the NEC and multiple power sources

Chapter (5): In this chapter, ATP transmission line technique and FEM methods are used to study the characteristics of the grounding electrodes that are installed in uniform and non-uniform soil including the effect of soil ionization with frequency, soil resistivity and permittivity variations. The critical breakdown field strength of two layers of soil with different values of the reflection factor is presented. The transient probabilistic grounding potential rise and the transient impedances of horizontal and vertical electrodes are calculated in non-uniform soils including the effect of soil ionization with frequency and soil resistivity variations. Finally, the conductivity and permittivity variations on the soil ionization electric field are considered in the used model.

Chapter (6): this chapter investigates the influence of direct lightning characteristics in terms of human body safety. Specifically, such investigation is carried out on the effect of resistivities of two-layer soils on human safety when lightning stroke hits the towers of the high voltage transmission lines (HVTLs). The merit of the proposed study is that the soil ionization phenomenon is taken into consideration. Further, the study focuses on the current passing through the human heart, when step and touch (contact) voltages are generated by grounding potential rise, caused by direct lightning strikes transmission tower and the produced potential rise that a person could be exposed.

Chapter (7): This chapter contains: Safety Criteria According to IEEE Standards - Step and Touch Voltage Criteria - Reduction factor- Simplified Equations for Mesh and Step Voltage - Application of step and mesh potential in Safe earthing system design- SAFETY CRITERIA-IEC 479-1 STD- Permissible Touch and Step Voltage IEC 479-1- Optimum Compression Ratio (OCR) - Test and verify- Approximated analysis to calculate voltage profile using apparent soil resistivity- Scaling factor- Test and verify the approximated method- Ground Potential Rise (GPR) of Faulty Substations having Equal and Unequal Spacing Grounding Grids Conductors

Chapter (8): This chapter contains the corrosion theory of grounding electrodes and basic electrochemistry in corrosion reactions. It contains also the forms of substation grounding grid corrosion (uniform corrosion, pitting corrosion, galvanic corrosion, microbial influenced corrosion, survey on corrosion rate of substation grounding grid, copper and steel corrosion rates, corrosion protection methods (coating, cathodic protection (CP)) . The chapter also contains the methods of applying cathodic protection in grounding grids, anode selection, anode spacing, and impressed current in the grounding grid cathodic protection. Finally, it contains the required information for design grounding system cathodic protection and sacrificial anode (galvanic) cathodic protection system design steps.

Chapter 1
Introduction to Grounding Grids at Faulty and Lightning Strokes Conditions

Osama El-Sayed Gouda
Faculty of Engineering, Cairo University, Egypt

ABSTRACT

This chapter gives a general background on the contents of the study presented in this thesis which deals with the grounding of electric substations. With the increasing demand for electricity, many power substations have been established in urban and non-urban areas. Substations are exposed to many faults, whether internal in operation or external such as transient lightning stroke. As a result of these faults that the power substation is exposed to, an electrical protection system is established to protect the equipment in the power substation, as well as the protection of people.

1.1 OVERVIEW

Grounding system is one of the most important items inside power plants and substation systems. The main purpose of power system substation grounding grids is to maintain reliable operation and provide protection for personnel and equipment during fault conditions. Grounding system allows controlling harmonics and drain to earth the fault currents. Poor design methods can lead to high erection costs and

DOI: 10.4018/979-8-3693-2969-6.ch001

unsafe conditions (Khodr H. M., Salloum G. A., and Miranda V., (2006)), (IEEE Std. 80. (2000)).

Grounding grids are considered an effective solution for grounding systems for all sites which must be protected against lightening strokes such as, telecommunication towers, petroleum fields, substations and plants (Ghoneim S. Sherif, (2007)). The optimal design of grounding systems for substations is to suitably arrange the conductors of grounding systems to equalize the leakage current distribution and the potential gradient of ground surface. At the same time to keep the maximum touch voltages within the safe limit. This would ensure making all grounding conductors sufficiently utilized, safe and economic design method (Gouda O. E. and Ibrahim H. M., (2010)).

Safety grounding system design is an important part of substation design. An appropriately designed grounding system ensures the adequate safety for people and protects the installation. With the continuous development of power systems world wide; the higher level voltages has been used. With these high levels of voltages, the fault current and the area of grounding system will accordingly increase. Consequently, the design of grounding system become more sensitive at fault conditions. Another extreme of grounding system design is the economic aspect. The grounding system in substations consists of many rods and conductors. Personnel safety is analyzed by two main factors called step and touch voltages. It is obvious that the safety factors are affected by geometrical grounding system and soil characteristics. The optimum design of grounding system is affected by optimum decision of the number and diameters of conductors and rods which will carry the fault currents. Moreover, the optimum spacing between conductors to accommodate safety factors, the laying depth of the grounding grid, surface potential gradient, step voltage and touch potential and mesh voltage (Nezhad N. K., Hadi Fallahi M., and Dozein M. G., (2013)). There are two important parameters to design the grounding systems of substations:

1- The equivalent electrical grounding resistance (Rg) of the system must be low enough to assure that fault currents are dissipated mainly through the grounding grid into the earth, while maximum potential difference between close points into the earth's surface must be kept under certain tolerances (step, touch, and mesh voltages) (IEEE Std. 80. (2000)).

2- The effect of voltage distribution is an important parameter to design the grounding systems of substations to ensure the safety of people above the substation ground. This can be achieved by equalizing the potential distribution of the ground surface and reducing step and touch potentials (Gouda O., Amer G. and EL-Saied T., (2006)). In general, the design procedure of a safe grounding system consists of five major steps:

i. Analysis of soil resistivity measurements
ii. Estimation of the allowable limits of touch/step voltages
iii. Fault current analysis for the estimation of the maximum grid current
iv. Grounding system design.
v. Evaluation of the safety performance of the designed grounding system (Datsios Z. G. and Mikropoulos P. N., (2012)).

In an equally spaced grounding grid, the fault current dispersion from peripheral conductors is around 4-10 times greater than the conductors at the middle portion of the grid due to protecting effect, resulting unbalanced potential distribution (Unde M. G. and Kushare B. E., (2014)).The potential is less at peripheral meshes than the center meshes which further results into higher touch voltage at the corner meshes than those in center meshes. However, use of same number of conductors with unequal spacing makes the potential distribution more uniform and thereby reduces the touch voltage, enhancing the safety level.

The area of grounding grid and number of grid conductors remains same for equal and unequal spacing with exponentially distributed conductors. In equally spaced grounding grid, the conductors are spaced at equal distance throughout the substation area. Whereas, in an unequally spaced grid, conductor spacing is large at center of the grid and it progressively decreases towards the periphery of the grid. This kind of grid configuration produces surface potential more uniform and thereby reduces the safety touch voltage, enhancing the safety level. Thus, unequal spacing minimizes the effect of protecting and makes the full consumption of grounding grid conductors. There is definite ratio at which grid conductors spacing increases or decreases is known as compression ratio (CR).

Once the area of substation grid and soil model are defined, every grid has an Optimal Compression Ratio (OCR) that makes the ground surface potential and touch voltage distributions mostly uniform. This OCR further makes the grid resistance, touch voltage and step voltage minimum. The grounding grid designed with an OCR, minimizes the touch voltage and thereby enhances the safety.

Optimization is a mathematical technique that concerns the finding of maxima or minima of functions in some feasible region. There is no business or industry which is not involved in solving optimization problems. A variety of optimization techniques compete for the best solution. Particle Swarm Optimization (PSO) is a relatively new and powerful method of optimization that has been empirically shown to perform well on many of these optimization problems. It is widely used to find the global optimum solution in a complex search space (Satyobroto Talukder, (2011)), The optimization of the design of substation grounding grids is an issue that has assumed an increasing importance, since it seeks to reduce the amount of conductive material buried, at the same time, it must ensure an effective and safe flow of the ground fault current through the soil.

Figure 1. Electrical utility system that use electricity require earth grounding, either directly or through a grounding system

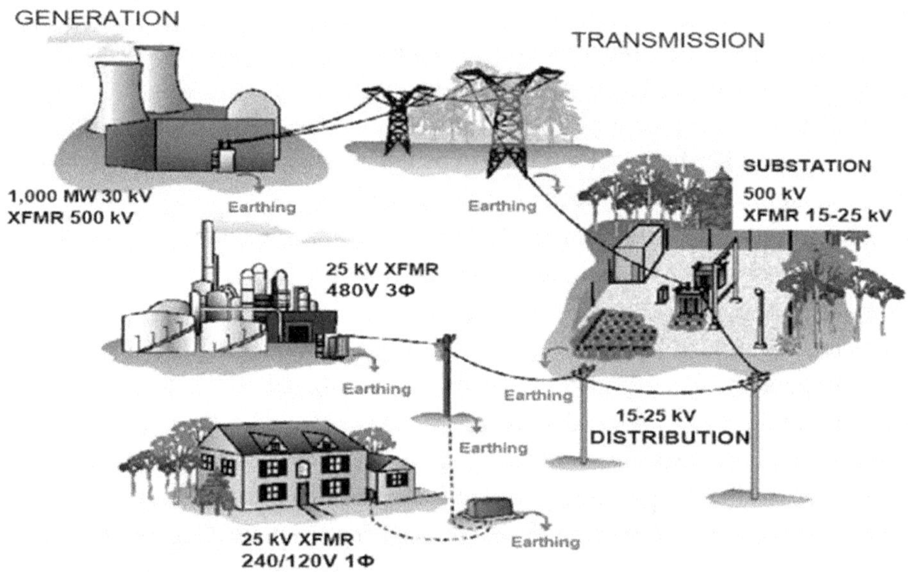

The optimization of a grounding system can have two distinct objectives: the grid is not safe so the aim will be to ensure that it meets the tolerable values for step and touch voltages, or the grid is already safe so the goal is to reduce the amount of conductive material buried in order to reduce the costs involved without compromising the safety.

The main target of this study is to apply particle swarm optimization algorithm to obtain optimal design of grounding grid system by obtaining optimal basic design quantities of the grounding grids: ground resistance Rg, touch voltage Vt, step voltage Vs and the cost of the grounding system design. These mentioned quantities depend on the grid parameters, which are its side lengths, radius of grid conductors, length of vertical rods, soil resistivity and crushed layer thickness.

Every Building, equipment, power plants, substations and/or facilities included in electrical utility system that use electricity require earth grounding, either directly or through a grounding system. (See Figure 1).

The grounding system is considered one of the main components of the substation's protection system. Therefore, electrical grounding systems are important in creating a safe environment for human operators and equipment under fault or transient conditions. Electrical installation must be grounded for the following reasons:

- It provides a low-impedance path between the electrical equipment and the ground.
- It provides a reference potential for electrical equipment.
- It prevents extreme overvoltage and potential gradients that might harm the human personnel working around it and damage the power equipment.

In general, any fault current in a power substation flows through the ground via a ground electrode system, which has impedance to the current flow. The voltage of the ground electrode system will rise. The potential difference caused by the excessive voltage increase could damage equipment and endanger the lives of humans and animals near the grounding system. Many standards like (IEEE Std. 80. (2000)).Provide detailed descriptions of grounding designs and procedures for power frequencies. Grounding systems behave differently under high frequency and transient conditions than they do under conventional power frequency conditions. This book presented the characteristics of the grounding electrodes in uniform and two-layer soils when subjected to lightning including the impacts of soil ionization with frequency, soil resistivity, and permittivity variations. Also, this thesis investigates the influence of direct lightning characteristics in terms of human body safety. Specifically, such investigation is carried out on the effect of resistivity of two-layer soils on human safety when lightning stroke hits the towers of the high voltage transmission lines (HVTLs).

1.2 DEFINITIONS OF GROUNDING PRINCIPLES

A grounding system should be installed in a manner that will limit the effect of ground potential gradients to such voltage and current levels that will not endanger the safety of people or equipment under normal and fault conditions, as well as assure continuity of service. In the beginning some related definitions are specified as follows: -

1. Ground

A conducting connection, whether intentional or accidental, by which an electric circuit or equipment is connected to the earth or to some conducting body of relatively large extent that serves in place of the earth.

2. Grounded

A system, circuit, or apparatus are usually provided with ground for the purpose of establishing a ground return circuit and for maintaining its potential at approximately the potential of earth.

3. Ground Return Circuit

A circuit in which the earth or an equivalent conducting body is utilized to complete the circuit and allow current circulation from or to its current source

4. Ground Electrode

A conductor embedded in the earth and used for collecting ground current from or dissipating ground current into the earth.

5. Ground Grid

A system of horizontal ground electrodes that consists of a number of interconnected, bare conductors buried in the earth, providing a common ground for electrical devices or metallic structures, usually in one specific location.

6. Ground Mat

A solid metallic plate or a system of closely spaced bare conductors that are connected to and often placed in shallow depths above a ground grid or elsewhere at the earth surface, in order to obtain an extra protective measure minimizing the danger of the exposure to high step or touch voltages in a critical operating area or places that are frequently used by people. Grounded metal gratings placed on or above the soil surface or wire mesh placed directly under the crushed rock, are common forms of a ground mat.

7. Ground Potentials Rise (GPR)

The maximum voltage that a station-grounding grid may attain relative to a distant grounding point assumed to be at the potential of remote earth.

8. Step Voltage

The difference in surface potential experienced by a person bridging a distance of one meter with his feet without contacting any other grounded object.

9. Touch Voltage

The potential difference between the ground potential rise (GPR) and the surface potential at the point where a person is standing, while at the same time having his hands in contact with a grounded structure.

10. Mesh Voltage

The maximum touch voltage to be found within a mesh of a ground grid

11. Harmonic Impedance

One of the most important properties in the high-frequency analysis of grounding system is harmonic impedance to ground $Z(jw)$. It can be computed as the input impedance of an open transmission line in frequency domain as:

$$Z(jw) = Z_o coth(\gamma l) \tag{1.1}$$

$$Z_0 = \sqrt{\frac{j\omega L_g}{\left(1/R_g + j\omega C_g\right)}} \tag{1.2}$$

$$\gamma = \sqrt{j\omega L_g\left(1/R_g + j\omega C_g\right)} \tag{1.3}$$

From above equations, it is noticed that the harmonic impedance depends on only geometry, electromagnetic properties of the ground system and electric soil parameter.

12. Transient Potential

.After solving above equation and obtain$Z(jw)$, it is enable to evaluate the time functions of the potential$v(t)$ as response to an arbitrary current pulse$i(t)$ by (*Grace* L. and Grceva S., (2009).)

$$v(t) = F^{-1}\{F[i(t)]Z(jw)\} \tag{1.4}$$

Where F and F^{-1} denote Fourier and inverse Fourier transforms, respectively. The transient voltage is a parameter of great practical important since it is able to reveal the maximum voltage which the grounding is submitted when it is subjected to impulsive phenomena. Moreover, as it can also evaluate the time that the ground is subjected to certain levels of voltage transient.

13. Impulse Impedance

The impulse impedanceZ_p is defined as the ratio between the peak voltage and peak current at point of injection and its concept is originally in the time domain Grcev Leonid, (2005).

$$Z_p = \frac{V_p}{I_p} \tag{1.5}$$

Impulse impedance is a very important parameter as it allows to quickly estimating the maximum voltage at the point of injection just multiplying the value of peak current by Z_p (Alipio Rafael and Visacor S., (2013))

14. Effective Length

The effective length is defined as the maximum length of the electrode from which increases in size do not cause significant decrease in the value of the impedance impulsive (Alipio Rafael and Visacor S., (2013)).

15. Decrement Factor

The decrement factor is defined as the ratio of maximum ground potential rise for frequency dependent soil parameters and maximum ground potential rise for constant soil parameter (Akbari Majed, et al., (2013)).

$$DF = \frac{max\{GPRforfrequencydependentsoilparameter\}}{max\{GPRforconstantsoilparameter\}} \tag{1.6}$$

16. Charge Simulation Method of One Layer Soil

In charge simulation method, the electrical field is simulated with a field that calculated by a number of charges. The values of factious charges must satisfy the boundary condition. When values of factious charges are computed, the potential and electrical field can easy computed at anywhere on the space. Nazar Hmalik explained the charges simulation method (MalikN. H., (February 1989)). If several factious charges of any type (ring, point, and line) are calculated in the space, the potential at any point *CC* can be computed by summation of individual potential of each charge. Let $\boldsymbol{Q}_j$ be values of ***nn***individual charges and $\boldsymbol{\varphi}_i$ be the potential at any point *CC* within the space. According to the theory of superposition the potential at any point *CC* is calculated from the following equation:

$$\varphi_i = \sum_{j=1}^{nn} P_{ij} Q_j \tag{1.7}$$

In the above equation $\boldsymbol{P}_{ij}$are the potential coefficients which can be calculated analytically for any types of charges by solving Poisson orLaplace'sequations, $\boldsymbol{\varphi}_i$is the potential at $\boldsymbol{i}^{th}$ contour point, and $\boldsymbol{Q}_j$is the charge at $\boldsymbol{j}^{th}$ point charge. In our simulation technique, point type of charges is used. A number of point charges placed on each conductor in the grid. Due to flatting of ground surface, the method of the image can be used with the charge simulation method, and hence the potential will be constant on the grounding grid and its symmetry grid (MalikN. H., (February 1989)). The potential coefficients will be:

$$P_{ij} = \frac{1}{4\pi\varepsilon}\left[\frac{1}{d_{ij}} + \frac{1}{d'_{ij}}\right] \tag{1.8}$$

Where $\boldsymbol{d}_{ij}$is the distance between contour point ***i***and point charge ***j***and $\boldsymbol{d'}_{ij}$is the distance between a contour point***i***and an image point charge***j***as shown in Figure 2.

Figure 2. Illustration of the charge simulation method

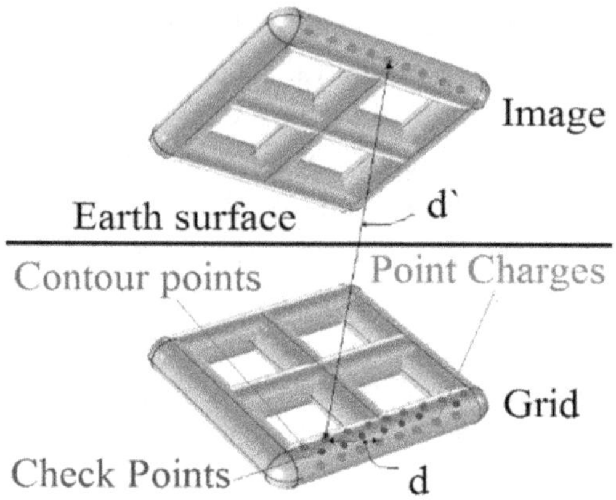

The positions of each contour point and charge point are determined in ***X, Y, Z*** coordinates, i.e., the distances between the charge point ***j*** and the contour point ***I*** are computed as (MalikN. H., (February 1989)).

$$d_{ij} = \sqrt{\left(X_j - X_i\right)^2 + \left(Y_j - Y_i\right)^2 + \left(Z_j - Z_i\right)^2} \tag{1.9}$$

WhereX_j, Y_j, $Z_j$are the dimensions of the point charge***j***, andX_i,Y_i, $Z_i$are the dimensions of the contour point ***i***. It is assumed $\varphi_i$equals to one volt at each contour point to solve the magnitude of charges in Equation (1.7). To check the accuracy of this method, checkpoints are used instead of the contour points. These checkpoints are located on the surface of the grid on places ratherthan those of the contour points, where their potentials also equal to one volt. Once the values of the charge points are computed the potential at any point outside the grid can be calculated.

17. Charge Simulation Method in Two-Layer Soil

In two-layer soil, the dipoles are realigned in the surface of different soils when the voltage applied. They appear as surface charges, these charges must satisfy the following condition (Mukherjee P. K. and Roy C. K., (February 1978))

i-The potential at any point on one side must be equal the other side.
ii- The flux at any point on each side must be equal.

Figure 3 shows fictitious charge points for field calculation by charge simulation method in two horizontal soils. Grid is replaced by a number of charges *NE* with same number of contour points on thegrid surface of which *NE*Acontour points are on the sideof soil 1 and *NE-NE*A=*NE*D contour points are on the side of soil 2. The different soil boundary is replaced by *NB* contourpoints with *NB* charges in the soil 2 and *NB* in soil 1.

Figure 3. Fictitious charge points for field calculation by charge simulation method in two horizontal soils

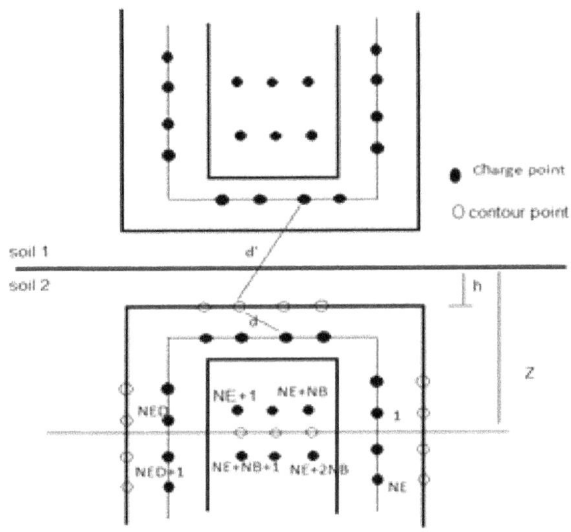

There are thus *NE* + 2*NB* unknown fictitious charges.

In grid 1….,.. *NE*

In soil (1) (*NE*+ 1)…….,..*NE*+*NB*

In soil (2) (*NE* + *NB* + 1) ….,..*NE* + 2*NB*

The equations can be formulated as follows:

In the soil1, the potential *VE*at the contour points is the integrated effect of all the *NE*charges on the grid and the *NB*charges in the soil 2. Therefore;

$$\sum_{j=1}^{NE} P_{ij} Q_j + \sum_{j=NE+NB+1}^{NE+2NB} P_{ij} Q_j = VE$$

For

$$i = NED + 1, \ldots\ldots NE \tag{1.10}$$

. In the soil 2, the potential ***VE***at contour points is the integrated effect of all the ***NE*** charges on the grid and the ***NB*** charges in the soil 1. Therefore;

$$\sum_{j=1}^{NE} P_{ij} Q_j + \sum_{j=NE+1}^{NE+NB} P_{ij} Q_j = VE \text{ (1.11) For} i = 1,2,\ldots\ldots NED$$

The potential at contour point on different soil boundary is unknown but the potential at each point on the different soil boundary must be the same whether it is computed from region of soil 1 or from region of soil 2.

$$\sum_{j=1}^{NE} P_{ij} Q_j + \sum_{j=NE+1}^{NE+NB} P_{ij} Q_j = \sum_{j=1}^{NE} P_{ij} Q_j + \sum_{j=NE+NB+1}^{NE+2NB} P_{ij} Q_j$$

OR

$$\sum_{j=NE+1}^{NE+NB} P_{ij} Q_j - \sum_{j=NE+NB+1}^{NE+2NB} P_{ij} Q_j = 0$$

For

$$i = NE + 1, \ldots\ldots NE + NB \tag{1.12}$$

Also, at the contour points on the different soil boundary, the normal field intensity in soil 2 must be ε_1 times that in the soil 1.

$$\left(\sigma_1 + jw\varepsilon_0\varepsilon_1\right)\left[\sum_{j=1}^{NE} f_{ij} Q_j + \sum_{j=NE+1}^{NE+NB} f_{ij} Q_j\right] = \left(\sigma_2 + jw\varepsilon_0\varepsilon_2\right)\left[\sum_{j=1}^{NE} f_{ij} Q_j + \sum_{j=NE+NB+1}^{NE+2NB} f_{ij} Q_j\right]$$

OR

$$\left(\left(\sigma_1 + jw\varepsilon_0\varepsilon_1\right) - \left(\sigma_2 + jw\varepsilon_0\varepsilon_2\right)\right)\sum_{j=1}^{NE} f_{ij} Q_j + \left(\sigma_1 + jw\varepsilon_0\varepsilon_1\right)\sum_{j=NE+1}^{NE+NB} f_{ij} Q_j - \left(\sigma_2 + jw\varepsilon_0\varepsilon_2\right)\sum_{j=NE+NB+1}^{NE+2NB} f_{ij} Q_j = 0$$

For

$$i = NE + 1, \ldots\ldots NE + NB \tag{1.13}$$

Where f_{ij} is the field coefficient in the normal direction to the soil boundary at the respective contour point, σ_1 is the soil conductivity of soil 1, σ_2 is soil conductivity of soil 2, ε_0 is permittivity of air, ε_1 is soil permittivity of soil 1, and ε_2 is soil permittivity of soil 2. After solving Equations (1.7) to (1.10) to determine the unknown fictitious charge source points, the potential on the earth surface can be calculated by using Equation (1.7). Also grounding resistance can be calculated using Equation (1.15).

$$c_1 = \frac{\sum_{j=1}^{n} Q_j}{V} \tag{1.14}$$

$$R_g \times c_1 = \rho_a \times \varepsilon_1 \tag{1.15}$$

Where, c_1 is the capacitance of the grounding grid (Farad), V is the GPR that is defined as 1 V [20-21], Q_j is the charge of point charge j that used for the calculation, ρ_a is the apparent soil resistivity.

18. Computed Apparent Resistivity in Two Layer Soil

In two soil horizontal layer, the apparent resistivity ρ_a can be calculated by the following formula (IEEE Std. 80. (2000))

$$\rho_a = \frac{\rho_1}{\left[1 + \left[\left(\frac{\rho_1}{\rho_2}\right) - 1\right]\left[1 - e^{\frac{1}{k(z+2h)}}\right]\right]}$$

For

$$\rho_2 < \rho_1 \tag{1.16}$$

$$\rho_a = \rho_2 \times \left[1 + \left[\left(\frac{\rho_2}{\rho_1}\right) - 1\right]\left[1 - e^{\frac{-1}{k(z+2h)}}\right]\right]$$

For

$$\rho_2 > \rho_1 \tag{1.17}$$

Where h is the burial depth of grounding grid, k is the reflection factor $k = [(\rho_2 - \rho_1)/(\rho_1 + \rho_2)]$, z is the top layer depth, ρ_1 is resistivity of soil 1, and ρ_2 is resistivity of soil 2 .

In (IEEE Std. 80 (2000)) Equation (1.17) is changed due to very large depth of upper soil layer; resistivity ρ_a given by Equation (1.17) tends to ρ_2. This is physically improper if the electrode lies down in the upper soil layer, as assumed in (J. Nahman, L Paunovic, (2006)). Therefore, Equation (1.17) is modified as follows:

$$\rho_a = \rho_1 \times \left[1 + \left[\left(\frac{\rho_2}{\rho_1}\right) - 1\right]\left[1 - e^{\frac{-1}{k(z+2h)}}\right]\right]$$

For

$$\rho_2 > \rho_1 \qquad (1.18)$$

19. Grounding Codes and Standards

Grounding needs vary according to function. The grounding requirements of a power system will vary from those of electrical equipment, lightning protection or for the proper function of electronic equipment. Proper installation of appropriate grounding systems requires knowledge of the needs and layout of the facility. Soil characteristics, grounding conductor materials grounding connections and terminations, are significant factors determining the design of a grounding system. Applicable standards and codes must be applied.

While many codes and standards contain minimum grounding and bonding requirements, the design and installation of electrical grounding systems is one of the most important aspects of any electrical distribution system. However, codes and standards are often misunderstood and grounding systems subsequently installed improperly. Standard used for ground calculations of High and medium voltage electricity AC substations is IEEE 80-2000: Guide for safety in AC substation grounding.

REFERENCES

Alipio, R., & Visacro, S. (2013). Frequency Dependence of Soil Parameters: Effect on the Lightning Response of Grounding Electrodes. *IEEE Transactions on Electromagnetic Compatibility*, 55(1), 132–139. 10.1109/TEMC.2012.2210227

Datsios, Z. G., & Mikropoulos, P. N. (2012), Safe grounding systems design for a photovoltaic power station. *8th Mediterranean Conference on Power Generation, Transmission, Distribution and Energy Conversion MEDPOWER*. 10.1049/cp.2012.2041

Gouda O., Amer G. & EL-Saied T. (2006). Computation of Potential Profile at a Surface above Energized Unequally Spaced Grounding Grid. The eleventh international middle east power systems conference, MEPCON', 2, 517–22.

Gouda, O. E., & Ibrahim, H. M. (2010), Earth Surface Potentials and GPR for Grids Having Different Conductors' Distributions Buried in different Layer Soil Structures, MEPCON'10, Cairo University, Egypt, Dec., Paper ID 229.

Grcev, L., & Grceva, S. (2009). on high-frequency circuit of horizontal grounding electrodes. *IEEE Transactions on Electromagnetic Compatibility*, 51(3), 873–875. 10.1109/TEMC.2009.2023330

IEEE Std. 80. (2000). IEEE Guide for Safety in AC Substation Grounding, IEEE Std, 80, 2000, 4th edition.

Khodr, H. M., Salloum, G. A., & Miranda, V. (2006). *Grounding System Design in Electrical Substation: An Optimization Approach*. IEEE PES Transmission and Distribution Conference and Exposition Latin America.

Leonid, G. (2005). on high-frequency circuit equivalents of a vertical ground rod. *IEEE Transactions on Power Delivery*, 20(2), 1598–1603. 10.1109/TPWRD.2004.838460

Majed, A., Keyhan, S., & Mohamad, R. A. (2013). The Effect of Frequency Dependence of Soil Electrical Parameters on the Lightning Performance of Grounding Systems. *IEEE Transactions on Electromagnetic Compatibility*, 55(4), 739–746. 10.1109/TEMC.2012.2222416

Malik, N. H. (1989, February). A review of charge simulation method and its application. *IEEE Transactions on Electrical Insulation*, 24(1), 3–20. 10.1109/14.19861

Mukherjee, P. K., & Roy, C. K. (1978). Computation of Fields in and around Insulators by Fictitious Point Charges. *IEEE Transactions on Electrical Insulation*, EI-13(1), 24–31. 10.1109/TEI.1978.298095

Nahman, L., & Paunovic, I. (2006). Resistance to earth of earthing grids buried in multi-layer soil. *Electrical Engineering*, 88(4), 281–287. 10.1007/s00202-004-0282-y

Nezhad, N. K., Hadi Fallahi, M., & Dozein, M. G. (2013). An Optimal Design of Substation Grounding Grid Considering Economic Aspects Using Particle Swarm Optimization. *Research Journal of Applied Sciences, Engineering and Technology*, 6(12), 2159–2165. 10.19026/rjaset.6.3840

Sherif, G. S. (2007). *Optimization of Grounding Grids Design with Evolutionary Strategies, Ph.* D., Univeristät Duisburg-Essen.

Talukder, S. (2011). *Mathematical Modeling and Applications of Particle Swarm Optimization, Master Degree*. Blekinge Institute of Technology.

Unde, M. G., & Kushare, B. E. (2014), Effect of compression ratio on touch voltage in layered soil, 2014 6th IEEE Power India International Conference (PIICON).

Chapter 2
Optimum Design of Grounding Grids Using OCR

Osama El-Sayed Gouda
Faculty of Engineering, Cairo University, Egypt

ABSTRACT

Grounding of power substations is to provide safety for personnel during normal and fault to ground conditions by limiting step and touch potentials, to dissipate lightning strokes and to stabilize voltage during transient conditions. The important goals of this chapter are to present approaches for the optimum design of ground grid and studying the earth surface potentials and ground potential rise (GPR) of substation grounding in case of uniform and non-uniform soils. The equalization of the current density and then decreasing of GPR is achieved. The compression ratio "CR" of unequally spaced grounding grid conductors as well as particle swarm optimization (PSO) technique are applied for the design of grounding grids of HV substations.

2.1. PROPOSED ALGORITHMS AND USED TOOLS

2.1.1. Compression Ratio Technique

The compression ratio "CR" of unequally spaced grounding grid conductors is a function of the grid length L, number of grid conductors N, and the maximum separation between the grid conductors. This relation changes depending on number of conductors N, if it is odd or even, as given in equations 2.1 and 2.2 (J. He, R. Zeng, Y. Gao, Y. Tu, W. Sun, J. Zou, and Z. Guan, (Jan. 2005)), (IEEE Std. 80, (2000)).

DOI: 10.4018/979-8-3693-2969-6.ch002

$$d_{max} = \frac{L(1 - CR)}{1 + CR - 2\,CR^{(N/2+1)}}\,N \text{ is even} \quad (2.1)$$

$$d_{max} = \frac{L(1 - CR)}{2(1 - CR^{(N-1)/2})}\,N \textit{ is odd} \quad (2.2)$$

When the grounding conductors are arranged according to an exponent regularity, the conductors span decreases gradually from the center of the grounding grid to its sides. The ith conductor span from the center can be obtained using equation (2.3)

$$d_i = d_{max} \cdot CR^i \quad (2.3)$$

Where,
CR is the compression ratio ≤ 1
d_{max} is the maximum spacing of center mesh
For optimal design of the HV substation grounding grid, the optimum compression ratio can be used (W.Sun, Jingling He, Yanqing Gao, R.Zeng, W.Wu and Qi Su, (2000)). Gouda O.E., (Gouda Osama El-Sayed, (2017)), (Gouda, O.E., (2011)) proved that the mesh conductors of unequally spaced grid having denser conductors at the edges gave the most safe and efficient design. However, one advantage of the use of unequally spaced grid is that it produces more uniform current density in the grid conductors and therefore remarkable decrease in voltage gradients of the earth's surface can be observed. An empirical expression is obtained to calculate OCR as follows (Gouda Osama E.; El Dein Adel Z. (2017)).

OCR=a_0+$a_1$exp (0.0001h) + $a_2$exp (bh). (2.4)

Where

$$b = -0.3503 - 9.6311\exp(-0.03666L). \quad (2.5)$$

$$a_0 = a_{01} + a_{02}k + a_{03}k^2 \quad (2.6)$$

$$a_1 = a_{11} + a_{12}k + a_{13}k^2 \quad (2.7)$$

$$a_2 = a_{21} + a_{22}k \quad (2.8)$$

Where k is the segments number of the conductor in any direction; h is the grid laying depth. The constants of equations 2.4- 2.8 as function of the grid length are illustrated in Table 1 (Gouda Osama E.; El Dein Adel Z. (2017)).

Table 1. Relationships between a_0, a_1, a_2 and **L** *(Gouda Osama E.; El Dein Adel Z. (2017))*

Coefficients	L≤100 m	100 m≤L≤175 m	175 m≤L≤250 m	L>250 m
a_{01}	0.44	0.38	-0.51	0.32
a_{02}	-77.43	-50.65	-33.18	-15.44
a_{03}	15.63	13.88	18.49	13.42
a_{11}	0.033	0.19	1.15	0.38
a_{12}	76.9	50.21	32.82	15.16
a_{13}	-15.56	-13.83	-18.44	-13.38
a_{21}	-0.067	-0.037	-0.029	-0.022
a_{22}	0.5	0.41	0.34	0.26

2.1.2 Particle Swarm Optimization (PSO)

The local best PSO method allows each particle to be affected by the best-fit particle that is chosen from its neighborhood, and it reflects a ring social topology. This social information exchanged within the neighborhood of the particle, denoting local knowledge of the environment (El-Fergany Attia, (2013)). Therefore, in a PSO technique, all particles are begun randomly and estimated to compute fitness together with finding the personal best and global best. After that a loop starts to find an optimum solution. In the loop, first the particles' velocity is updated by the personal and global bests, and then each particle's position is updated by the current velocity. The loop ends with a stopping criterion predetermined in advance. Basically, two PSO algorithms, namely the Global Best (Gbest) and Local Best (Lbest) PSO, have been developed which differ in the size of their neighborhoods (Chun-Yao Lee and Yi-Xing Shen, (2009)).

2.2 VOLTAGE PROFILE COMPUTATION BY USING IMAGE METHOD

Infinite Series Potential Method (I.S.M), can be used in the calculations of the potential difference in volts in three dimensions (x, y, z) using equation (2.9) .

$$v(x, y, z) = \frac{I\rho\,[lnf(x, y, z)]}{4\pi l} \tag{2.9}$$

Where

$$f(x, y, z) = \frac{\sqrt{(x+0.5l)^2 + y^2 + z^2} + x + 0.5l}{\sqrt{(x-0.5l)^2 + y^2 + z^2} + x - 0.5l} \tag{2.10}$$

Where I is the grounding fault current in Amperes, L is the length of the grid side in meters and ρ is the soil resistivity in Ω.m. If there is a grid of total conductor length (L_C) and buried in uniform soil, the total potential differences (V_{Total}) caused by this grid; is the sum of the potential caused at this point by the grid conductors and their images (Thabet A, (2002)).

$$V_{Total} = \frac{I\rho\sum_{x,y}[\ln f(x, y, z) + \ln f(x, y, z + 2h)]}{4\pi L_C} \tag{2.11}$$

If the grid is buried in double layer earth structure, the total potential differences (V_{Total}) caused by the grid and all its images at any point will be:

$$V_{Total} = \frac{I\rho_1\sum[\ln f(x, y, z)]}{4\pi L_C} + \frac{I\rho_1}{4\pi L_C}\sum\Big[\ln f(x, y, z + 2h) + \sum_{q=1}^{\infty}\big(K^q \ln f(x, y, z + q2b_1 + (q-1)2h\big)$$

$$+ 2K^q \ln f(x, y, z + q2b_1 + q2h) + K^q \ln f\big(x, y, z + q2b_1 + (q+1)2h\big)\big)\Big] \tag{2.12}$$

Where q is the number of images, r is the radius of grid conductor, K is the reflection factor between upper and lower soil layers, h is the burial grid depth, b_1 is the difference between upper layer depth and grid burial depth and ρ_l is the first layer soil resistivity.

2.3 SAFETY CRITERIA CONSTRAINS AND GROUNDING RESISTANCE

The safety standard includes that the actual grid mesh and step potentials must be lesser or even equal to the values of the corresponding acceptable potentials which are borne by humans without being subjected to electric shocks (IEEE Std. 80, (2000));

$$E_{mesh-Actual} \le E_{mesh-acceptable} \text{ and } E_{step-Actual} \le E_{step-acceptable} \tag{2.13}$$

The recommended limits of the grounding potential rise, GPR value is usually 5000V for the most substations grounding grids.

$GPR_{-acceptaple} \leq GPR_{max}$ (2. 14)

The grounding resistance of the designed grid has to be equal or less than the target grid resistance $R_g \leq R_{target}$ (2.15)

For the highest voltage and large substations, the grounding resistance is usually less than 0.5 Ω. In smaller substations, the acceptable value usually depends on the local conditions (IEEE Std. 80, (2000)). The conductor's separation in x and y directions is proportional to the grounding grid dimensions as follows. (IEEE Std. 80, (2000))

$$D_{x,min} \leq D_x \leq D_{x,max} D_{y,min} \leq D_y \leq D_{y,max} \tag{2.16}$$

In case of unequally spaced grid the space between each two conductors depends on the designed compressed ratio. The burial depth of the grid influences the ground potential rise and the grid resistance. According to standards, the grid burial depth is recommended to be between 0.5-2.5 m.

$$h_{min} \leq h \leq h_{max} \tag{2.17}$$

Finally, in this article mathematical model is formulated in general constrained optimized problem.$minimise\{f(x, v) = Volume\}$

$$S.t \begin{cases} g(x, v) = 0 \\ h(x, v) \leq 0 \end{cases} \tag{2.18}$$

$f(x, v)$ is defined as the objective function, $g(x, v)$ and $h(x, v)$ are defined as the set of equality and inequality constraints, respectively, x is known as state variables and v is the vector of variables. The control variables depend on the designer vision.

2.4. TOTAL VOLUME OF THE GROUNDING GRID CONDUCTORS

The increase of the grid meshes, and the use of rods produce more favorable surface voltage distribution for personal safety and also reduce the resistance grid and grounding potential rise which promotes the electrical network reliability and the protection system devices. However, attention must be paid to the copper grid volume. To have the optimal design of grounding grid with minimum cost, the objective volume function must be optimized while satisfying the safety considerations and predetermined value of grounding resistance. The copper volume of the grid conductors is evaluated as given in equation (2.19)

Objective Volume

$$= \text{minimise}\left\{\left(\pi \times \left(\frac{d}{2}\right)^2 \times (NL_c + Nw_c) \times L\right) + \left(\pi \times \left(\frac{d_r}{2}\right)^2 \times N_r \times L_r\right)\right\} \quad (2.19)$$

Where d grounding grid diameter of the conductor (m)
NL_c conductors number in the length direction
Nw_c conductors number in the width direction
d_r the ground rod diameter (m)
N_r Number of rods
L_r Rod length
L the grid side length

2.5. OPTIMAL DESIGN OF GROUNDING GRID

2.5.1 Design of the Grounding Grid by the Use of Optimum Compression Ratio

To obtain the relation between the ground potential rise, maximum touch voltage, step voltage and grid resistance and compressed ratios, Software is designed by the authors and calculations are done assuming the line to ground fault current was 10 kA of 220 kV substation and the soil resistivity was 100 Ω.m.

The grid material is copper having 0.01 m diameter. fault duration t_f is 1 second, shock duration t_s is 1 second, the decrement factor D_f is 1, S_f is 0.6,

$$S_f = \frac{I_g}{3I_0} \cong 0.6 \quad (2.20) \text{ (IEEE Std. 80, (2000))}$$

Where I_g is the grid current and I_0 is the zero sequance current
ρ_s(Crushed Rock Resistivity) is 2500 Ω.m, and h_s (Thickness of Crushed Rock) is 0.102 m. In each case the grid number of meshes was 64, (105×105 m²)

.Fig.1 clarifies the relation between the compression ratios versus the GPR, mesh voltage, step voltage and the grounding grid resistance for different values of compression ratios (CR).

Summary of the relations between GPR, step and mesh voltages and grounding grid resistance as a function of the compressed ratios are illustrated in Table 2. It is concluded from this table and the results given in Figure 2.1 that the GPR, step and mesh voltages and grounding grid resistance reach their lowest value when the compression ratio for this grid reaches 0.493. These findings are in agreement with that obtained by O.E.Gouda (Gouda Osama E.; El Dein Adel Z. (2017))

Figure 1. The potential rise of the grid (GPR)

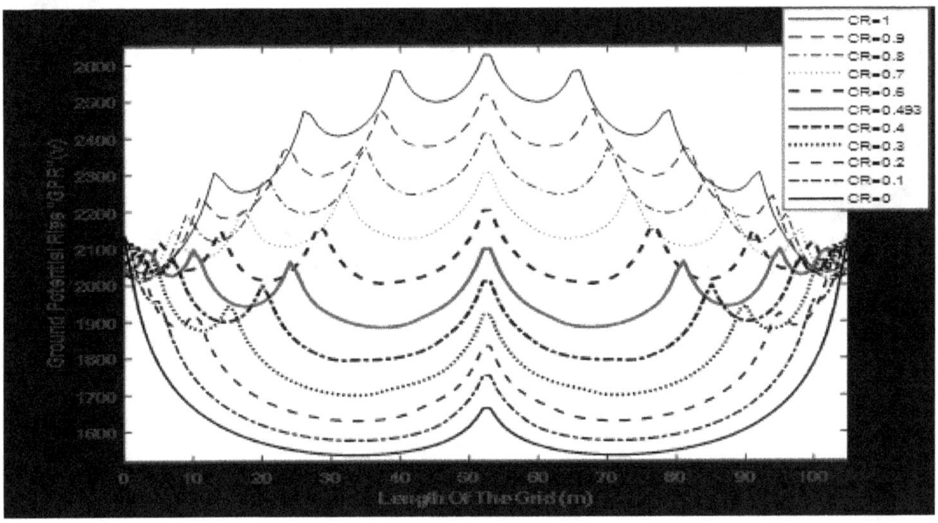

Figure 2. Touch potentials of grounded structures

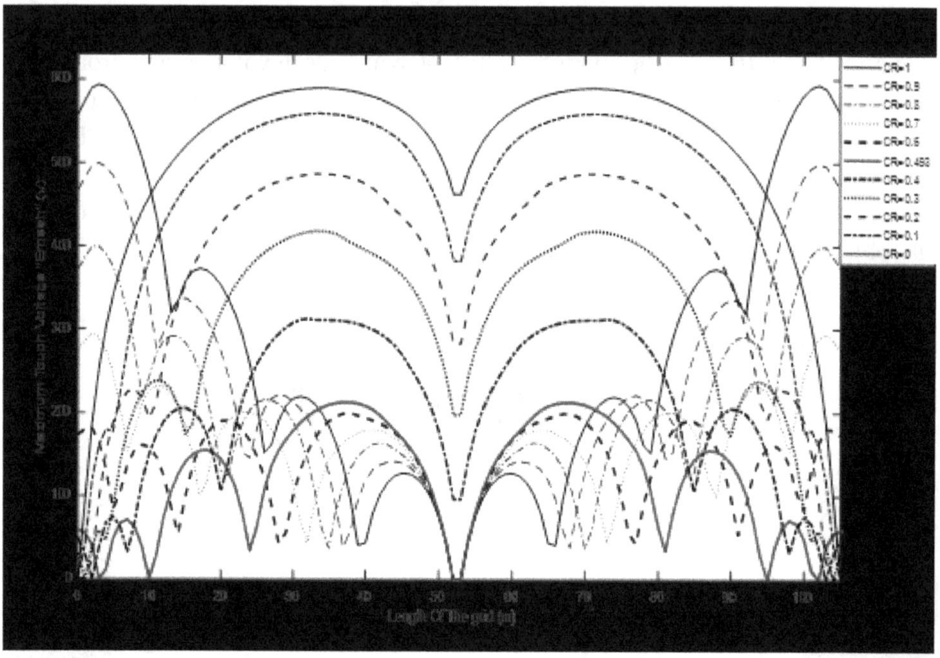

Figure 3. Step potential along the center of the square grounding grids

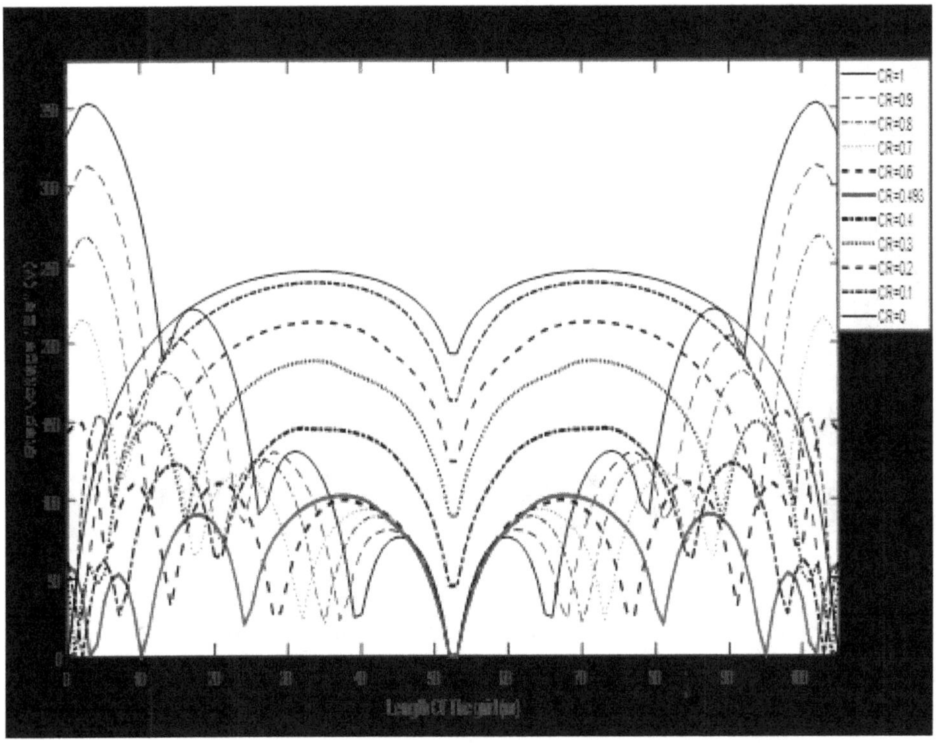

Figure 4. Grid resistance, for different values of compression ratios (CR)

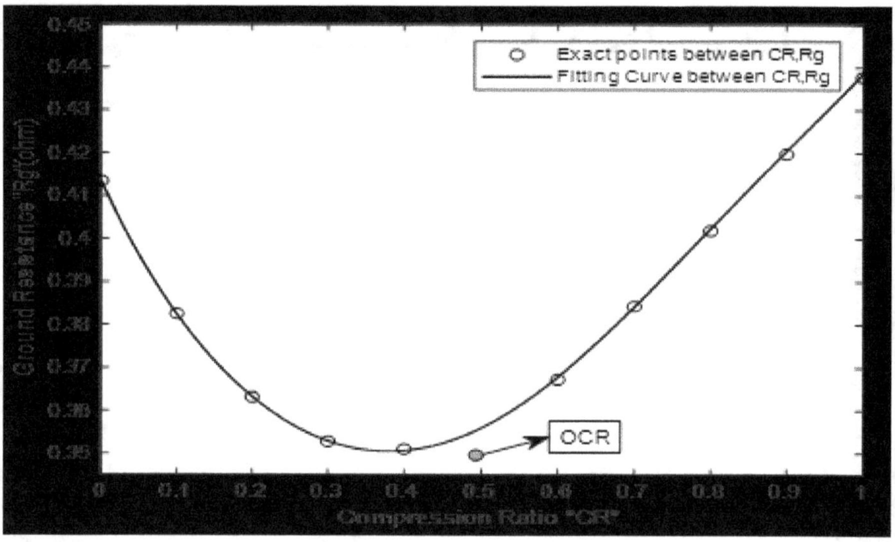

Figure 4 clarifies the relations between the GPR, mesh voltage, step voltage, and grid resistance as functions of conductor compression ratio and number of meshes.

Table 2. Summary of the GPR, Step and mesh potentials and grid resistance verses the compressed ratio, N=64

Compression Ratio (CR)	GPR (Volt)	Emesh (Volt)	Estep (Volt)	Ground Resistance (Ω)
CR=0	2480.5	601.49	245.7	0.4134
CR=0.1	2295.3	559.26	238.45	0.3825
CR=0.2	2178.6	486.98	213.08	0.3631
CR=0.3	2116.6	417.96	188.28	0.3528
CR=0.4	2106	313.57	145.4	0.351
CR=0.493	2097.6	218.66	102.38	0.3496
CR=0.6	2203.2	266.56	**149.34**	0.3672
CR=0.7	2305.9	326.45	215.82	0.3843
CR=0.8	2411.5	401.7	267.5	0.4019
CR=0.9	2518.8	499.5	13.115	0.4198
CR=1	2626.4	593.7	353.2	0.4377

To build up useful charts for the optimal design of grounding grids for high voltage substations, the used Software has been modified for obtaining the OCR of the grids with different number of meshes and to investigate the influences of the compact ratio on the GPR, mesh and step potentials. The meshes number is changed between 200 and 1600 while the grid area was kept constant at (105×105m^2). From Figure 3, it is found that the OCR value is not constant for all the grounding grids even if they have the same area. Each grid has its own OCR depending on its number of meshes. It is remarkable that these relationships usually have an obvious 'U' shape. It is noticed also from Figure 4 that increasing of the grid meshes number increases its OCR. As example OCR is 0.493 for N=64, 0.58 for N=100 and 0.65 for N=256 after that number of meshes tends to be constant. This can be understood from the perspective that the increase in the number of meshes and therefore the conductors number tends to the network to be similar to a metal plate.

Samples of 3D voltage profile distribution for different compressed ratios values are given in Figure 5, 6, and 7. It is observed that at CR=0.493 the ground potential is more uniformly distributed

Figure 5. The relation between the meshes number and GPR

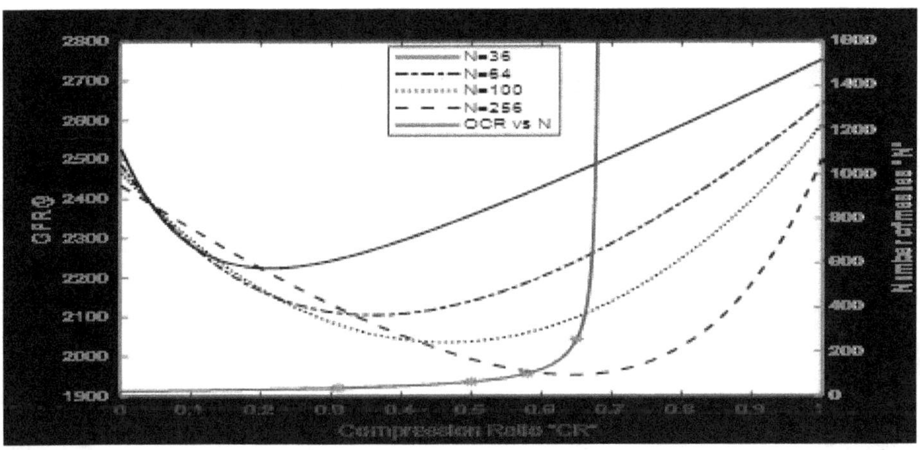

Figure 6. Mesh voltage

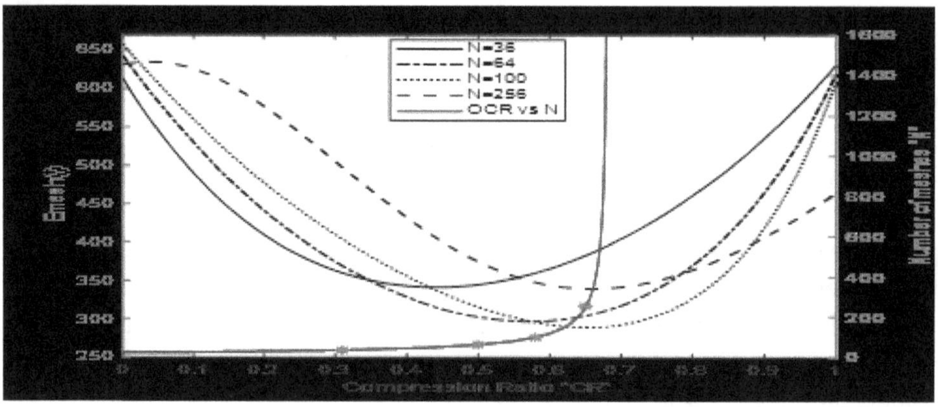

Figure 7. Step voltage

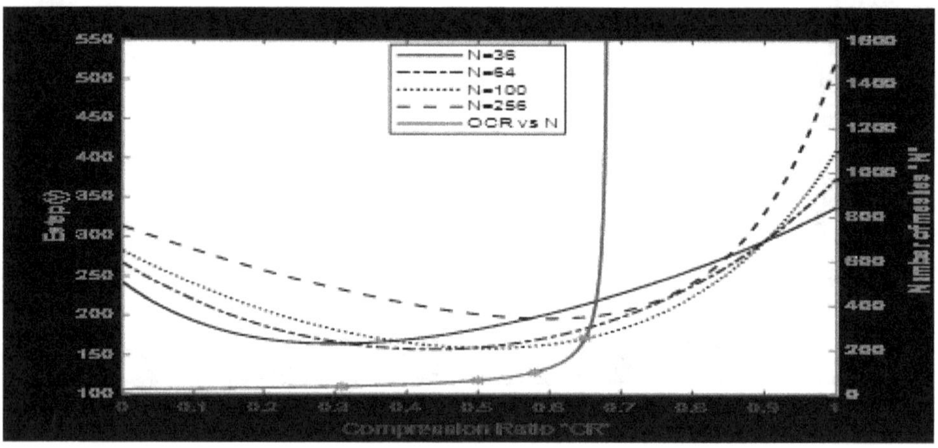

Figure 8. The relation between CR and ground resistance

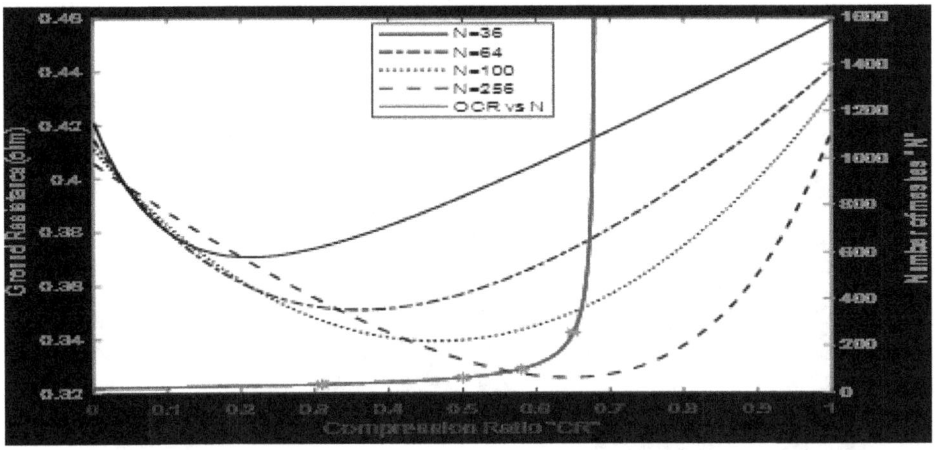

Figure 9. Voltage profile along all grid conductors in case of unequally spaced grid having denser conductors at the edges of ($A=105\times105\ m^2$, $N=64$) in case of $CR=0.2$

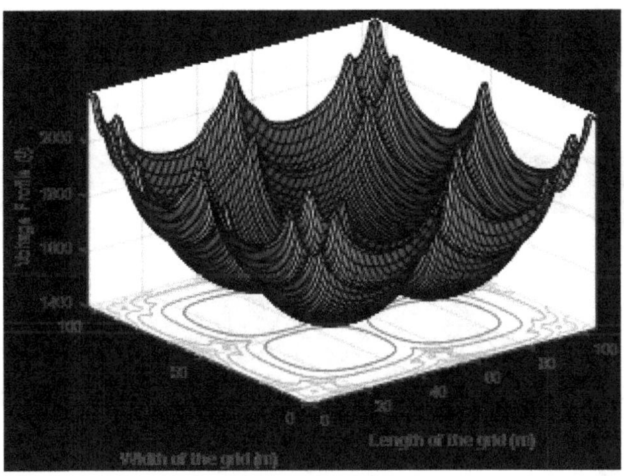

Figure 10. Voltage profile along all grid conductors in case of unequally spaced grid having denser conductors at the edges of ($A=105\times105$ m^2, $N=64$) in case of $CR=0.493$

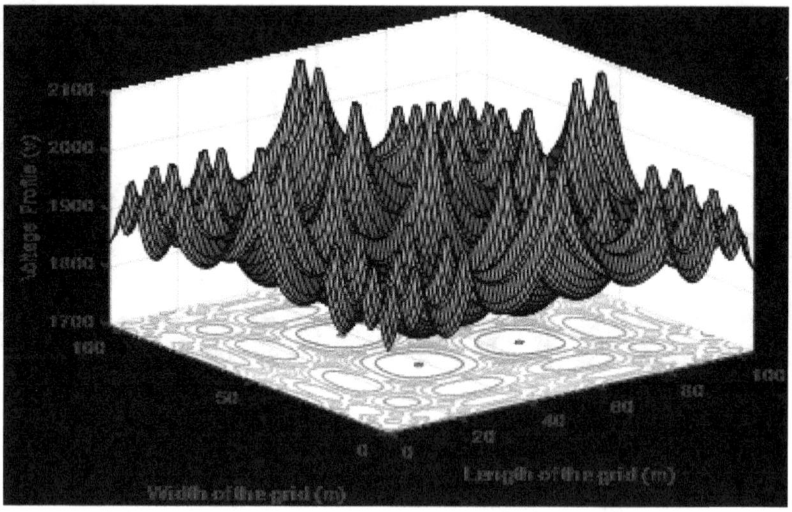

Figure 11. Voltage profile along all grid conductors in case of unequally spaced grid having denser conductors at the edges of ($A=105\times105$ m^2, $N=64$) in case of $CR=0.9$

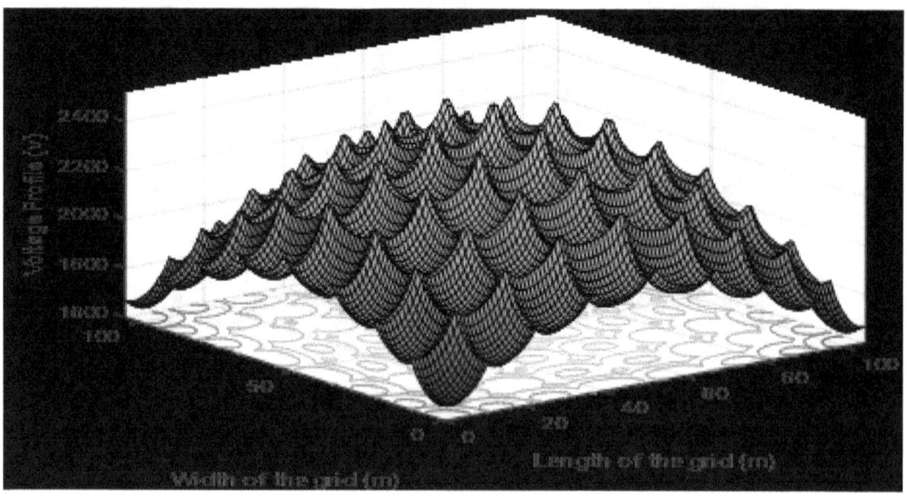

2.5.2 Optimum Design of Ground Grid Using Particle Swarm Optimization (PSO)

In this part of the article the step potential, touch potential, ground potential rise (GPR), and grounding resistance, based on the particle swarm optimization (PSO) are investigated to obtain the optimal grounding grid design, which includes number of meshes, and its laying depth. In the grid design the factors of safety and minimum used copper have been considered as explained in items E and D illustrated above under the topic the proposed algorithm used tools. The used copper volume is estimated based on the particle swarm optimization (Chun-Yao Lee and Yi-Xing Shen, (2009)).

The field data given in Table 3 are used for the design of grounding grid by using PSO algorithm. The optimized parameters in this study are the laying depth, spacing between conductors and the number of conductors on each side. Figure 12 illustrates the PSO algorithm flowchart of the grounding grid optimization (Chun-Yao Lee and Yi-Xing Shen, (2009)).

Table 3. The field data of the grounding system

Grid parameter	Field data
A (Grid area, m^2)	105×105 m^2
ρ (Soil resistivity, Ω.m)	100 Ω.m
I_f(Fault current, A)	10000 Amp
D_f(Decrement factor)	1
S_f (Current division factor)	0.6
t_f (Duration of fault current)	1 sec
t_s (Duration of shock)	1 sec
ρ_s(Crushed Rock Resistivity)	2500 Ω.m
h_s (Thickness of Crushed Rock)	0.102 m
Diameter of the conductors	0.01 m
$E_{touch-tolerable}$	572.72 v
$E_{step-tolerable}$	1819.9 v

The obtained designed parameters are illustrated in Table 3 and it is compared with the parameters of designed grounding grid equally spaced between conductors by using IEEE Std 80-2000.

Table 4. Grounding grid parameters using IEEE STD. 80-2000 and PSO without rods, equal spacing between conductors

Ground grid design without using rods	Using IEEE Std 80-2000	Using PSO
h (laying depth, m)	1	0.9
I_f(Fault current, A)	10000	10000
D (Spacing between each conductor and the other, m)	15	21
ρ (Soil resistivity, Ω.m)	100	100
NLc (Number of conductors in each side)	8	6
Volume (m^3)	0.1319	0.0990
Mass (kg)	1179.18	885.06
Fault current (Ampere)	10000	10000
GPR (v)	2553.9	2695.2
Estep (v)	303.35	266.01
Emesh(v)	538.92	565.51
Ground Resistance (Ω)	0.4257	0.4492

From Table 4, it is noticed that the design of the grounding grid using PSO algorithm gives saving in copper volume and mass of the grounding grid by about 24.9% and decreases the step potential by 12.3%, but unfortunately an increase in GPR and the mesh voltage by 5.5% and 4.9% respectively is noticed . It should be noted that these values give a safe design for the grounding grid. Table 4 gives similar results for grounding grid containing rods. In this case the optimization is done on the laying depth, spacing between conductors, number of rods and the number of conductors on each grid side. The voltage profile of the grounding grid design using IEEE Std 80-2000 and using PSO in the 3-D and the GPR in 2-D are given in Figure 12.

2.5.3 Hybridization of CR Method and PSO Algorithm

Hybridization of the CR method and PSO algorithm is done to obtain the most economical and safe grounding design grid. In this technique the compressed ratio CR is considered as one of the variables that investigated by PSO besides the spacing between conductors and the number of conductors on each side. In the case under study a square grounding grids of 105×105 m^2 area and buried in 100 Ω.m uniform soil resistivity and the fault current emanating from this grid into the soil is 10000 (Amp) it is found that the optimized value of CR is 0.23.

Figure 12. PSO algorithm of the grounding grid optimization flowchart

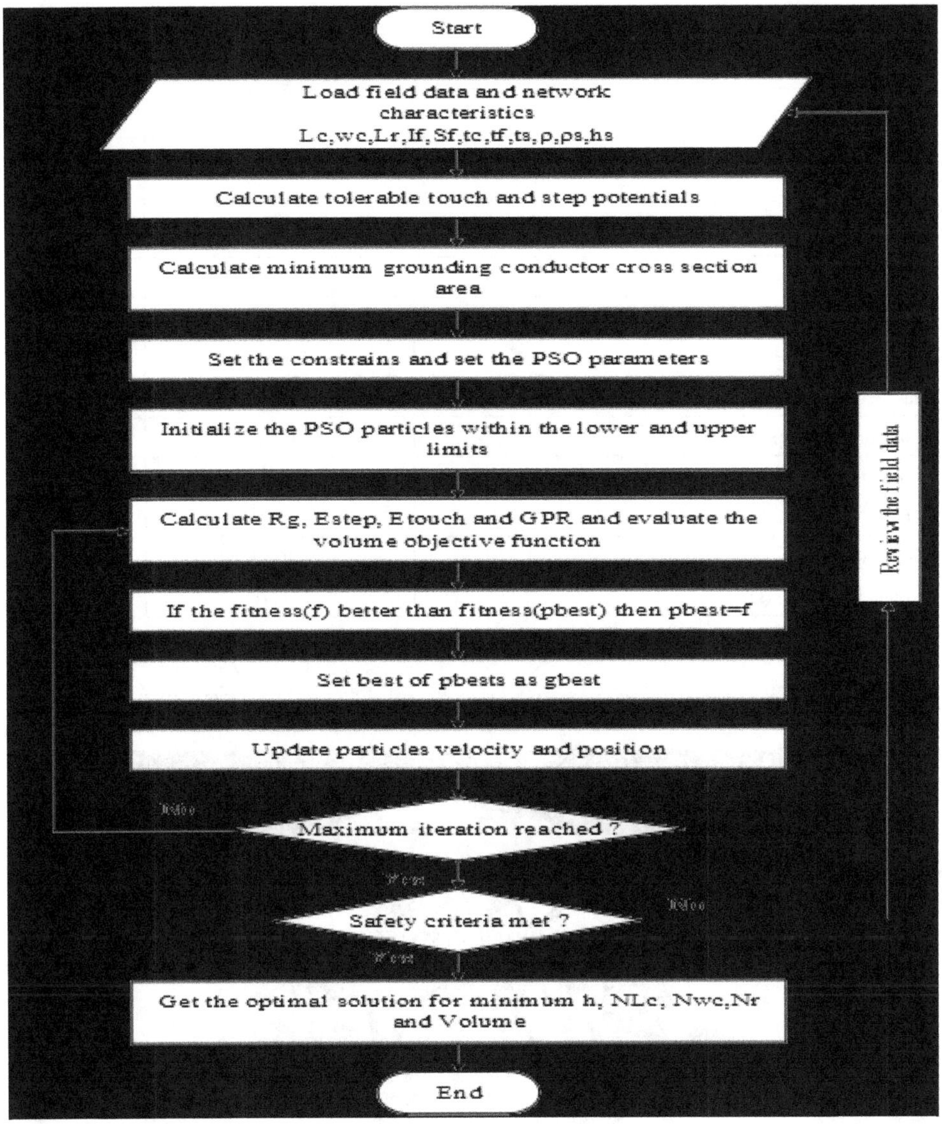

Table 5. Grounding grid parameters using IEEE STD. 80-2000 and PSO with using rods, equal spacing between conductors

Ground grid design with rods	Using IEEE Std 80-2000	Using PSO
h (laying depth, m)	1.5	1.1
I_f(Fault current, A)	10000	10000
D (Spacing between each conductor and the other, m)	26.25	35
NLc(Number of conductors in each side)	5	4
Nr (Number of rods)	17	16
Lr (Rod length, m)	3	3
Dr (Rod diameter, m)	0.03	0.03
Volume (m^3)	0.1185	0.0999
Mass (kg)	1059.39	893.106
Fault current (Ampere)	10000	10000
GPR (v)	2566	2755.5
Estep (v)	215.53	185.03
Emesh(v)	476.87	559.11
Ground Resistance (Ω)	0.4277	0.4593

Figure 13. Voltage profile of the ground system design using IEEE Std 80-2000

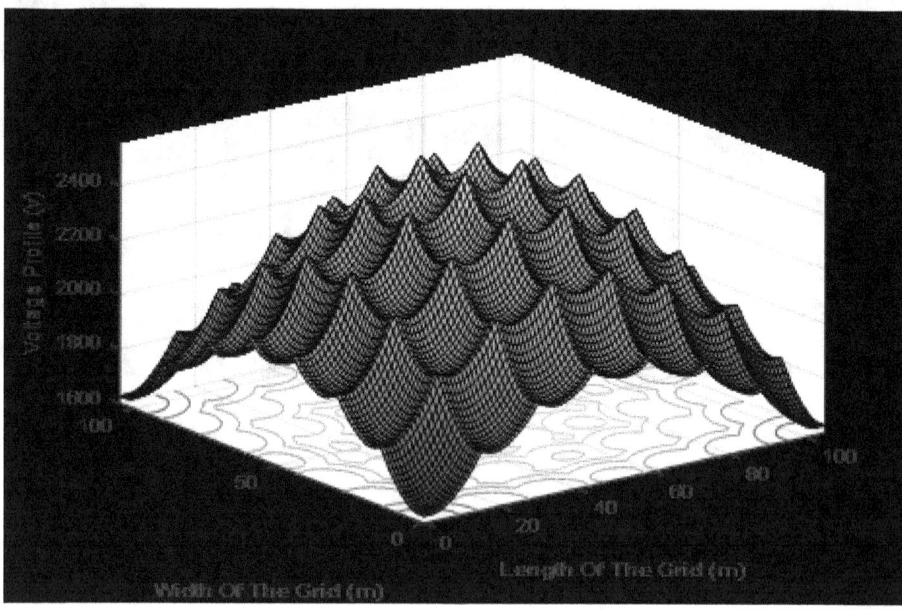

Figure 14. Voltage profile of the ground system design using PSO in 3-D

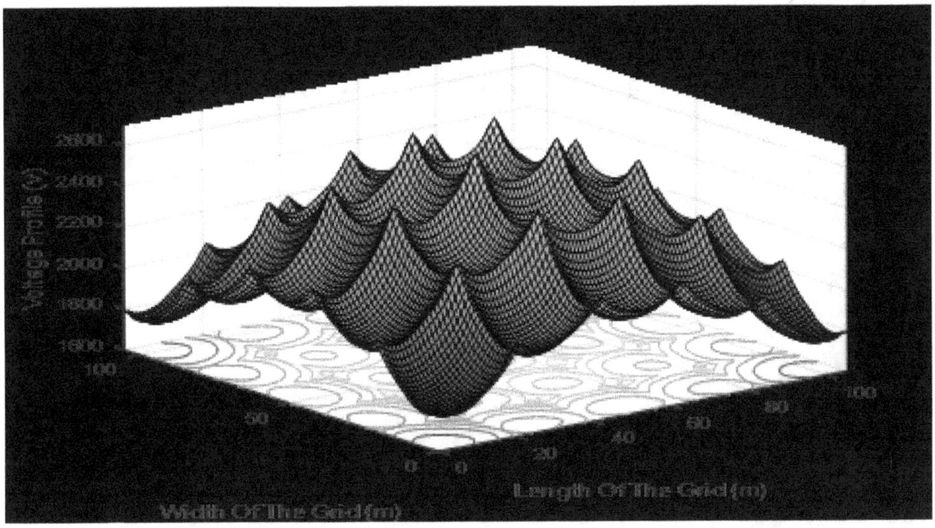

Figure 15. GPR of grounding grid using IEEE Std 80-2000 and PSO in 2-D

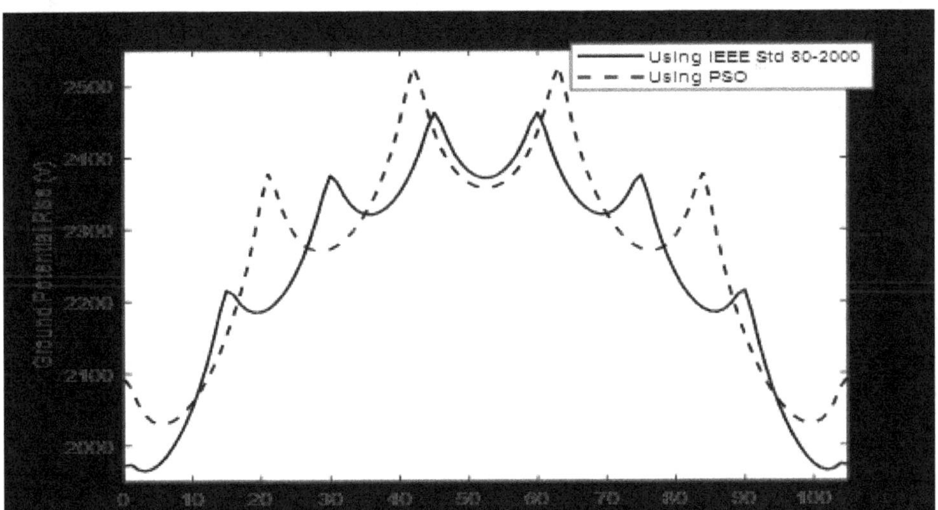

Figure 16. The voltage profile the grounding grid when using PSO combined with CR, the grid conductors are unequally spaced and CR= 0.23

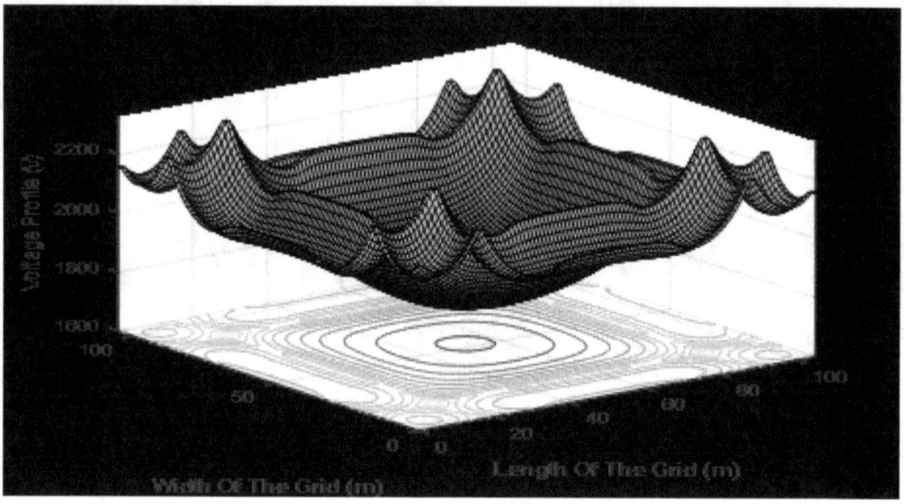

Figures 13 -16 show the voltage profile of the ground grid design by using IEEE Std 80-2000 as given in Figure 13, by using PSO in 3-D as shown in Figure 14 and GPR of grounding grid by using IEEE Std 80-2000 and PSO in 2-D as given in Figure 15. The grid conductors in case of using the new technique are unequally spaced. Table 6 gives design of the grounding grid using OCR and PSO; the grid conductors are unequally spaced. Figure 16 gives the voltage profile of the grounding grid when using hybridization of CR method and PSO algorithm, the grid conductors are unequally spaced and CR= 0.23

Table 6. Design of the grounding grid using OCR and PSO, the grid conductors are unequally spaced

Ground grid design	**using CR**	**using OCR**	**New technique**
CR	0.9	0.119	0.23
h	1.5	1.5	1.5
D (Spacing between each conductor and the other, m), NLc	[17.5 – 35- 35- 17.5], 5	[17.5 – 35- 35- 17.5], 5	[26.25- 52.5 - 26.25], 4
Volume (m^3)	0.0825	0.0825	0.0660
Mass (kg)	737.55	737.55	590
GPR (v)	2656.6	2252.4	2327.2
Estep (v)	204.55	109.08	128.14

continued on following page

Table 6. Continued

Ground grid design	using CR	using OCR	New technique
Emesh(v)	472.56	295.43	429.15
Resistance (Ω)	0.4428	0.3754	0.3879

Figure 17. GPR

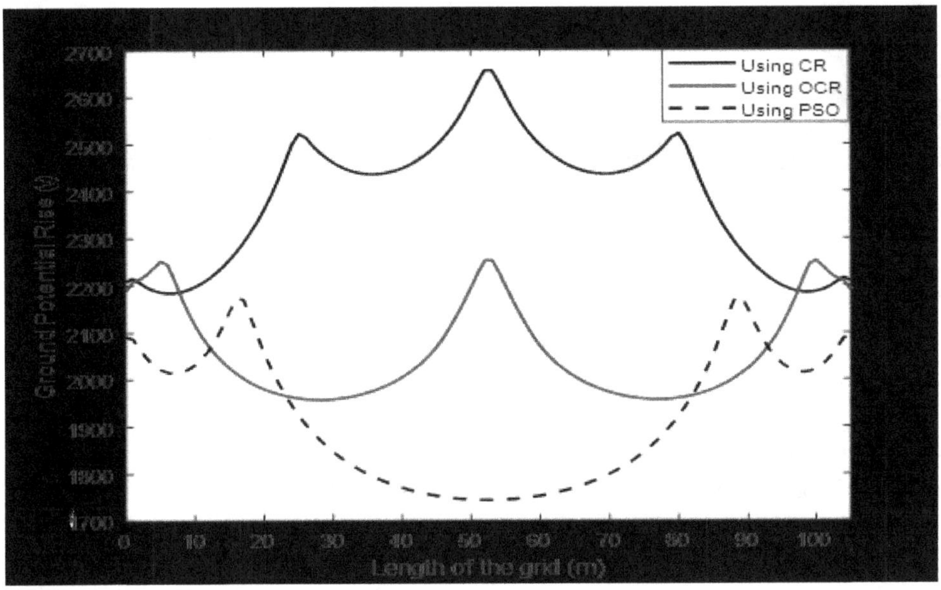

Figure 18. Mesh voltage

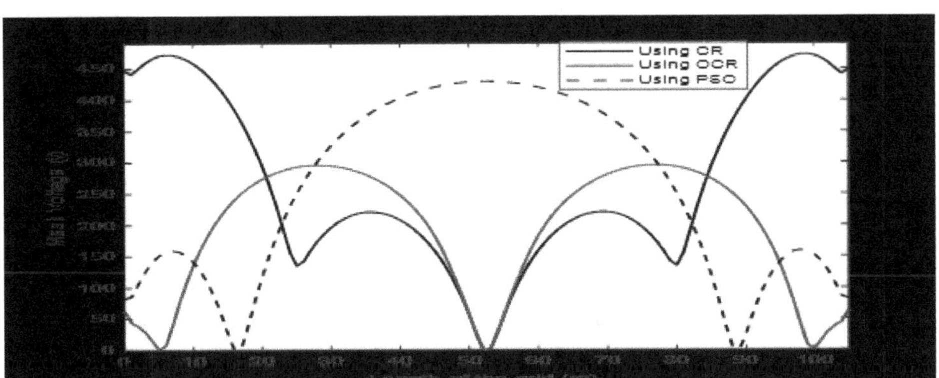

Figure 19. Step voltage of the designed grids given in Table 6

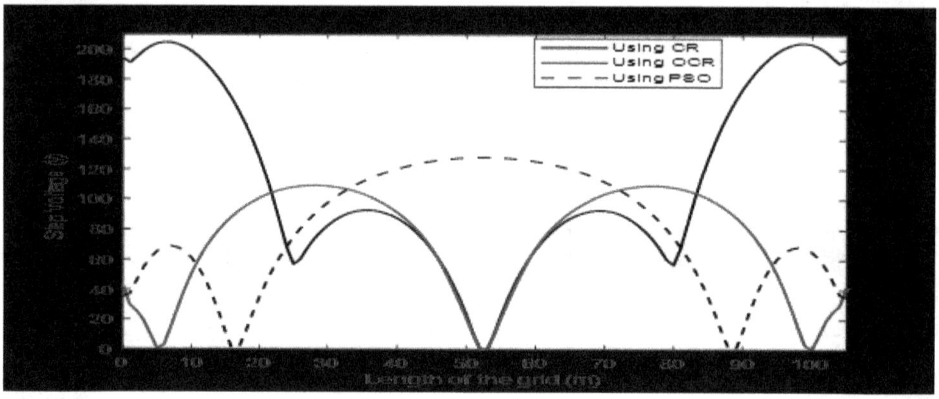

Figure 17 gives the ground potential rise; Figure 18 shows the mesh voltage and Figure 19 gives the step voltage of the designed grids.

The same results are given in Table 6. Using the new technique in the grounding grid design indicates that 4 conductors only on each side of the grid give economical and safe grounding grid. From this table it is noticed that 19.27% saving in copper volume of the grounding grid is done by using the new technique and slight increase is observed in the grid resistance, ground potential rise and mesh voltage comparing with using OCR and CR= 0.9 methods in the design of the grounding grid, but with the acceptable safe design.

2.6 EFFECT OF THE CRUSHED ROCK ON THE STEP AND TOUCH VOLTAGES

Among the methods that can be used to achieve safety in grounding grids is placing crushed rock on the grid surface. This reduces the dimensions and lengths of the grid. In the current part of the article the study of the effect of crushed rock on the step and touch voltages is done.

If the underlying soil has a lower resistivity than the surface material, only some grid current will go upward into the thin layer of the surface material, and the surface voltage will be very nearly the same as that without the surface material. The current through the body will be lowered considerably with the addition of the surface material because of the greater contact resistance between the earth and the feet. However, this resistance may be considerably less than that of a surface layer thick enough to assume uniform resistivity in all directions.

Crashed rock layers of high resistivity material such as gravel are often spread on the earth's surface above the ground grid to increase the contact resistance between the soil and the feet of persons in the substation. The safety criteria for touch and step voltages were changed to provide accommodations lower limits for a 50 kg or 70 kg body weight, reflecting the outcome of more recent studies by (IEEE Std. 80, (2000)); and a special corrective factor was added, to calculate the effect of a finite thickness of the surface material. The reduction of touch and step voltage criteria according to IEEE 80 depends on the relative values of the soil and the surface material resistivity, and on the thickness of the surface material.

According to IEEE 80 (IEEE Std. 80, (2000)); touch and step voltages values of the corresponding acceptable potentials which are borne by humans without being subjected to electric shocks can be calculated by using the following equations: For body weight in the range of 50 kg

$E_{Step} = (1000 + 6C_s \cdot \rho_s) \cdot \frac{0.116}{\sqrt{t_s}}$ (2.20), and for body weight in the range of 70 kg

$$E_{Step} = (1000 + 6C_s \cdot \rho_s) \cdot \frac{0.157}{\sqrt{t_s}} \quad (2.\ 21)$$

In addition, the touch voltage criteria are for body weight in the range of 50 kg

$E_{touch} = (1000 + 1.5C_s \cdot \rho_s) \cdot \frac{0.116}{\sqrt{t_s}}$ (2. 22), and for body weight of 70 kg

$$E_{touch} = (1000 + 1.5C_s \cdot \rho_s) \cdot \frac{0.157}{\sqrt{t_s}} \quad (2,\ 23)$$

The surface layer de-rating factor C_s can be calculated by

$$C_s = \frac{1}{0.96}[1 + 2\sum_{n=1}^{\infty} \frac{k^n}{\sqrt{1 + (2nh_s/0.08)^2}}] \quad (2.24)$$

Where K is the reflection factor and defined as

$$K = \frac{\rho_s - \rho_1}{\rho_s + \rho_1} \quad (2.\ 25)$$

Equation 24 can be approximated as given in equation (2.26)

$$C_s = 1 - \frac{0.09\left(1 - \frac{\rho_1}{\rho_s}\right)}{2h_s + 0.09} \quad (2.26)$$

Where:

ρ_s: The surface material resistivity in Ω·m

ρ_1: The resistivity of the earth beneath the surface material in Ω·m
h_s : The thickness of the surface material in m
t_s : The time to clear the fault, it is considered 1 second

Table 7. The surface layer De-Rating factor versus $\frac{\rho_1}{\rho_s}$ and h_s

$\frac{\rho_1}{\rho_s}$ **with constant** h_s **= 0.0 5 m**	**100/500**	**100/1000**	**100/1500**	**100/2000**	**100/5000**
The surface layer de-rating factor C_s	0.696	0.573	0.557	0.552	0.535
h_s, m with constant $\frac{\rho_1}{\rho_s}$ = 100/1000	0.05	0.1	0.15	0.2	0.3
The surface layer de-rating factor C_s	0.573	0.72	0.792	0.834	0.882

From Table 7, it is observed that the coefficient C_s decreases with the increase in the resistivity of the crushed rock layer, which eventually leads to an increase in the values of the step and touch voltages. It is also noted that the increase in the thickness of the crushed rock layer helps in a noticeable increase of the constant C_s , which is reflected in an accelerated rise in the touch and step voltages.

Figure 20a illustrates the 3-D relation between the soil resistivity, crashed rock resistivity, touch, and step voltages which are borne by humans without being subjected to electric shocks, Es is the step voltage and Et is the touch voltage.

Figure 20b shows the 3-D relation between crashed rock thickness and resistivity and touch and step voltages (ρ =100 Ω.m, laying depth =0.5m, Es is the step voltage and Et is the touch voltage)

Figure 20. 3-D relations between the soil resistivity, crashed rock resistivity, touch, and step voltages

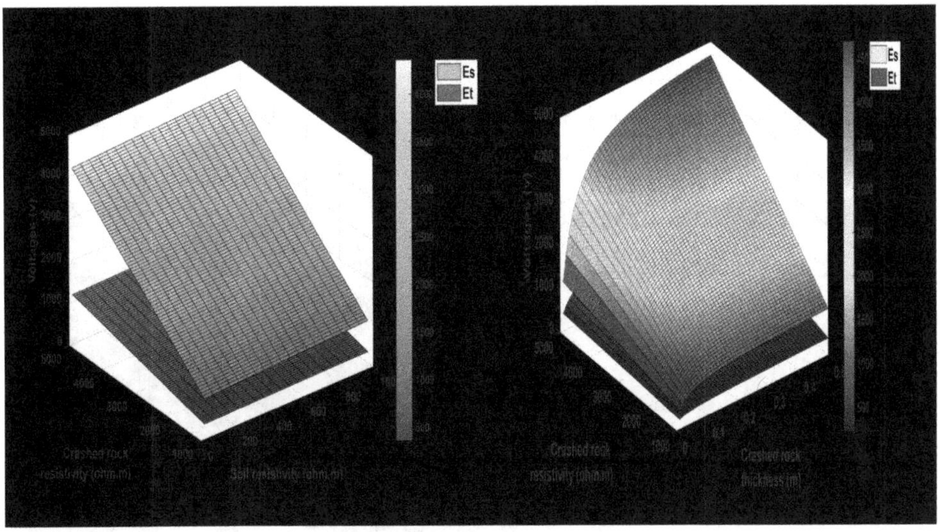

The effect of crashed rock resistivity on touch and step voltages which are borne by humans without being subjected to electric shocks is investigated while the soil resistivity is changed between 200 to 1000 Ω·m and crashed rock resistivity is changed between 1000 to 5000 Ω·m and thickness of crashed rock h_s is fixed at 0.2 m. The calculations are done according to IEEE80 standard and by using MATLAB program.

2.7 VERIFICATION OF THE CALCULATED RESULTS

2.7.1 Experimental Set-Up and Testing Measurements

In this section the validation of the calculated results is done by comparing some of calculations with the obtained experimental results. Fig. 2.9 shows the components of experimental setup, which contains, glass tank has dimensions of 1.5 m × 1.5 m × 0.5 m. The inner surface of the tank is covered by a conducting sheath that is connected to earth. Tap water with NaCl salt is used as an electrolyte, representing homogenous soil. Its electric resistivity is planned to be 100 Ω.m. Electrolytic resistivity can be controlled by changing the salinity of the tap water (Osama El-Sayed Gouda, (2017)). The water conductivity is measured by means of DC probe. The electrolytic temperature inside the tank was about 27°C during the measurements. A power supply ranges from zero to 220V is used to change the applied voltage up to appropriate value. Two voltmeters are used; one for measuring the input voltage to the model and the other is connected to a probe for measuring the voltage distribution on the grid surface. One ammeter is used to measure the grid current simulates the short circuit current, as shown in Figure 21.

Figure 21. Homogenous and nonhomogeneous experimental setup simulation (a) and (b)

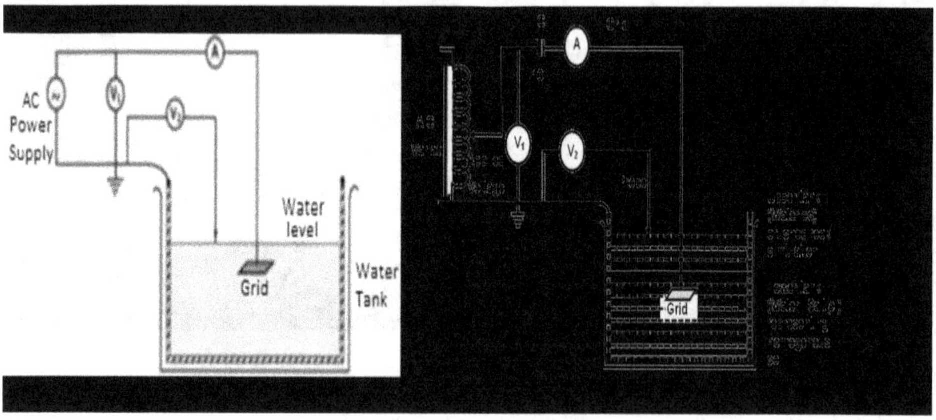

The CR of grounding grid is varied from 0 to 1 as mentioned in Table 1. In the experimental simulation five grids are selected with scale factor of 1/100, to be experimentally tested for the verification of the obtained calculations. The simulated grids are A, B, C, D and E with 105 × 105 cm dimensions of each grid. Grid A with CR=1 is equally spaced conductors whereas grid B is exponentially spaced with CR-0.5 (0.493). In this grid type the conductors are densely populated at the periphery of the grid and sparse at the center. The third grid (C) has CR=0, it contains only four meshes because all other conductors are clustered at the periphery of the grid, (Thabet, 20p2). Grids with CR=0.2 and CR=0.8 are also simulated. Samples of grid configurations with different compression ratios are shown in Figure 22.

Figure 22. Grounding grids with various compression ratios "CR"

In the researchers' opinion, the impacts of the diameter of grid conductors on the grid performance are negligible. The conductors of the simulated grids are made from copper with 2.5 mm diameter. The scale factor of the grid laying depth during the tests was 1/100.

The fault current passed through the ground grids is simulated by 5 A. This value is obtained by controlling the input voltage to the grid using variac. The used scaling factor is 1: 2000 of the faults current 10 kA.

2.7.2 Experimental Results

The surface potential rise, mesh voltage, step voltage and grid resistance are measured using a probe as shown in Figure 22. The measurements are done on grid models A, B, C, D and E simulating 105 × 105m^2, the number of grid meshes depends on the compressed ratio of each grid. The homogenous soil is simulated by water with resistivity of 100 Ω m. The measured surface potential rise, step potential value, meshes voltage and grid resistance are multiplied by the scaling factor of current and compared with the calculated values. The measured results are illustrated in Table 6. From the tabulated results, it is observed that the measured values are slightly higher than the calculated values. This may be due to the boundary impacts of the tank walls, as each grid pattern dimension exceeds 1/5 of the flat of the tank dimensions that represents the soil.

Table 8. Calculated surface potential rises, Mesh voltage, step voltage and grid resistance compared with measured values

CR	0	0.23 (0.2)	0.493 (0.5)	0.8	1
Calculated Surface potential rise in volts	2480.5	2178.6	2097.6	2411.5	2626.4
Measured Surface potential rise in volts	2490	2180	2100	2420	2630
Calculated Mesh voltage in volts	601.49	486.98	218.66	401.7	593.7
Measured Mesh voltage in volts	610	490	220	410	595
Calculated step voltage in volts	245.7	213.08	102.38	267.5	353.2
Measured step voltage in volts	250	215	110	270	360
Calculated grid resistance (Ω)	0.4134	0.3631	0.3496	0.4019	0.4377
Measured grid resistance (Ω)	0.43	0.37	0.36	0.42	0.44

To investigate the impact of crashed rock on the grounding system performance, new testing arrangement is illustrated Figure 24. In this figure the upper tank is added to the lower one to simulate the crashed rock layer. Capacitor (10 F) is connected between the variac and the grid model to prevent any flow of DC current due to the use of dissimilar materials.

The top ends of the lower tank are coated with plastic to obtain electrical isolation between the metallic bodies of the two tanks. Only the electrical contact between the two mediums (representing the soil at which the grid is buried and the crashed rock) occurs through the metallic base of the upper tank. The lower tank is connected to the earth. The scale factor of the grid laying depth and the crashed rock layer thickness is 1/100

The surface potential rise and mesh voltages are measured using a probe as illustrated in Figure 22. The measurements are done on grid size 105×105 m^2 .The measurements are done using grid models A, B, C, D and E. The number of conductor's grid depends on the compressed ratio. The soil at which the grid is buried is simulated by water with resistivity of 100 Ω m. The crashed rock is simulated by water in an upper tank with thickness layer of 20 cm and 1000 Ω m. The measurement values are given in Table 9. Comparing the measured values in Tables 8 and 9, it is noticed that the presence of crashed rock increased the measured surface potential rise by range from 15.96% to 28%. The same thing is happening to mesh voltage with crashed rock, where it is increased in percentages ranging from 12.24% to 14.6% when the crashed rock resistivity was 1000 Ω m with 0.2 thicknesses.

Table 9. Measured surface potential rise and mesh voltage with crashed rock

CR	**0**	**0.23 (0.2)**	**0.493 (0.5)**	**0.8**	**1**
Measured Surface potential rise with crashed rock in volts	2900	2655	2500	3100	3050
% increase	16.4	21.7	19.04	28	15.96
Measured mesh voltage with crashed rock in volts	680	550	248	470	670
% increase	13.1	12.24	12,7	14.6	12.6

REFERENCES

Attia, E.-F. (2013). Accelerated Particle Swarm Optimization-based Approach to the Optimal Design of Substation Grounding Grid. *Przegląd Elektrotechniczny*.

Gouda, O. E. (2011). *Optimum design of grounding system of high voltage substations*. LAMBERT Academic Publishing.

Gouda Osama, E., & El Dein Adel, Z. (2017). Ground potential rise of faulty substations having equal and unequal spacing grounding grids conductors. *IET Generation, Transmission & Distribution*, 11(1), 18–26. 10.1049/iet-gtd.2016.0263

Gouda Osama El-Sayed. (2017). *Design Parameters of Electrical Network Grounding Systems, book*. Advances in Computer and Electrical Engineering. IGI Global.

He, Zeng, Gao, Tu, Sun, Zou, & Guan. (2005), Optimal Design of Grounding System Considering the Influence of Seasonal Frozen Soil Layer, IEEE Trans. Power Del., 20(1).

IEEE Std. 80. (2000). IEEE Guide for Safety in AC Substation Grounding, IEEE Std, 80, 2000, 4th edition.

Lee, C.-Y., & Shen, Y.-X. (2009). Optimal Planning of Ground Grid Based on Particle Swam Algorithm, World Academy of Science, Engineering and Technology. *International Journal of Electrical, Computer, Energetic, Electronic and Communication Engineering, Vol*, 3(12).

Sun, W., He, J., Gao, Y., Zeng, R., Wu, W., & Su, Q. (2000). Optimal Design Analysis of Grounding Grids for Substations Built in Non-uniform Soil. *IEEE Transactions on Power Apparatus and Systems*, 1455–1460.

Thabet, A. (2002), Grounding systems of electric substations in non-uniform earth structure with new analysis, M. Sc. Thesis, High Institute of Energy, Aswan.

Chapter 3
Performance of Grounding System Under Lightning Strokes

Osama El-Sayed Gouda
Faculty of Engineering, Cairo University, Egypt

ABSTRACT

In this chapter, one of the most important types of faults of the electrical networks is presented, which is the lightning phenomenon. This chapter presents an explanation of the phenomenon of lightning, the types of lightning, the types of lightning's effect on the electrical network, difference between direct and indirect lightning strokes, as well as the problems associated with the phenomenon of lightning and how to solve these problems. Also, this chapter presents a study of the effect of transient lightning stroke on the grounding electrodes.

3.1. LIGHTNING

3.1.1 Lightning Phenomenon

Lightning is a natural phenomenon caused by electrical discharge, which emits electromagnetic radiation during thunderstorms. The lighting is occasionally associated with thunder due to the electric current passing through the lightning channel (http://en.citizendium.org/wiki/Lightning, 05. (2009))

DOI: 10.4018/979-8-3693-2969-6.ch003

3.1.2 Lightning Process

From Figure 1, it is seen that the area of negative charge at the base of the thundercloud builds up, and it induces a region of positive charge to develop on the ground below. As a result of this, a potential difference or voltage is created across the cloud-to-ground gap. Once the voltage reaches a certain strength, the air between the base of the cloud and the ground develops electrical conductivity. At first, a channel, known as a stepped leader, is formed. Although invisible to the naked eye, this allows electrons to move from the cloud to the ground. It is called a stepped leader because it travels in 50 to 100 m sections, with a slight pause in between, to the ground, as it becomes near to the ground, positively charged streamer fires upwards from the ground to connect with it. Streamers are most often initiated from tall objects on the ground. Once connected, electrons from the cloud can flow to the ground and positive charges can flow from the ground to the cloud. It is this flow of charge that is the visible lightning stroke. After the first discharge, it is possible for another leader to form down the channel. Once again, a visible lightning stroke is seen. This can happen 3–4 times in quick succession. All of this happens in a time interval of about 200 milliseconds (https://www.sciencelearn.org.nz)

Figure 1. Process of lightning (https://www.sciencelearn.org.nz)

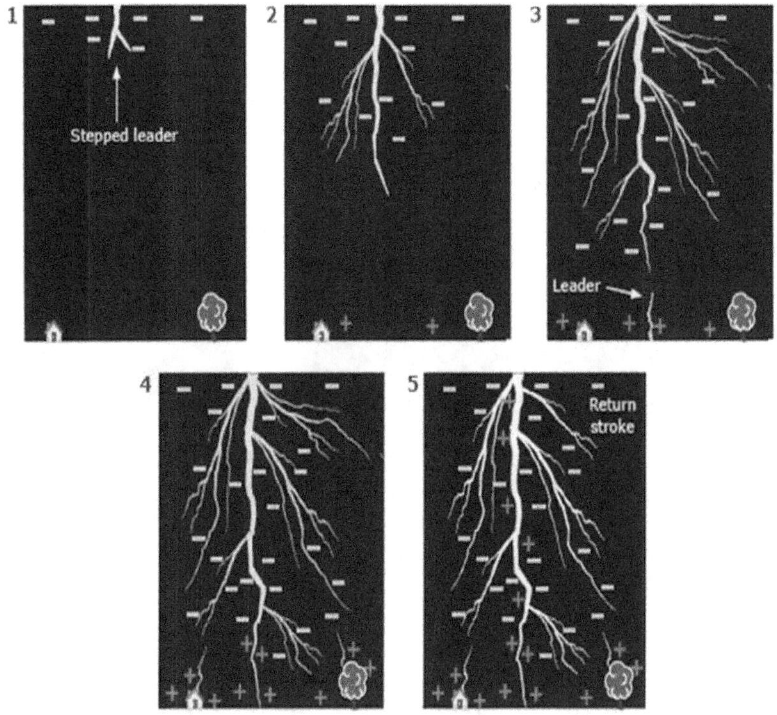

3.1.3 Types of Lightning

Lightning discharges can be classified into three main types see Figure (2) (Thanasaksiri T., (2004)), (Mcdermott Tom, (2004)), (IEEE Standards 1410, (2004))

i) Intra-cloud discharge.
ii) Cloud-to-cloud discharge.
iii) Cloud-to-ground discharge.

Figure 2. Type of lightning (https://www.nvent.com/en-eg/erico/solutions/lightning-protection)

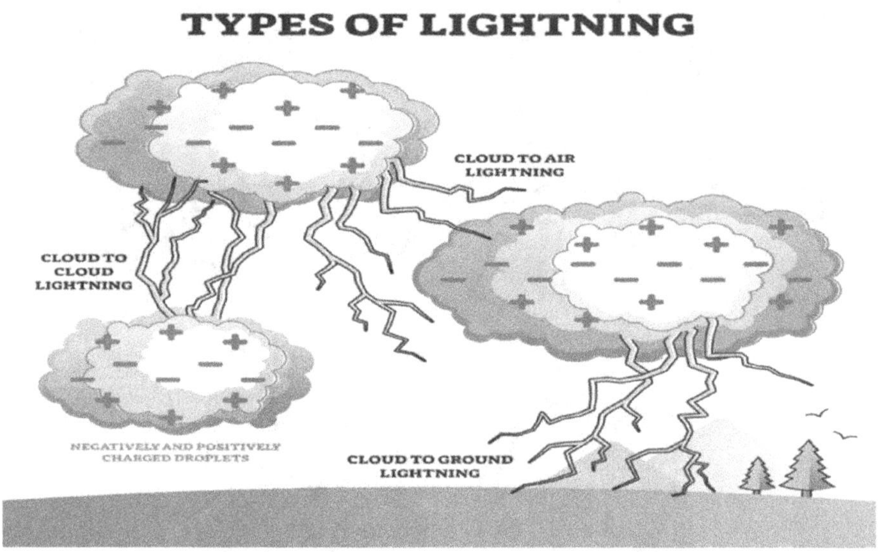

3.1.4 Difference Between Direct and Indirect Lightning Strokes

Lightning strikes can generate direct or indirect overvoltage, which might result in more frequent transmission line outages. When there are indirect strokes, the surrounding lines' return stroke current creates an electromagnetic field that results in overvoltage caused by lightning (S. Viscaro, (2007)).

Figure 3. Difference between direct and indirect lightning stroke (Dong-Hyuk Ha, Choi Jong-Kee, and Yoon Yong-Beum, (2019))

A) Direct Stroke

A lightning direct stroke can be defined as a lightning stroke that directly hits any part of the electrical network. In most cases in power distribution lines, the insulation flashover occurs although the return stroke current is small due to the high generated overvoltage which is very high compared with the overhead distribution lines insulation level (Viscaro S., (2007)).

B) Indirect Stroke

An indirect lightning stroke can be defined as a lightning stroke that does not directly hit any part of the electrical network; however, the induced overvoltage is generated and traveled over the network. This type of stroke is responsible for many lightning outages of low-insulation lines. Most of the flashes hit near the line produce voltages less than 300 kV (IEEE Standards 1410, (2004)). Although the indirect strokes generate induced overvoltage with small amplitude compared with generated overvoltage by direct stroke, they frequently occur affecting the performance of overhead distribution lines (Viscaro S., (2007)), (Viscaro S., (2005)).

3.1.5 Lightning Problems

As is known, lightning stroke has a large amount of current. So, over voltages will generate. However, these over voltages can appear in different manners such as overhead line over voltages and ground potential rises. These over voltages are discussed in the following subsections.

3.1.5.1 Transmission Line Over Voltages

Induced overvoltage due to nearby lightning stroke
Back-flashover
Shielding failure or direct stroke to the phase conductor

3.1.5.2 Transient Ground Potential Rise (TGPR)

Transient ground potential rise (TGPR) can be defined as a rapid increase of ground potential while dissipating a surge into the earth. It causes serious concern for personnel safety and electrical network protection.

3.1.5.3 Transient Earth Surface Potential (TESP)

Transient Earth surface Potential Rise is a phenomenon that occurs when a large amount of current enters the earth through the grounding system. Not only has the potential of the grounding system risen, but also the surrounding soil.

3.2 PERFORMANCE OF GROUNDING ELECTRODE UNDER FREQUENCY AND TRANSIENT EFFECT

As is already known, the main goal of the grounding system is to keep electrical equipment and people safe by sending the high currents of the network faults to the ground. So, grounding resistance, step voltage, touch voltage, and earth surface potential are all considered when evaluating any grounding system design. Many researchers have studied the performance of grounding systems in power frequency and the factors affecting it (Schwarz S. J., (1954)). This section presents the performance of grounding electrode under lightning frequency content, transient effect, and soil ionization effect.

The function of the grounding system is to protect the electrical networks, devices and people from any fault occurring. The behavior of grounding electrodes against high-frequency lightning is completely different from that of the steady state or power frequency faults. The lightning strokes usually have large amounts of current that change the electrical performance of the grounding electrodes surrounding by the soil and causes the soil ionization. One of the most severe faults that occurs in the electrical power network is a lightning current stroke. This lightning stroke fault has heavy current with high frequency content that changes the achievement of the grounding electrodes (Grcev Leonid, (2009)), (Grcev L., Arnautovski V., (1998))

3.2.1 Effect of Transient Lightning Stroke on the Grounding Electrodes

This section will review the harmonic impedance of grounding electrode including the frequency effect, the transient ground potential rise, and transient impedance and review experimental results comparing with simulation results.

A. Harmonic Impedance

Leonid Grcev (Grcev Leonid, (2009)) carried out study on vertical grounding electrode to investigate the harmonic effects on the grounding impedance. The length of grounding electrode was 12 m long with 0.7cm radius in earth with 100-ohm meter resistivity and relative permittivity of 10. Figure (2.4) shows the magnitude of the harmonic impedance versus the frequency. It is seen that the harmonic impedance is frequency independent and equal to the low frequency resistance to ground in the low frequency (LF) range up to a frequency of about 100 kHz. In the high frequency (HF) range it exhibits inductive behavior, and its value becomes large (Grcev Leonid, (2009)).

Figure 4. Amplitude of the harmonic grounding impedance of vertical rod (Grcev Leonid, (2009))

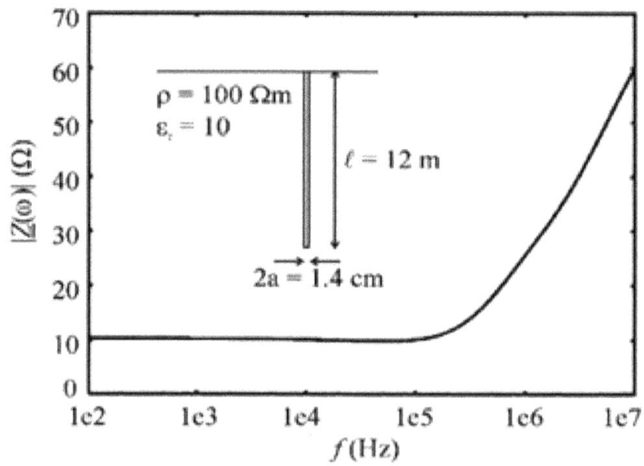

B. Transient Ground Potential Rise

Figure 5 shows the first stroke lightning current pulse and subsequent stroke injected in vertical grounding electrode, and the response to this excitation, such as the potential at the feed point and the transient impedance. As it is seen from Figure (3.5) the transient potential is decreased with the increase in the time of lightning stroke. Also, it is noted that transient potential in first lightning stroke is larger than the transient potential of subsequent (Grcev Leonid, (2009))

C. Transient Impedance

Figure 2.5 shows transient grounding impedance z(t) in both cases first lightning stroke and subsequent lightning stroke. As it is noticed in Figure (2.5) z(t) rises fast to some large value in the initial surge period, after which it converges to the stationary condition characterized by values that may be approximated with R(t). This can be used to distinguish two periods of the transient response:

1) The initial surge period [when z (t) $\neq$ R (t)]; 2) The stationary period [when z (t) $\approx$ R (t)]

Such division of the transient period is important, since the surge period is very short, from one to a few or few tens of microseconds, and the rest is the stationary period when the behavior may be approximated as dominantly resistive (Grcev Leonid, (2009)).

D. Experimental Verification

Figure 6 shows the oscillograms of the recorded current pulse injected at the end point of horizontal grounding electrode and transient voltage to remote ground at the same point during field measurements performed by references (Grcev Leonid, (2009)) and (Rochereau H., (1988)). The grounding electrode has length 14 m, radius 116 mm buried at 0.6 m depth in soil with resistivity 70 Ω.m and relative permittivity 15. These experimental results are compared with simulation results based on the electromagnetic field theory approach (Grcev L., Arnautovski V., (1998)) where a good agreement is noticed.

Figure 5. Lightning current pulse, transient potential at feed point and ground impedance of ground rod. Response to typical first stroke current pulse (Grcev Leonid, (2009))

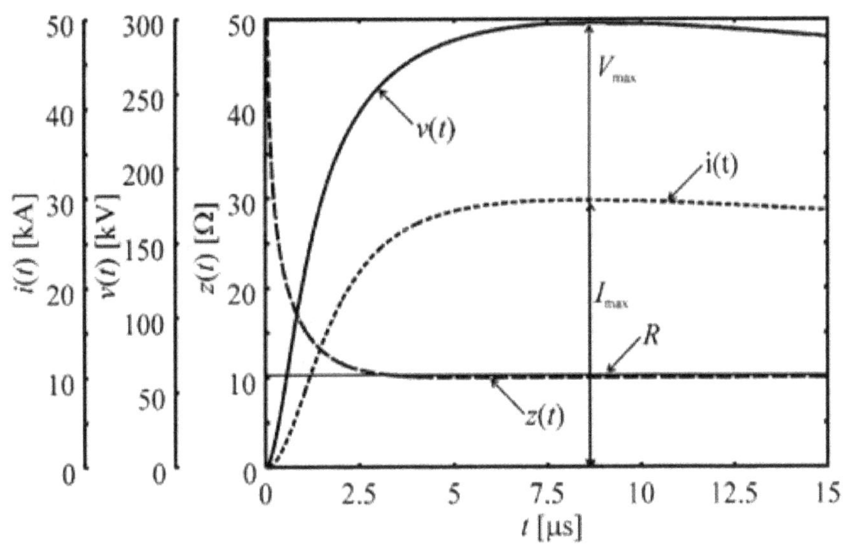

Figure 6. Lightning current pulse, transient potential at feed point and ground impedance of ground rod. Response to typical subsequent stroke current pulse (Grcev Leonid, (2009))

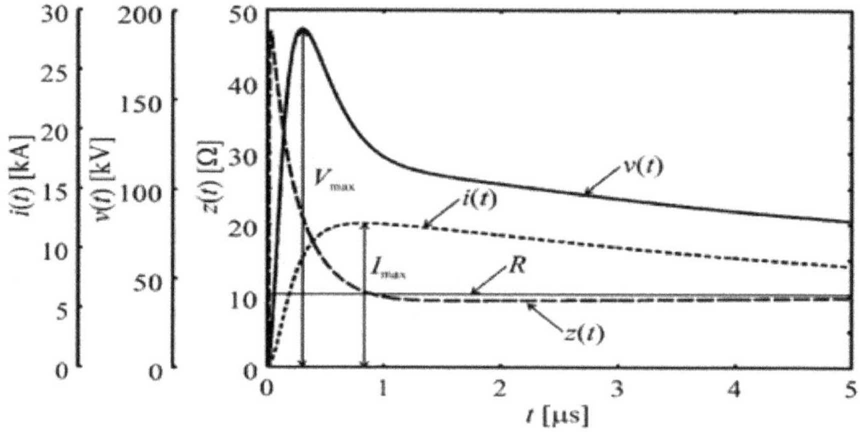

3.2.2 Effect of Frequency Content on the Soil Resistivity

Many researchers studied the effect of frequency on electrical parameters (resistivity and permittivity) of soil (Cavka Damir, Mora Nicolas, and Rachidi Farhad, (2014)) and (Navinesshani Permal, et al., (2020)). Table (1) summarizes models which have been proposed by the researchers for representation effect of frequency content on soil electrical parameters of soil (the electrical resistivity and permittivity). All models are based on laboratory and field measurements (Cavka Damir, Mora Nicolas, and Rachidi Farhad, (2014)). This section reviews the effect of lightning frequency content on electric parameter of soil (resistivity and permittivity). Lightning frequency affects the performance of grounding electrode such as harmonic impedance, transient voltage, transient impedance, and effective length of grounding electrode.

Figure 7. Measurement and simulation of transient voltages to remote ground at the beginning point of horizontal electrode (Grcev Leonid, (2009))

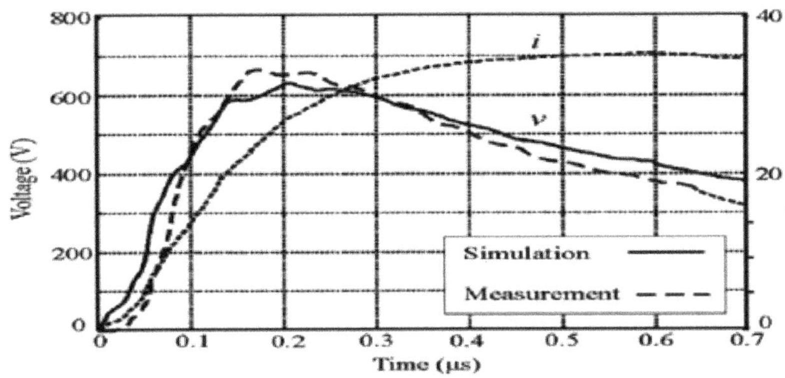

A. Soil Electrical Parameters

This section gives a comparison between the six considered soil models presented. Figures (3.7) and (3.8) present the comparison between soil models results for different values of low-frequency (100 Hz) resistivity 93.46 and 1806 Ω·m (Cavka Damir, Mora Nicolas, and Rachidi Farhad, (2014)). The adopted values correspond to the experimental data of Bigelow and Eberle (Bigelow R. C. and Eberle W. R., (1972)) and He *et al.* (He Jinliang, Zeng Rong, and Zhang Bo, (2013)), which are also plotted in the same figure. Overall, it can be seen that:

The soil models by Scott (S) (J., H., Scott, (1966, May)), (Smith–Longmire (SL) (K., S., Smith, C., L., Longmire, (1975)), Messier (M) (M., A., Messier, (1980)), and (Visacro–Alipio (VA)) predict overall similar results, which are in reasonable agreement with the independent experimental data of Bigelow–Eberle (Bigelow R. C. and Eberle W. R., (1972)) and He et al (He Jinliang, Zeng Rong, and Zhang Bo, (2013)). Differences between the soil models are more significant at high frequencies and for low-resistivity soils.

In general, predictions of the models by Portela and Visacro–Portela differ significantly from those of other considered models.

Table 1. Main characteristics of experimental datasets used for the elaboration of curve-fit expressions considered in this study (Grcev Leonid, (2009))

Curve Fitting Expression	Frequency Range	Measurement Type	No. of Samples
Scott (S)	100 Hz – 1 MHz	Field And Laboratory	105 (With Different Water Content)
Smith and Longmire (SL)	100 Hz – 200 MHz	Based On Scott's and Wilkenfeld's (5 Samples Data)	
Messier (M)	100 Hz – 1 MHz	Based On Scott's Data	
Visacro and Portela	40Hz – 2 MHz	Laboratory	Unknown
Portela (P)	100 Hz – 2 MHz	Laboratory	68
Visacro And Alipio (VA)	100 Hz – 4 MHz	Field	31

Figure 8. Comparison of different soil models (ρ_{LF}= 93.5 Ω · m) (Cavka Damir, Mora Nicolas, and Rachidi Farhad, (2014))

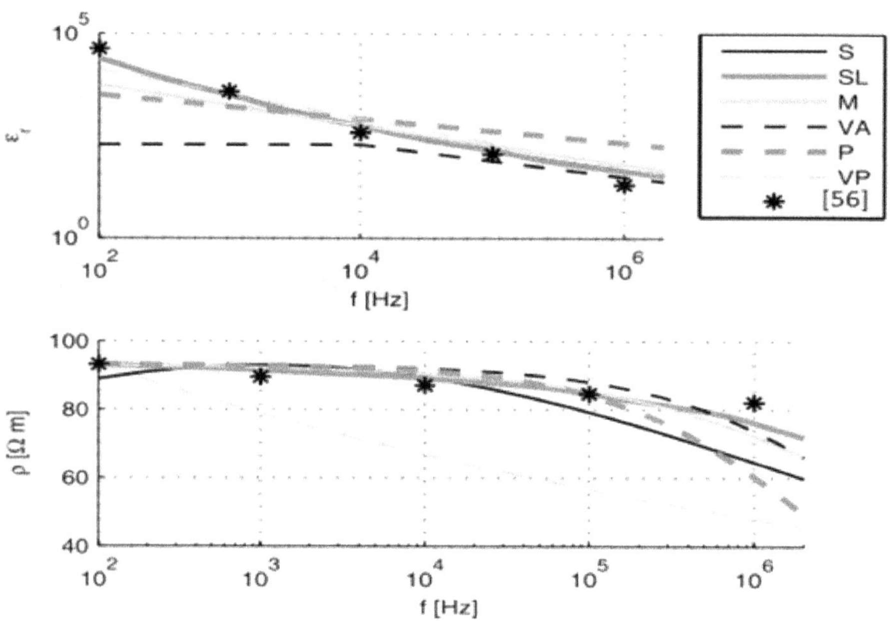

Figure 9. Comparison of different soil models (ρ_{LF}= 1806 $\Omega \cdot m$) (Cavka Damir, Mora Nicolas, and Rachidi Farhad, (2014))

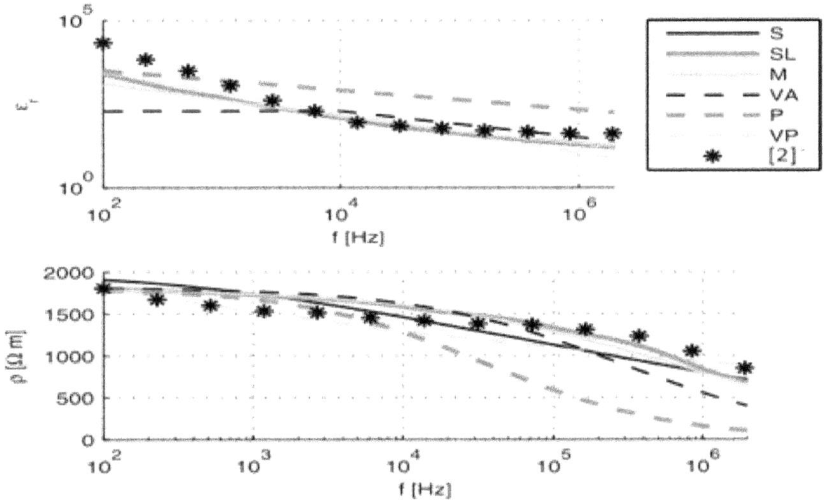

B. Harmonic Impedance

Figure 10 shows harmonic impedance of horizontal grounding electrode. In this case a horizontal grounding electrode and buried at a depth d = 1 m is simulated.

Figure 10. Input impedance for various soil models (L = 10 m; ρ = 100 Ω·m) (Cavka Damir, Mora Nicolas, and Rachidi Farhad, (2014))

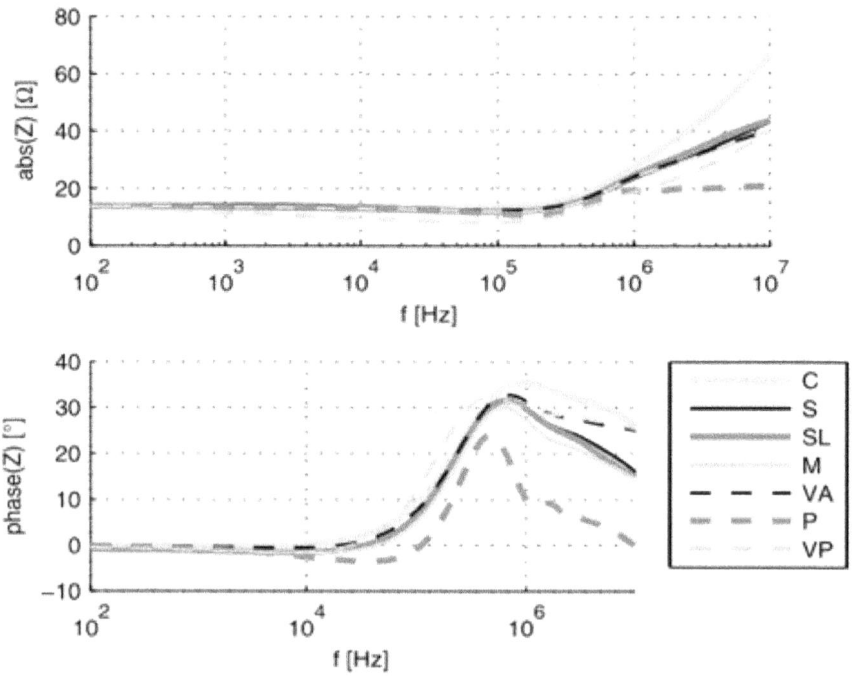

Figure 10 shows the frequency spectrum (magnitude and phase) of the harmonic grounding impedance of the horizontal electrode for various soil models. This figure shows the differences between the predictions of the frequency dependent models and those obtained assuming constant soil parameters (labeled C in all the figures). The results show that the differences between various soil models increase for higher values of the low-frequency resistivity. Also, differences start to appear earlier in the frequency spectrum with the increase of resistivity (Cavka Damir, Mora Nicolas, and Rachidi Farhad, (2014))

C. Transient Ground Potential Rise

Figures 11 and 12 show the transient voltage response in time domain (transient GPR), for first and subsequent return strokes, respectively. It can be noted that the frequency dependence of the soil parameters results in a decrease of the potential of the grounding electrode with respect to the case where the soil parameters are assumed to be constant (Cavka Damir, Mora Nicolas, and Rachidi Farhad, (2014)).

Figure 11. Transient GPR of a horizontal buried electrode for various soil models at (ρ = 100 Ω·m), first return stroke (Cavka Damir, Mora Nicolas, and Rachidi Farhad, (2014))

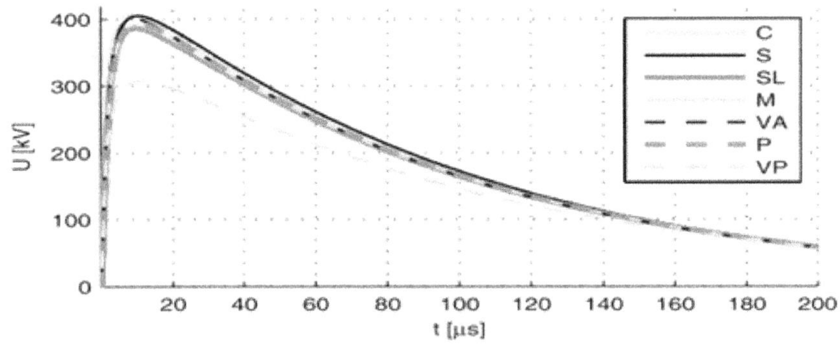

Figure 12. Transient GPR of a horizontal buried electrode for various soil models at (ρ = 100 Ω·m), (b) subsequent return stroke

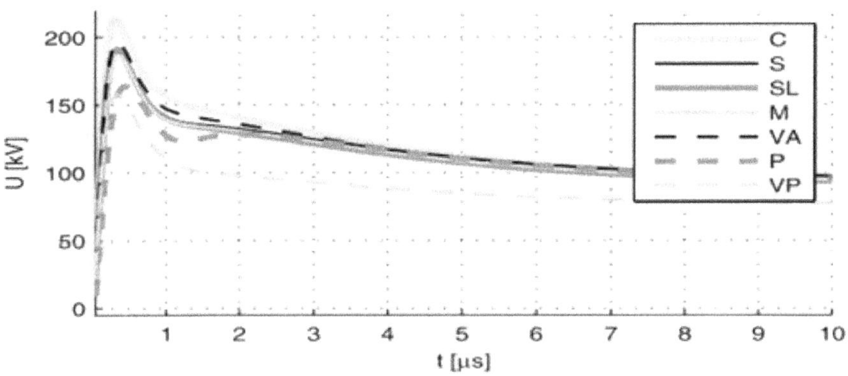

D. Transient Impedance

Figures 13 and 14 show the transient impedance of horizontal grounding electrode when subjected to first and subsequent return stroke current, calculated by various soil models. From these figures, it is seen also that when soil electrical parameters are constant the transient impedance is higher than transient impedance of soil electrical parameters depends on frequency (Yassin S., (2017)).

Figure 13. Transient impedance of horizontal grounding electrode, for first stroke (Yassin S., (2017))

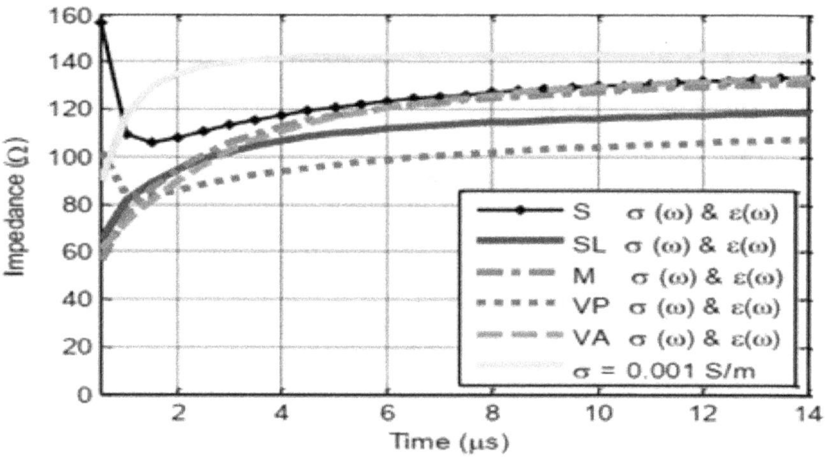

Figure 14. Transient impedance of horizontal grounding electrode for the second stroke (Yassin S., (2017))

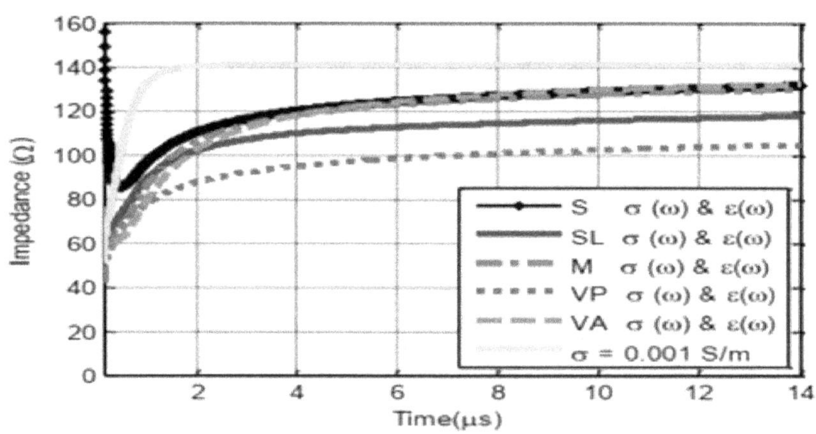

3.2.3 Effect of Soil Ionization

Soil ionization is a nonlinear phenomenon that is created by injecting high currents of lightning stroke into grounding systems. In this section an Overview brief about soil ionization effect and nonlinear phenomena. As early as the nineteenth century, Towne (Towne H.M., (1929)) in 1928 published his work about the non-linearly behavior of soil when the driven rod is subjected to discharge currents with a rise time of 20–30 ms and a peak current of up to 1500 A. It was observed that the V–I curves formed "loops," and the resistances under impulse conditions were lower than 60-cycle values. He ascribed the difference to an "arc of sparks," which expanded the conducting contacts between the earth electrodes and the conducting soils. The soil breakdown phenomenon occurs when the high impulse current discharges into the earth through the grounding conductors (Visacor S., (2007)). Due to the soil breakdown, the soil's resistance and consequently the grounding electrode's peak voltage response decrease. Therefore, the soil breakdown improves the efficiency of the grounding systems. Two main processes have been advanced to explain the increase of soil conduction during high impulse current discharges, namely, (1) thermal heating process and (2) soil ionization process (Mehrdad Mokhtari, Abdul-Malek Zulkurnain, and Gharehpetian Gevork B., (2016)).

A) Thermal Heating Process

In the thermal heating process, the discharged current increases the temperature of the existing water filling among the soil grains. Due to the heating process, the resistivity of heated water decreases (Mehrdad Mokhtari, Abdul-Malek Zulkurnain, and Gharehpetian Gevork B., (2016)).

B) Soil Ionization Process

The air voids and soil grains operate as a mixed dielectric in a soil medium. This is because the conductivity and permittivity of air spaces (air and air) and soil grains (soil and soil) vary. As a result, the displacement vectors at the interface between the air void and soil grain are equal, that is $E_{soil}\varepsilon_{soil} = E_{air}\varepsilon_{air}$ When current discharges into the earth, the resistive paths of the air void and soil grain act as a resistive voltage divider. Because the permittivity of air is substantially higher than that of soil grain, the electric field in the air gap exceeds the breakdown level [64]. Ionized air produces an arc inside the soil as a result of air breakdown. In soil, the arc is fully developed at the current peak. The volume at which the arc is conducting the current in the soil is called the arc channel from zero to peak, arc resistance is a function of current; by which increasing the current will reduce the arc resistance.

The minimum value of the arc resistance is obtained at the current peak (Liew A.C., Darveniza M., (1974)). For the duration of the current decay time, the resistance of the arc is mainly dependent on the energy balance between the arc channel and soil rather than the current magnitude. By decreasing the energy stored in the arc channel, the resistance of the arc increases until the arc is extinguished in the soil. Soil dynamic resistance for any instance of time is obtained as equivalent parallel soil resistance and arc resistance. The soil dynamic resistance changes for the duration of the impulse current since the resistance of the arc is changed. Due to the energy between the arc channel and soil, dynamic soil resistance would show a hysteresis characteristic (Mehrdad Mokhtari, Abdul-Malek Zulkurnain, and Gharehpetian Gevork B., (2016))

The effect of soil ionization is simulated by many methods such as the Finite Difference Time Domain (FDTD) method (Ala, G., et al., (2008), Finite Element (FEM) method (Anton Habjanic, Trlep. Mladen, (2006)) and the Transmission Line (TL) model (Grcev Leonid, (2005), (Navinesshani Permal, et al., (2020)). In this section, the soil ionizations models suggested by CIGRE (CIGRE Working Group 33-01 (Lightning) of Study Committee 33, (1991)), (Bellaschi et al. P.L. Bellaschi, R.E. Armington, A.E. Snowden, (1942)), Nor et al. (Nor, et al., (2012)), and (Liew and Darveniza Wang J., Liew A. C., and Darveniza M., (2005)) are summarized. Table (2.2) summarizes the suggested models' equation, advantages, and disadvantages. This classification is not rigorous as indicated in (Navinesshani Permal, et al., (2020)) but is generally adopted in the literature for fundamental understanding.

Table 2. The equations, advantages, and disadvantages of the suggested models (Navinesshani Permal, et al., (2020))

Soil Ionization model	The equation for ground resistance with soil ionization	Advantages	Disadvantages
CIGRE	$R_i = \frac{R}{\sqrt{1 + i(t)/I_g}}$ Where R is a grounding electrode resistance in [Ω], and $i(t)$ is the impulse current in [kA]. I_g is the limit current at which the soil ionization happens in [kA].	• simple • Easily used in circuit approaches.	• Only valid for a maximum of 30m long electrodes. • In computations, it does not consider the energy balance concept. • cannot show the hysteresis characteristic of grounding electrode resistance.

continued on following page

Table 2. Continued

Soil Ionization model	The equation for ground resistance with soil ionization	Advantages	Disadvantages
Bellaschi	Considers the geometry of the ionized zone as the grounding electrode' new geometry because it considers the arc's resistance to zero. Consequently, the length and radius of the ionized region are used as the effective radius, *ai* and effective length *li* of conductor. The following relation is considered for calculating the effective radius a_i,. $$a_i = \frac{a\, i_{(t)}}{I_c}$$ where *a*, is the conductor radius in [m], i_t is the magnitude of impulse current in [kA], and I_c is the critical current in [kA] where soil ionization occurs at the electrode surface. The relation of a_i to the current is used to calculate the conductor effective length, $l_i$as $$\frac{i_{(t)}}{I_c} = \frac{a_i l_i}{al}$$	• can be used for a multi-electrode system. • Simple and applicable to use in circuit approaches.	• does not consider the energy balance concept in the calculations
Nor	$$R(t) = R_1(i)\left(\frac{R_2(t)}{R_1(t)} + e^{t/\tau_i}\right)$$ Where $\tau_i$is the time constant of ionization. Due to thermal effect and soil capacitive effect, resistance R_1reflects the conduction of the soil in the preionization branch. R_2resistance represents the conduction performance in the post-ionization branch after the ionization is completely expanded	• simple	• does not consider the deionization time. • cannot accurately predict the resistance characteristics for the decay time duration of the impulse current. • Difficult to obtain experimental data to obtain the resistances in the pre- and post-ionization branches. • cannot generate an accurate grounding electrode resistance and its voltage response, unless the experimental data is used to calibrate the model. • does not consider the energy balance concept because the model does not consider the deionization time

continued on following page

Table 2. Continued

Soil Ionization model	The equation for ground resistance with soil ionization	Advantages	Disadvantages
Liew Darveniza	$\rho = \rho_i + (\rho_o - \rho_i) \cdot (1 - e^{t/\tau_2}) \cdot (1 - J/J_c)^2$ Where ρ_i is the lowest soil resistivity value, $\tau_2 = 4.5\ \mu s$, the approximate recovery time. ρ_o is the soil resistivity at the instance of current discharge. t, is the time measured from the ionization onset, J is the current density and J_c is the critical current density. $R = \frac{\rho}{2\pi l} \ln \frac{a+l}{a}$ Where ρ is the soil resistivity in Ω.m, l is the electrode length in [m], and a is the conductor radius in [m].	• considers the concept of energy balance. • exhibits the hysteresis characteristic of the grounding electrode resistance under impulse condition	• More complex. • In the equivalent circuit, the electrode inductance and soil capacitance are ignored. • cause large errors in electrode voltage response under fast and slow fronted currents when the inductive and capacitive characteristics are important.

REFERENCES

Ala, G., Viola, F., Buccheri, P. L., & Romano, P. (2008). Finite difference time domain simulation of earth electrodes soil ionisation under lightning surge condition. *IET Science, Measurement & Technology*, 2(3), 134–145. 10.1049/iet-smt:20070001

Bellaschi, P. L., Armington, R. E., & Snowden, A. E. (1942). Impulse and 60-cycle characteristics of driven grounds-II. *Transactions of the American Institute of Electrical Engineers*, 61(6), 349–363. 10.1109/T-AIEE.1942.5058539

CIGRE Working Group 33-01 (Lightning) of Study Committee 33, (1991), Over voltages and insulation coordination, Guide to procedures for estimating the lightning performance of transmission lines, Paris.

Damir, C., Nicolas, M., & Farhad, R. (2014). A Comparison of Frequency-Dependent Soil Models: Application to the Analysis of Grounding Systems. *IEEE Transactions on Electromagnetic Compatibility*, 56(1), 177–187. 10.1109/TEMC.2013.2271913

Dong-Hyuk Ha, Choi Jong-Kee, & Yoon Yong-Beum, (2019), Parametric Analysis of Lightning Induced Voltage on Distribution Lines using EMTP, 33(3), 19-26.

Grcev, L., & Arnautovski, V. (1998), Comparison between simulation and measurement of frequency dependent and transient characteristics of power transmission line grounding, *Proc. Int. Conf. Lightning Protection*, 524–529.

Habjanic, A., & Trlep, M. (2006). The simulation of the soil ionization phenomenon around the grounding system by the finite element method. *IEEE Transactions on Magnetics*, 42(4), 867–870. 10.1109/TMAG.2006.871625

He, J., Rong, Z., & Bo, Z. (2013). *Methodology and Technology for Power System Grounding*. John Wiley & Sons.

IEEE Standards 1410, (2004), IEEE Guide for Improving the Lightning Performance of Electric Power Overhead Distribution Lines. https://www.nvent.com/en-eg/erico/solutions/lightning-protection

Kosztaluk, R., Loboda, M., & Mukhedkar, D. (1981). Experimental study of transient ground impedances. *IEEE Power Engineering Review*, 1(11), 44–51. 10.1109/MPER.1981.5511930

Leonid, G. (2005). On high-frequency circuit equivalents of a vertical ground rod,||. *IEEE Transactions on Power Delivery*, 20(2), 1598–1603. 10.1109/TPWRD.2004.838460

Leonid, G. (2009). Impulse efficiency of ground electrodes. *IEEE Transactions on Power Delivery*, 24(1), 441–450. 10.1109/TPWRD.2008.923396

Liew, A. C., & Darveniza, M. (1974). Dynamic model of impulse characteristics of concentrated earths. *Proceedings of the Institution of Electrical Engineers*, 121(2), 123–135. 10.1049/piee.1974.0022

Manna, T. K., & Chowdhuri, P. (2007). Generalized equation of soil critical electric field EC based on impulse tests and measured soil electrical parameters. *IET Generation, Transmission & Distribution*, 1(5), 811–817. 10.1049/iet-gtd:20060559

Messier, M. A. (1980), The propagation of an electromagnetic impulse through soil: Influence of frequency dependent parameters, Mission Res. Corp., Santa Barbara, CA, USA, Tech. Rep. MRC-N-415.

Mokhtari, M., Zulkurnain, A.-M., & Gharehpetian, G. B. (2016), A critical review on soil ionization modeling for grounding electrodes, Archives of Electrical Engineering, 449-461.

Nor, . (2012). Comparison between utility sub-station and imitative earthing systems when subjected under lightning response. *International Journal of Electrical Power & Energy Systems*, 43(1), 156–161. 10.1016/j.ijepes.2012.04.042

Permal, N., Osman, M., Kadir, M. Z. A. A., & Ariffin, A. M. (2020). Review of substation grounding system behavior under high frequency and transient faults in uniform soil. *IEEE Access : Practical Innovations, Open Solutions*, 8, 142468–142482. 10.1109/ACCESS.2020.3013657

Rochereau, H. (1988). *Response of earth electrodes when fast fronted currents are flowing out," Response of Earth Electrodes When Fast Fronted Currents are Flowing Out, ser. EDF, B*. Bulletin de la Direction des Études et Recherches.

Schwarz, S. J. (1954), Analytical expression for resistance of grounding systems, AIEE Transactions on Power Apparatus and Systems, 73(13), 1011–1016.

Scott, J. H. (1966, May), Electrical and magnetic properties of rock and soil, Theoretical Notes, Note 18, U.S. Geological Survey. Available: https://www.ece.unm.edu/summa/notes/Theoretical.html

Smith K. S., Longmire C., L., (1975), A universal impedance for soils, Defense Nuclear Agency, Alexandria, VA, USA, Topical Report for Period Jul.

Thanasaksiri, T. (2004), Improving the Lightning Performance of Overhead Distribution Lines. TENCON 2004, IEEE Region 10 Conference, 21-24 Nov., 2004, Chiang Mai, Thailand.

Tom, M. (2004). Lightning Protection Design Workstation. *Electric Power Research Institute, LPDW*, V5, 2004.

Towne, H. M. (1929). Impulse characteristics of driven grounds. *General Electric Review*, 1929, 605–609.

Visacor, S. (2007). A Comprehensive Approach to the Grounding Response to Lightning Currents,‖. *IEEE Transactions on Power Delivery*, 22(1), 381–386. 10.1109/TPWRD.2006.876707

Visacor, S. (2012), How the frequency dependence of soil parameters affect the lightning response of grounding electrodes, *International Conference on Lightning Protection*, 1-4.

Viscaro, S. (2005). *Lightning: an Engineering Approach*. ArtLiber EditSao Paulo.

Viscaro, S. (2007). Direct Strokes to Transmission Lines: Considerations on the mechanisms of Overvoltage Formation and their Influence on the Lightning Performance of Lines. *Journal of Lightning Research*, 1, 60–68.

Wang, J., Liew, A. C., & Darveniza, M. (2005). Extension of dynamic model of impulse behavior of concentrated grounds at high currents. *IEEE Transactions on Power Delivery*, 20(3), 2160–2165. 10.1109/TPWRD.2004.839645

Yassin, S. (2017), Performance of Grounding Grids at Normal and Abnormal Conditions, Master thesis, South Valley University, Qena, Egypt.

Chapter 4
Grounding Grid Performance Under Fault Current

Osama El-Sayed Gouda
Faculty of Engineering, Cairo University, Egypt

ABSTRACT

This chapter discusses the following: TT system (earthed neutral), automatic disconnection for TT system, TN system (exposed conductive parts connected to the neutral), TN-C system, TN-S system- TN-C-S system, IT system (isolated or impedance-earthed neutral), IT system (impedance-earthed neutral), grounding transformer connection and design, grounding of industrial and commercial generators, Zigzag grounding transformer earthed systems (solid grounding, resistance grounding, reactance grounding, resonant grounding, ground-fault neutralizer), location of system grounding points selection, and grounding locations specified by the NEC and multiple power sources.

4.1 GROUND FAULT SCENARIO

As a condition of a grounding system study, a fault scenario is assumed to occur that provides ground current through the system under investigation, energizing the bonded and grounded components. According to symmetrical components concept, the ground current exists for specific fault types with a zero-sequence component which passes through the grounding system and through the earth path. As it is

DOI: 10.4018/979-8-3693-2969-6.ch004

known the zero sequence components are present in the cases of single-line-to-ground (SLG) and double-line-to-ground (DLG) faults:

As it is shown in the illustration given in Figure 1 the SLG fault current flows through the grid causes ground potential rise (GPR) and this voltage could result in hazardous touch and step shock hazards. As expected, the faulted site experiences GPR due to the impedance of the grounding system and the fault current returning to its remote source.

'There are three primary aspects of fault data that contribute to the grounding system evaluation (Easy *power, Grounding, (on line))*:

- The magnitude of fault current
- Time of fault duration
- Power factor of fault or X/R ratio

Each fault scenario may have unique factors that influence the fault data as it is incorporated into a grounding study.

Figure 1. Electric shocks due to SLG fault of a T.L. (Easy power, Grounding, (on line))

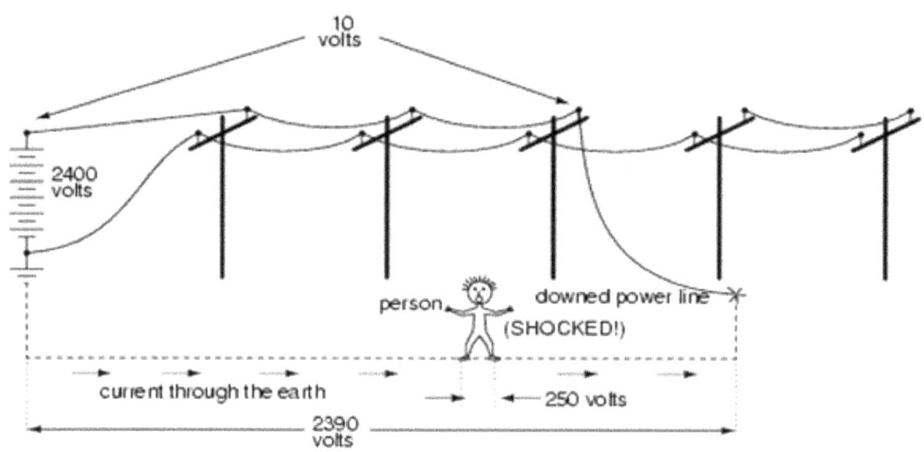

A. The Magnitude of Fault Current

The fault current magnitude depends on the power system interconnection and voltage levels associated with the fault. Grounding studies can consider faults at various voltage levels of the site under analysis to determine the maximum ground current. The maximum ground current will produce the largest GPR, which may be the worst-case for evaluating touch and step voltages. The largest touch and step voltages from the maximum ground fault current may not be the worst-case condition, as the duration of the fault affects the applicable voltage limits *(*Easy *power, Grounding, (on line)).*

The zero sequence reactance is typically lower than the positive and negative sequence generator reactance which determines three phase and line to line fault currents, the first cycle (symmetrical) line to ground bolted fault current at the terminals of a solidly grounded generator is given by the equation:

$$I_G = \frac{3E}{x1 + x2 + x0} \tag{4.1}$$

X1 is the direct axis sub transient reactance, X2 is the negative sequence reactance and X0 is the zero sequence reactance.

B. Time of Fault Duration

The time duration for a fault current to pass depends on the operation time of the protective devices and circuit breakers of the power systems. Standard specifications like IEEE Std 80 advised the use of primary and backup protection systems within specific times, which helps in the design and behavior of grounding networks during faults.

C. Power Factor of the Fault Current or X/R Ratio

A power system's inductive and resistive impedances may significantly affect the fault current characteristics. Power systems with greater X/R ratio produce a greater DC offset in the event of a fault. A classic visual representation of the DC offset on an AC curve is shown below.

The DC offset may increase the peak current considerably and increases the cumulative fault current through various paths. Regarding a grounding analysis for personnel safety, the cumulative current through the human body could result in fatal touch and step shock hazards and should consider this DC offset's effect. IEEE Std 80 provides a calculation method for considering the cumulative effect of the X/R ratio, described as the multiplying (Df).

The multiplying factor incorporates the X/R ratio, the frequency, and duration of the fault. Essentially this X/R multiplying factor is a method to equate the DC offset as an average fault current, thus resulting in greater touch and step voltage hazards. This multiplying factor equals 1.23 for a fault cleared in 0.1 seconds with an X/R that is equal to 20. The multiplying factors for the values of X/R for different switchgear contact-separation times, in ANSI Standard C37.5–1979are as given in graphs of Figure 2.

D. Split of the Fault Current

During ground fault events, a part of the fault current will take alternative paths that affect the GPR values in the area surrounding the grounding system. Determining the split of the fault current provides the percentage of fault current that goes through the grid producing a GPR and the portion that takes alternative paths reducing the maximum GPR.

Considering the fault current split allows for a more accurate analysis and more efficient grounding system design. Alternative paths often include a transmission line's overhead wires, distribution neutral wires, and cable shielding and armor. There are several methods for calculating the fault current split, but a common simplified approach is to calculate the equivalent impedance of the alternative paths and enter those Req and Xeq values into the XGSLab split factor tool. Another simplified approach is provided in the IEEE Std 80 Annex C graphical curves, where an engineer can quickly approximate the fault current split with the number of transmission and distribution lines connected to the grid under analysis. More thorough investigation is performed by software, such as NETS, that use the phase component method, which can model a complicated meshed power system such as the two stations with an interconnected transmission line and cables feeding a central site.

The NETS model below shows that of the 2.22 kA phase A fault, only 691 A will produce a GPR (through e) with the remaining taking the cable shields paths (D,E,F). Figure 2 gives the Multiplying factors for three-phase and line-to-earth faults according to IEEE (IEEE Standard C37.5-1979)

Figure 2. Multiplying factors for three-phase and line-to-earth faults (total current rating basis) (IEEE Standard C37.5-1979; reproduced by permission of the IEEE)

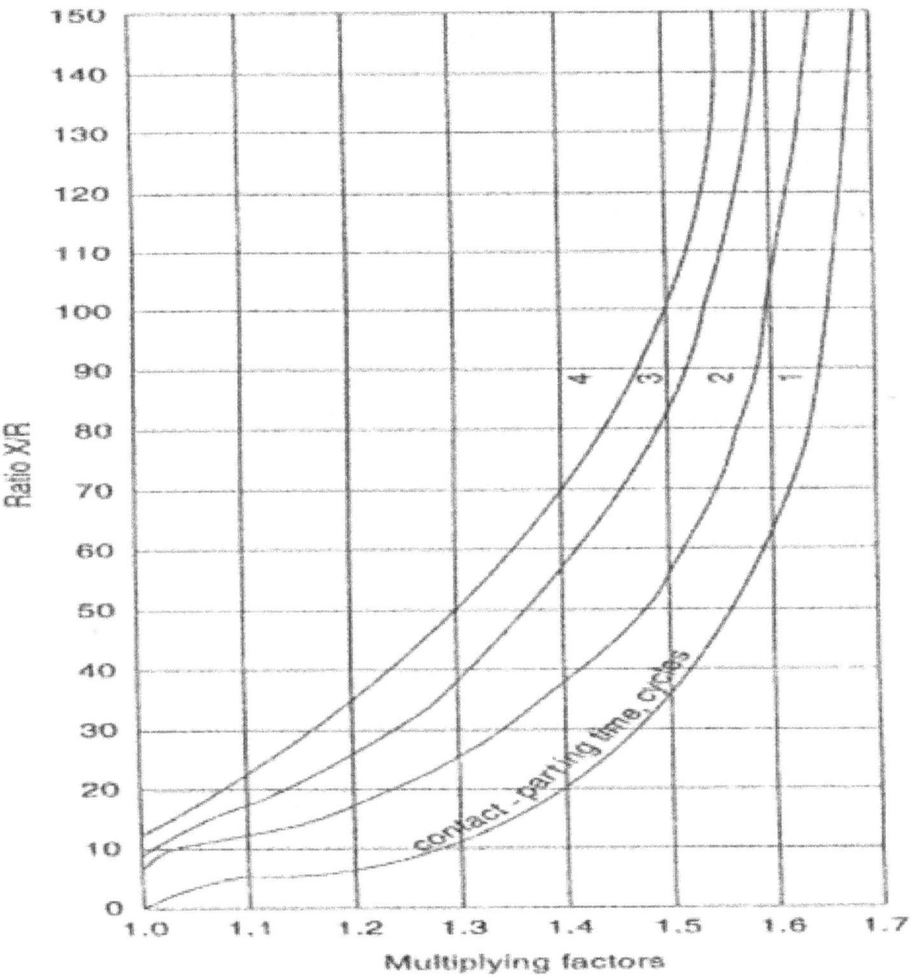

Simplified approaches, especially referencing Annex C of IEEE, require engineering judgment to determine if the assumptions included in the methods are applicable to the grounding system under investigation.

With many engineering studies, the quality of data inputs dramatically affects the quality of the output. Understanding the multiple aspects of the fault current data for a grounding system analysis can guide more accurate and efficient designs to reduce personnel and public shock hazards.

Dc Offset is the difference between the symmetrical current wave and the actual current wave during a power system transient condition, see Figure 3. Mathematically, the actual fault current can be broken into two parts, a symmetrical alternating component and a unidirectional (dc) component. The unidirectional component can be of either polarity, but will not change polarity, and will decrease at some predetermined rate.

Figure 3. The difference between the symmetrical current wave and the actual current wave during a power system transient condition (Easy power, Grounding, (on line))

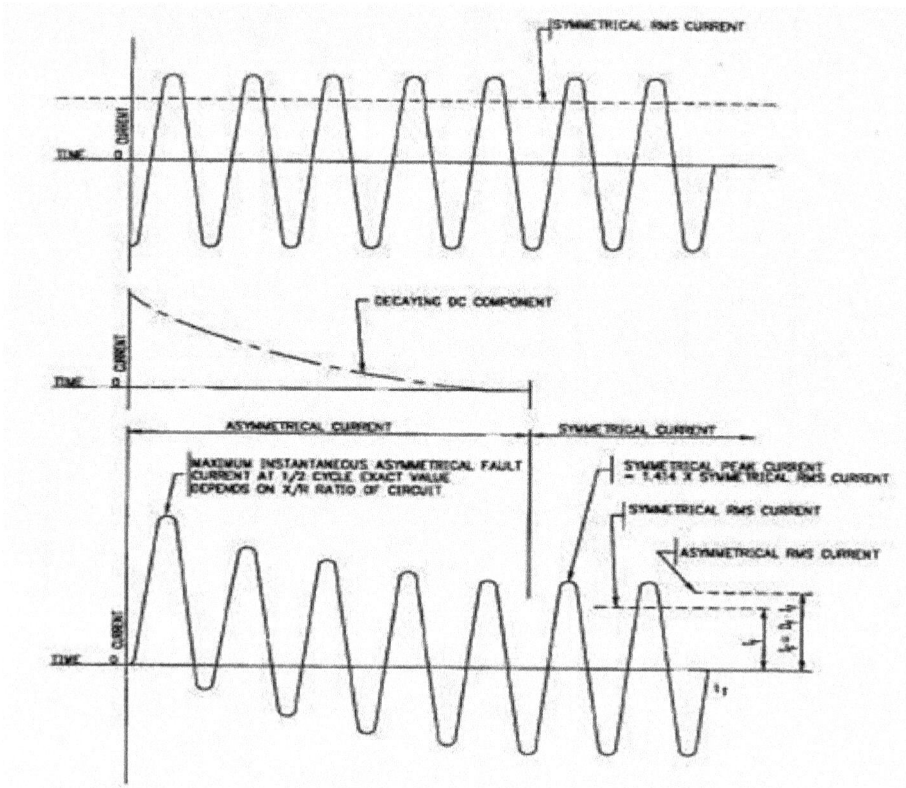

Relationship between actual values of fault current and values of I_F, I_f, and D_f for fault duration t_f

4.2 GROUNDING CONNECTIONS ACCORDING TO STANDARDS

International standards (IEC 60364) clearly define the various elements of grounding connections. The different grounding schemes (often referred to as the type of power system or system grounding arrangements) described characterize the method of grounding the installation downstream of the secondary winding of a MV/LV transformer and the means used for grounding the exposed conductive-parts of the LV installation supplied from it. The following terms are definitions of standardized grounding schemes.

4.2.1 TT System (Earthed Neutral)

One point at the supply source is connected directly to earth. All exposed- and extraneous-conductive-parts are connected to a separate earth electrode at the installation. This electrode may or may not be electrically independent of the source electrode. The two zones of influence overlap without affecting the operation of protective devices (Gouda Osama El-Sayed, (2017)).

TT system is technique for the protection of persons. The exposed conductive parts are earthed, and residual current devices (RCDs) are used. It is also used in the interruption for the first insulation fault.

The TT system is the simplest solution to design used in installations supplied directly by the public LV distribution network. It does not require continuous monitoring during operation (a periodic check on the RCDs may be necessary). The protection is ensured by special devices, the residual current devices (RCD), which also prevent the risk of fire when they are set to ≤ 500 mA. Each insulation fault results in an interruption in the supply of power, however the outage is limited to the faulty circuit by installing the RCDs in series (selective RCDs) or in parallel (circuit selection). Loads or parts of the installation which, during normal operation, cause high leakage currents, require special measures to avoid nuisance tripping, i.e. supply the loads with a separation transformer or use specific RCDs. (Figures 4 and 5) give the characteristics of TT system as reported by Wiki Electrical Installation and Schneider (2016)

Figure 4. Characteristics of TT system

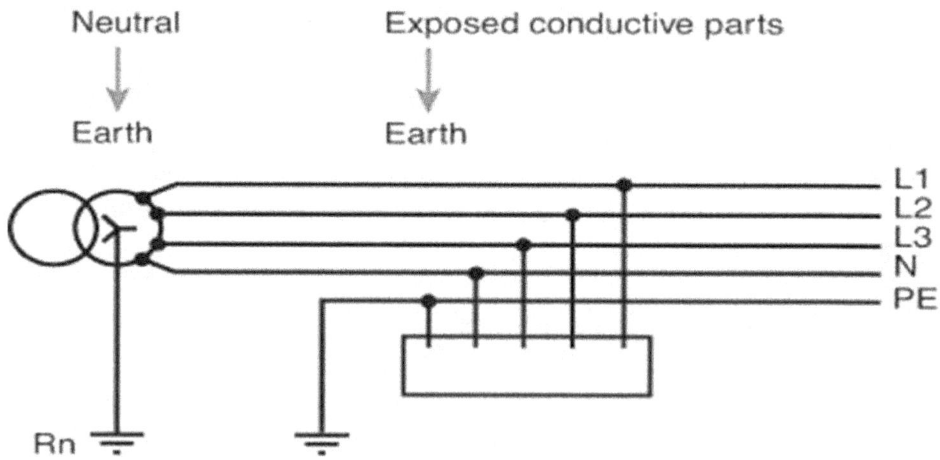

Figure 5. Main characteristics of TT system

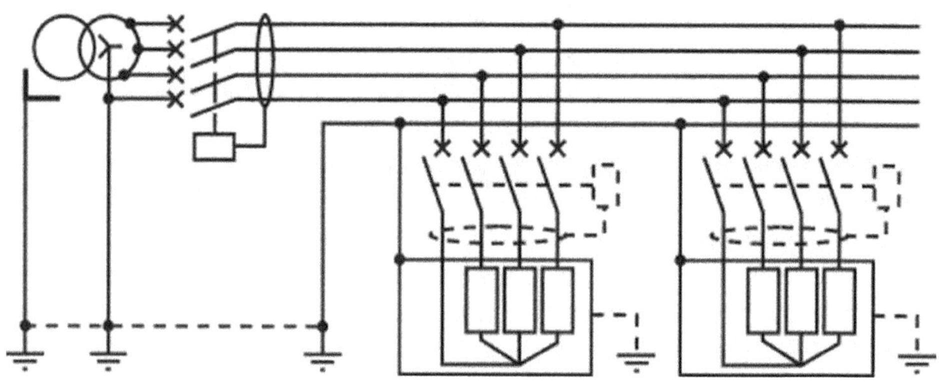

Automatic disconnection for TT system is achieved by RCD having a sensitivity of $\mathbf{I_{\Delta n}}$ is less than or equal 50/ $R_{A,}$ where R_A is the resistance of the earth electrode for the installation, $I_{\Delta n}$ is the rated residual operating current of the RCD. For temporary supplies (to work sites,) and agricultural and horticultural premises, the value of 50 V is replaced by 25 V. An example of TT system is given in (Figures 4 and 5), in this figure, the resistance of the earth electrode of substation neutral R_n is 10 Ω. The resistance of the earth electrode of the installation R_A is 20 Ω. The earth-fault loop current I_d = 7.7 A and the fault voltage $U_f = I_d \times R_A$ = 154 V and therefore dangerous, but $I_{\Delta n}$ = 50/20 = 2.5 A so that a standard 300 mA RCD will operate in about 30 ms without intentional time delay and will clear the fault where a fault voltage exceeding appears on an exposed-conductive-part. The upper limit of resistance for an installation grounding electrode which must not be exceeded, for given sensitivity levels of RCDs at U_L voltage limits of 50 V and 25 V. Figure 6 gives an example of TT system as reported by (Wiki Electrical Installation and Schneider (2016))

Figure 6. Example of TT system

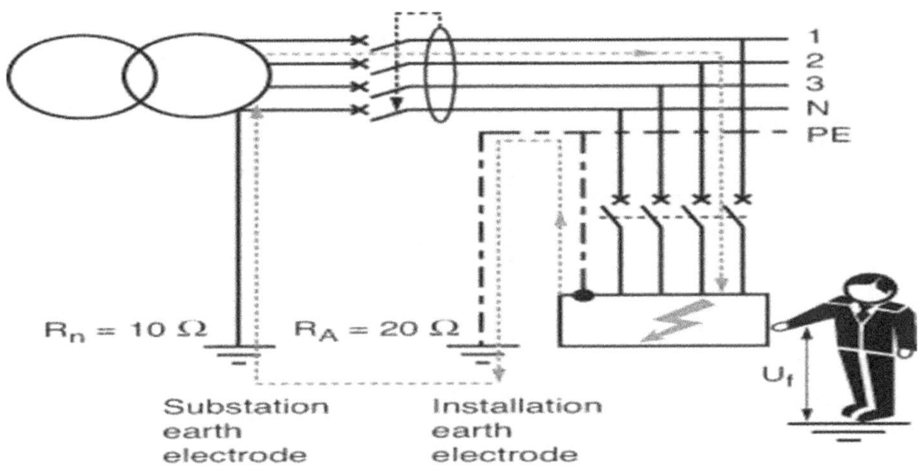

4.2.2 TN System (Exposed Conductive Parts Connected to the Neutral)

The several versions of TN systems are shown below:

4.2.2.1 TN-C System

The neutral conductor is also used as a protective conductor and is referred to as a PEN (Protective Earth and Neutral) conductor. This system is not permitted for conductors of less than 10 mm^2 or for portable equipment. The TN-C system requires an effective equipotential environment within the installation with dispersed earth electrodes spaced as regularly as possible since the PEN conductor is both the neutral conductor and at the same time carries phase unbalance currents as well as 3rd order harmonic currents and their multiples. The PEN conductor must therefore be connected to a number of earth electrodes in the installation. Caution: In the TN-C system, the "protective conductor" function has priority over the "neutral function". In particular, a PEN conductor must always be connected to the grounding terminal of a load and a jumper is used to connect this terminal to the neutral terminal. Example of TN-C system is given in (Figure 7) as reported by (Wiki Electrical Installation and Schneider (2016))

Figure 7. Example of TN-C system

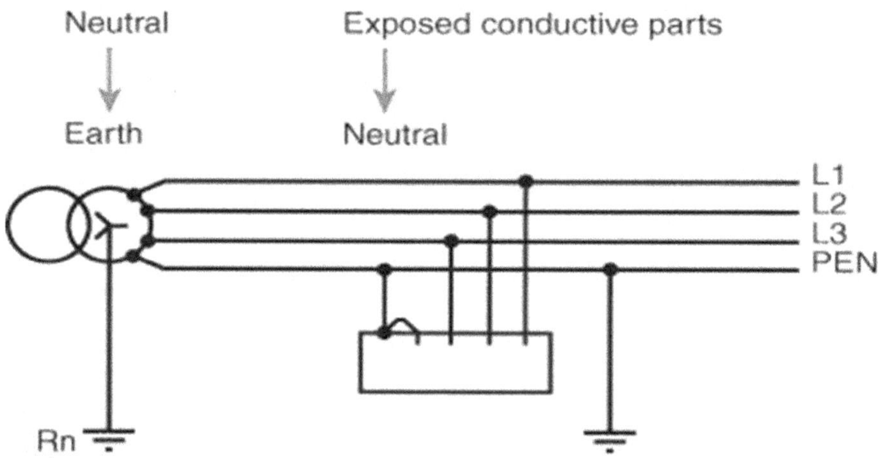

4.2.2.2 TN-S System

The TN-S system (5 wires) is obligatory for circuits with cross-sectional areas less than 10 mm^2 for portable equipment. The protective conductor and the neutral conductor are separate. On underground cable systems where lead-sheathed cables exist, the protective conductor is generally the lead sheath. The use of separate PE and N conductors (5 wires) is obligatory for circuits with cross-sectional areas less than 10 mm^2 for portable equipment. Example of TN-S system is drawn in (Figure 8) as reported by (Wiki Electrical Installation and Schneider (2016))

Figure 8. Example of TN-S system

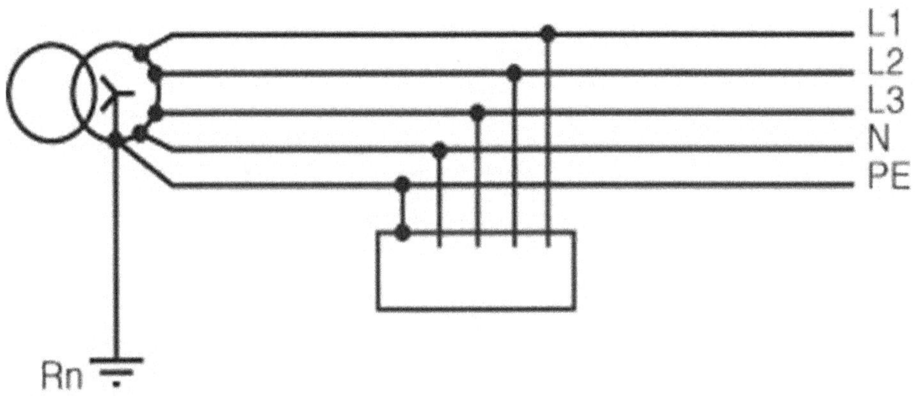

4.2.2.3 TN-C-S System

The TN-C and TN-S systems can be used in the same installation. In the TN-C-S system, the TN-C (4 wires) system must never be used downstream of the TN-S (5 wires) system, since any accidental interruption in the neutral on the upstream part would lead to an interruption in the protective conductor in the downstream part and therefore a danger. Example of TN-C-S system is given in (Figure 9) and connection of the PEN conductor in the TN-C system is shown in (Figure 10) as reported by (Wiki Electrical Installation and Schneider (2016))

Figure 9. Example of TN-C-S system

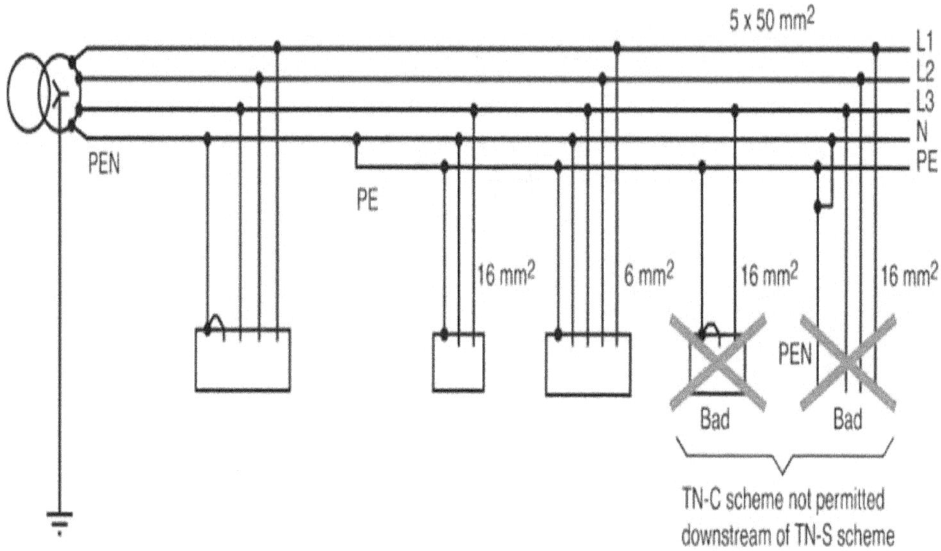

Figure 10. Connection of the PEN conductor in the TN-C system

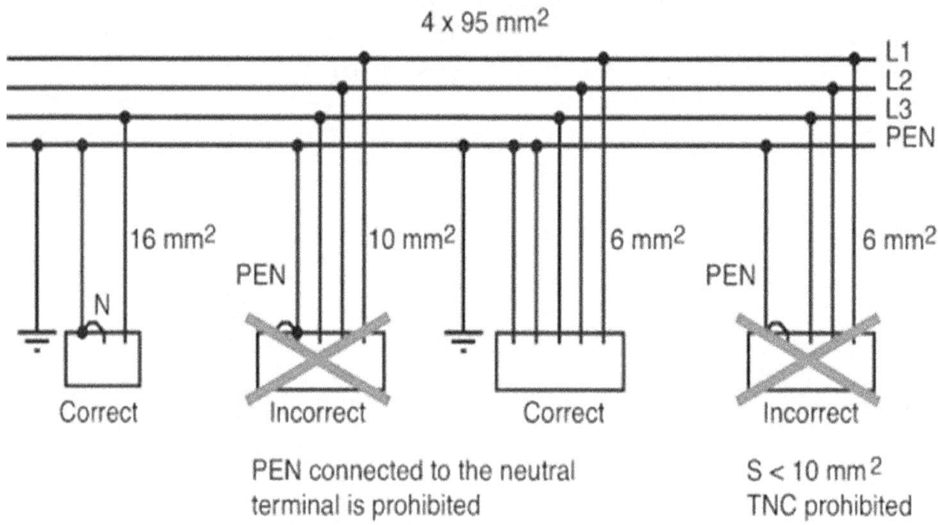

4.2.2.4 Characteristics of TN System

The TN system is used for the protection of persons, interconnection and grounding of exposed conductive parts and the neutral are mandatory and interruption for the first fault using over current protection (circuit- breakers or fuses). The operating technique is for the interruption for the first insulation fault.

4.2.3 Methods of Determining Levels of Short-Circuit Current

In TN-earthed systems, short-circuit to earth will, in principle; always provide sufficient current to operate an over current device. The source and supply mains impedances are much lower than those of the installation circuits, so that any restrictions in the magnitude of earth-fault currents will be mainly caused by the installation conductors, i.e., long wire loops lead to appliances greatly increase the "fault-loop" impedance, with a corresponding reduction of short-circuit current (Gouda Osama El-Sayed, (2017)).

Three practical methods are:

a) The method of impedances which based on the summation of all the impedances (positive-phase-sequence only) around the fault loop, for each circuit
b) The method of composition which is an estimation of short-circuit current at the remote end of a loop, when the short-circuit current level at the near end of the loop is known
c) The conventional method that is based on calculating the minimum levels of earth-fault currents, together with the use of tables of values for obtaining rapid results

These methods are only reliable for the case in which the cables that make up the earth-fault-current loop are in close proximity (to each other) and not separated. Three methods of calculation are commonly used:

a) The method of impedances, based on the trigonometric addition of the system resistances and inductive reactance
b) The method of composition
c) The conventional method, based on an assumed voltage drop and the use of prepared tables

4.2.3.1 Method of Impedances

This method summates the positive-sequence impedances of each item (cable, conductor, transformer, etc.) included in the earth-fault loop circuit from which the short-circuit earth-fault current is calculated, using the formula:

$$I = Uo / ((\Sigma R)^2 + (\Sigma X)^2)^{0.5} \qquad (4.1)$$

Where
$(\Sigma R)^2$ = (the sum of all resistances in the loop)^2 at the design stage of a project.
And $(\Sigma X)^2$ = (the sum of all inductive reactance in the loop)^2
and Uo = nominal system phase-to-neutral voltage. The application of the method is not always easy, because it supposes knowledge of all parameter values and characteristics of the elements in the loop. In many cases, a national guide can supply typical values for estimation purposes.

4.2.3.2 Method of Composition

This method permits the determination of the short-circuit current at the end of a loop from the known value of short-circuit at the sending end, by means of the approximate formula:

$$Isc = I\,(Uo/ (U + Zs\ Isc)) \qquad (4.2)$$

Where
Isc = upstream short-circuit current, ***I*** = end-of-loop short-circuits current
Uo = nominal system phase voltage and ***Zs*** = impedance of loop
Note: In this method the individual impedances are added arithmetically as opposed to the previous "method of impedances" procedure.
In this method the resultant calculated current is less than that it would actually flow. If the over current settings are based on this calculated value, then operation of the relay, or fuse, is assured.

4.2.3.3 Conventional Method

The maximum length of any circuit of a TN-earthed installation is:

$$Lmax = (0.8\ Uo\ Sph) / (\rho\ (1+ m)\ Ia) \qquad (4.3)$$

Where: $m= Sph / SPE$, SPE = cross-sectional area of the protective conductor concerned in mm^2.

Lmax = maximum length in meters, Sph = cross-sectional area of the phase conductors of the circuit concerned in mm^2 Uo = phase volts = 230 V for a 230/400 V system, ρ = resistivity at normal working temperature in ohm-mm^2/meter = 22.5 10^{-3} for copper; = 36 10^{-3} for aluminum. Ia = trip current setting for the instantaneous operation of a circuit-breaker, or Ia = the current which assures operation of the protective fuse concerned, in the specified time (Gouda Osama El-Sayed, (2017)).

This method is generally considered to be sufficiently accurate to fix the upper limit of cable lengths. The principle of this method is based on the assumption that the voltage at the origin of the circuit concerned (i.e., at the point at which the circuit protective device is located) remains at 80% or more of the nominal phase to neutral voltage. 80% value is used, together with the circuit loop impedance, to compute the short-circuit current. This coefficient takes account of all voltage drops upstream of the point considered. In LV cables, when all conductors of a 3-phase 4-wire circuit are in close proximity (which is the normal case), the inductive reactance internal to and between conductors is negligibly small compared to the cable resistance.

This approximation is considered to be valid for cable sizes up to 120 mm^2. (Figure 11) gives how to calculate L max for a TN-earthed system, using the conventional method.

Figure 11. Calculation of Lmax for TN-earthed system, using the conventional method

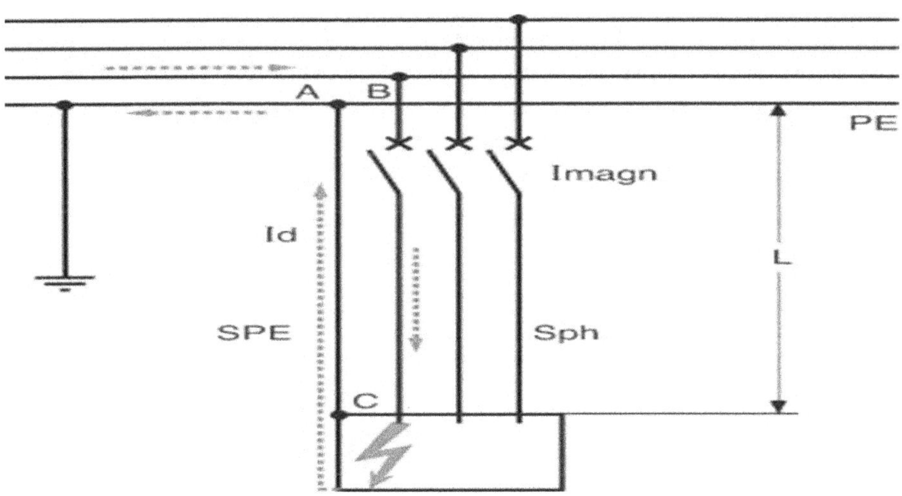

4.2.4. IT System (Isolated or Impedance-Earthed Neutral)

In IT system there is no connection between the neutral point of the supply source and safety grounding.

Figure 12. Example of IT system (isolated neutral)

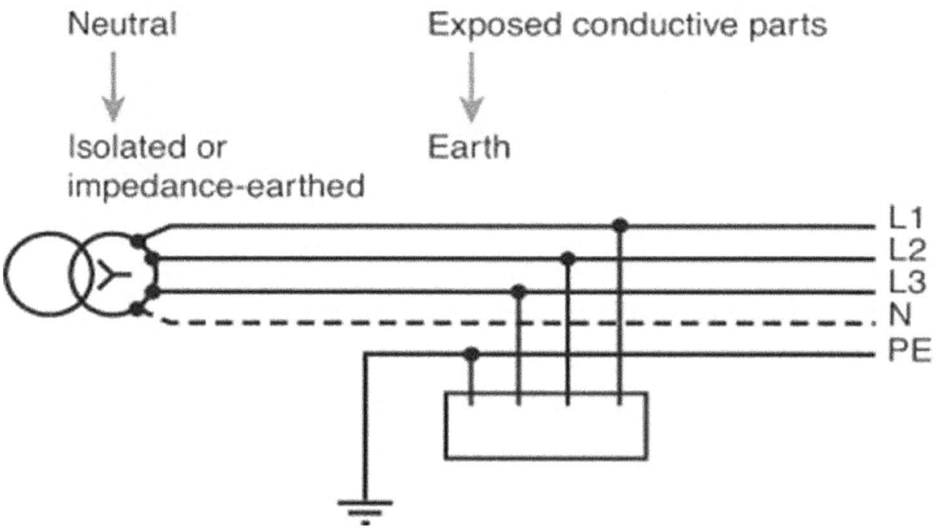

In IT system it is recommended to connect the conductive-parts of the installation with an earth electrode. In practice all circuits have leakage impedance to earth since no insulation is perfect. In parallel with this (distributed) resistive leakage path, there is the distributed capacitive current path, the two paths together constituting the normal leakage impedance to earth. Figure 12 gives an example of IT system (isolated neutral) as reported by (Wiki Electrical Installation and Schneider (2016))

Figure 13. IT system (isolated neutral)

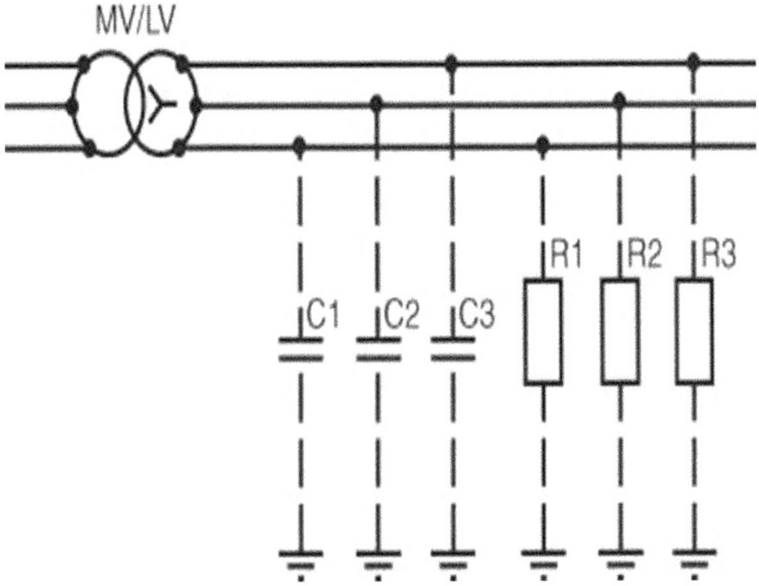

In a LV 3-phase 3-wire system, 1 km of cable will have leakage impedance due to C1, C2, C3 and R1, R2 and R3 equivalent to neutral earth impedance Zct of 3,000 to 4,000 Ω, without counting the filtering capacitances of electronic devices. See Figure 13.

Figure 14. Impedance equivalent to leakage impedances in an IT system

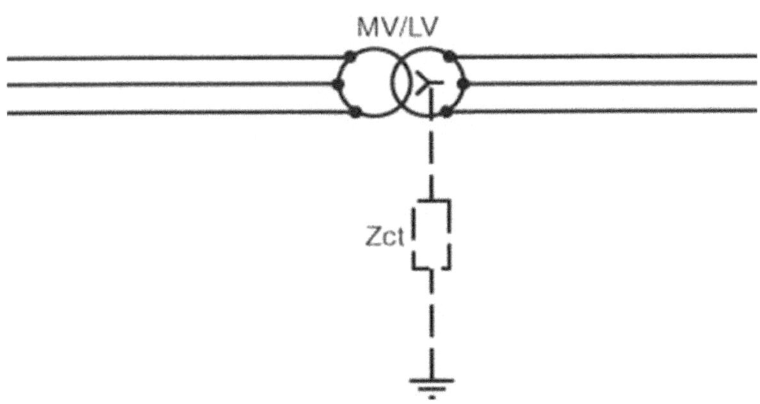

Figure 14 illustrates the impedance equivalent to leakage impedances in an IT system. The Protection technique of IT system is as follows (Gouda Osama El-Sayed, (2017)):

- Interconnection and grounding of exposed conductive parts
- Indication of the first fault by an insulation monitoring device (IMD)
- Interruption for the second fault using overcurrent protection (circuit-breakers or fuses). The operating technique of IT system is as follows:
- Monitoring of the first insulation fault.
- Mandatory location and clearing of the fault.
- Interruption for two simultaneous insulation faults.

Figure 15. IT system main characteristics

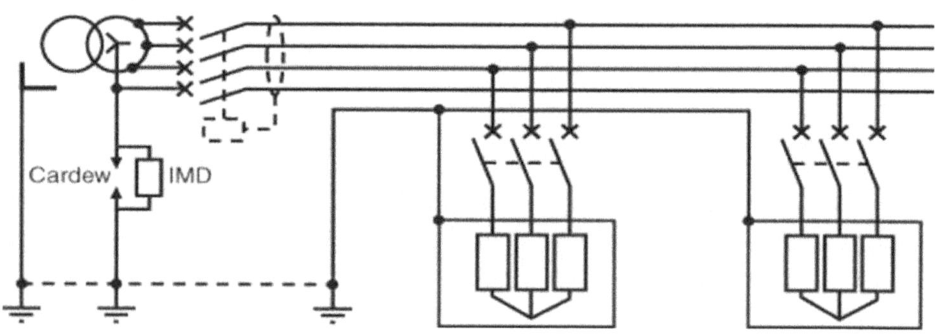

- Solution offering the best continuity of service during operation
- Indication of the first insulation fault, followed by mandatory location and clearing, ensures systematic prevention of supply outages
- Generally used in installations supplied by a private MV/LV or LV/LV transformer
- Requires maintenance personnel for monitoring and operation
- Requires a high level of insulation in the network (implies breaking up the network if it is very large and the use of circuit-separation transformers to supply loads with high leakage currents)
- The check on effective tripping for two simultaneous faults must be carried out by calculations during the design stage, followed by mandatory measurements during commissioning on each group of interconnected exposed conductive parts

Figure 15 gives the IT system main characteristics and Table 1 gives the IT system - Preliminary conditions.

Table 1. IT system - preliminary conditions

Minimum functions required	Components and devices
Protection against over voltages at power frequency	Voltage limiter
Neutral grounding resistor (for impedance grounding variation)	Resistor
Overall earth-fault monitor with alarm for first fault condition	Permanent insulation monitor PIM with alarm feature
Automatic fault clearance on second fault and protection of the neutral conductor against overcurrent	Four-pole circuit-breakers (if the neutral is distributed) all 4 poles trip
Location of first fault	With device for fault-location on live system, or by successive opening of circuits

Figure 16 shows the position of each protection device of IT system.

Figure 16. Positions of essential protection devices

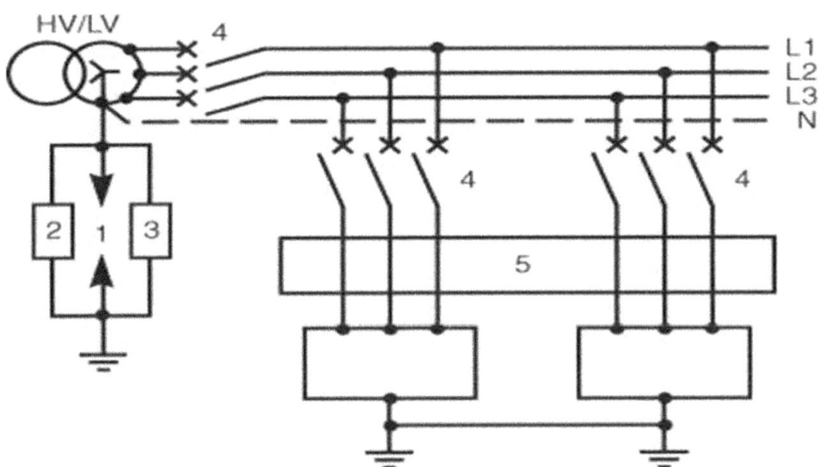

4.3 LOW VOLTAGE GROUNDING

The most dangerous circumstance is when a break occurs on the neutral service wire entering consumer's premises, for if the appliance is switched on, the full phase voltage will appear on the metalwork, although the appliance will not operate.

The danger can be minimized by connecting the framework of the appliance to the earth impedance on the consumer's premises as well, the resistance of which should be as low as possible, as the lower it is, so the lower will be the voltage appearing on the framework should a break occur on the neutral (Gouda Osama El-Sayed, (2017)).

The following points should be taken into consideration during the design of a grounding system:

1- The size of the neutral distribution should not be less than that of the phase wire.
2- The neutral should be connected to earth at consumer's terminals and the earth electrode consisting of single copper or mild- steel rod or pipe with 1.8 meter long
3- Water mains should not be used for grounding
4- An earth conductor should be used throughout each installation and should be connected to the neutral conductor and grounding lead at the meter board.
5- No switch or fuse should be used in the neutral wire throughout the installation
6- The resistance of the earth conductor from an appliance to the meter board should not exceed 0.5 ohm and this should be verified in each installation by means of a current of at least 10 A flowing for five minutes before protective multiple grounding is used.
7- Two separate grounding systems should be provided for grounding in the following cases:

a) The high voltage neutral network and any metal work of low or high voltage systems
b) The low voltage system neutral point and the network of the low voltage system. This should be so situated that the low voltage earth electrode is not within the resistance area of high voltage earth electrode. If this precaution is not taken then a breakdown on the high voltage system causing current to flow through the high voltage electrode would result in the potential rise of the high voltage earth electrode to be communicated to the neutral of the low voltage and all apparatus frames connected to the network.

8- Earth continuity conductor through the factory, connected to each item of plant to be protected. This conductor should then be continued, preferably underground to the substation, and connected directly to the neutral terminal of the low voltage winding of the substation transformer. This means the fault current return path is provided independent of the neutral wire, which will still be required to carry the out-of-balance current of the supply network. The earth return wire will carry no load current under normal circumstances but will carry all fault currents and provide short circuit conditions in the case of earth faults, thus the operation of protective equipment is insured even when a very large capacity plant is used.

4.4 METHODS OF SYSTEM NEUTRAL GROUNDING

Most grounded systems employ some method of grounding the system neutral at one or more points. These methods can be divided into two general categories: solid grounding and impedance grounding. Impedance grounding may be further divided into several subcategories: reactance grounding, resistance grounding, and ground-fault neutralizer grounding. Each method, as named, refers to the nature of the external circuit from system neutral to ground rather than to the degree of grounding. In each case the impedance of the generator or transformer who's neutral is grounded is in series with the external circuit. Thus, a solidly grounded generator or transformer may or may not furnish effective grounding to the system, depending on the system source impedance.

4.4.1. Isolated System

Many standards have discussed grounding of an electrical distribution system. The electrical distribution system may or may not include a reference to ground. In those instances, allowed by the National Electrical Code, some designers view the opportunity to use an ungrounded system as a way of providing higher reliability than would a grounded system. The thinking is that since a ground would not create a fault, an ungrounded system would therefore provide higher system reliability.

That must be realized, however, is that even if no current carrying conductor is intentionally grounded, current still flows from any or all phase conductors to ground. This is done through phase to ground capacitance. While not enough current flows to cause damaging fault currents, it does provide enough current to create a unique overvoltage problem due to the particular way that intermittent ground faults operate an ungrounded system. This capacitance becomes part of our circuit that already includes cable inductance and resistance.

In an ungrounded system, there is no intentional connection between the system conductors and ground. However, there always exists a capacitive coupling between one system conductor and another, and also between system conductors and ground. Consequently, the so-called ungrounded system is in reality a capacitance grounded system, by virtue of the distributed capacitance from the system conductors to ground. Since the capacitance between phases has little effect on the grounding characteristics of the system, it will be disregarded. For simplicity, the distributed capacitive reactance to ground is assumed to be balanced. In an un-faulted condition, with balanced three-phase voltages applied to the lines, the capacitive charging current, *I*co, in phase will be equal and displaced 120° from one another. The phase voltages to ground will also be equal and displaced 120° from one another.. Since the neutral of the distributed capacitances is at earth potential, it follows that the neutral of the transformer is also at earth potential, being held there by the capacitance to ground (Gouda Osama El-Sayed, (2017)).

If the phase conductor C for example, faults to ground, current flow through that capacitance to ground will cease, since no potential difference across it now exists.

The voltage across the remaining two distributed capacitors to ground will, however, increase from line to neutral to line to line. The capacitive charging current, *I*co, in the two healthy phases will therefore increase by the square root of 3 and the line-to-ground voltages are no longer 120°, but 60° apart.

Hence, the vectorial sum of the capacitive charging current to ground is no longer zero, but is 3 *I*co or three times the original charging current per phase. The fault current, *I*g, flowing from the faulted conductor to ground, leads the original line-to-neutral voltage by approximately 90°.

In an ungrounded system, it is possible for destructive transient over voltages to occur throughout the system during re-striking ground faults. These over voltages, which can be several times normal in magnitude, result from a resonant condition being established between the inductive reactance of the system and the distributed capacitance to ground.

In the isolated system, it is possible for destructive transient over voltages to occur throughout the system during re-striking ground faults. These over voltages, which can be several times normal in magnitude, result from a resonant condition being established between the inductive reactance of the system and the distributed capacitance to ground. Experience has proved that these over voltages may cause failure of insulation at multiple locations in the system, particularly at motors. Transient over voltages from re-striking ground faults are the main reason why ungrounded systems are no longer recommended and grounded systems of some form are the predominant choice. To reduce transient over voltages during re-striking ground faults, one should ground the system using either solid or impedance grounding. Various detection schemes are used to detect the presence of a single line-to-ground fault.

4.4.2 Earthed Systems

4.4.2.1 Solid Grounding

Solid grounding refers to the connection of a system conductor as given in (Figure 14), usually the neutral of a generator, power transformer, or grounding transformer directly to ground, without any intentional intervening impedance. However, both the impedance of the source and the unintentional impedance in the connection to ground must be considered when evaluating the grounding. To assess the benefits of a solid connection to ground, it is necessary to determine the degree of grounding provided in the system. A good guide in answering this question is the magnitude of ground-fault current as compared to the system three-phase fault current. The higher the ground-fault current in relation to the three-phase fault current, the greater the degree of grounding in the system. Effectively grounded systems will have a line-to ground short-circuit current of at least 60% of the three-phase, short-circuit value.

This is rarely a problem in typical industrial and commercial power systems. A Sufficiently low resistance to earth may be difficult to achieve, but the "ground" to which faults occur will be the bonded conductive electrical enclosures.

The zero-sequence impedance of most generators used in these systems is much lower than the positive sequence impedance of these generators. The zero-sequence impedance of a delta-wye transformer will not exceed the transformer's positive-sequence impedance. There are, however, conditions under which relatively high zero-sequence impedance may occur. One of these conditions is a power system fed by several generators and/or transformers in parallel. If the neutral of only one source is grounded, it is possible for the zero-sequence impedance of the grounded source to exceed the effective positive-sequence impedance of the several sources in parallel.

Another such condition may occur where power is distributed to remote facilities by an overhead line without a metallic ground-return path. In this case, the return path for ground-fault current is through the earth, and even though both the neutral of the source and the non-conducting parts at the load may be grounded with well-made electrodes, the ground-return path includes the impedance of both of these ground electrodes. This impedance may be significant. Another significant source of zero-sequence impedance is the large line-to-ground spacing of the overhead line.

Solid grounding is generally recommended for the following (Gouda Osama El-Sayed, (2017)):

a) Low-voltage systems (600 V and below) where automatic isolation of a faulted circuit can be tolerated or where capability is lacking to isolate a ground fault in a high-resistance grounded system.
b) Medium- or high-voltage systems (above 15 kV) in order to permit the use of equipment with insulation levels to ground rated for less than line to line voltage.
c) Medium- or high-voltage applications where the desire for a higher magnitude of ground-fault current in order to be able to provide selective ground-fault detection on lengthy distribution feeders outweighs concerns about arc flash and potential gradients as personnel hazards in a workplace setting.

4.4.2.2 Resistance Grounding

In a resistance-grounded system, the neutral of the transformer or generator is connected to ground through a resistor. A typical resistance-grounded neutral system is shown in (Figure 17). As commonly installed, the resistance has a considerably higher Ohmic magnitude than the system reactance at the resistor location. Consequently, the line-to ground fault current is primarily limited by the resistor itself.

Figure 17. Resistance-grounded systems

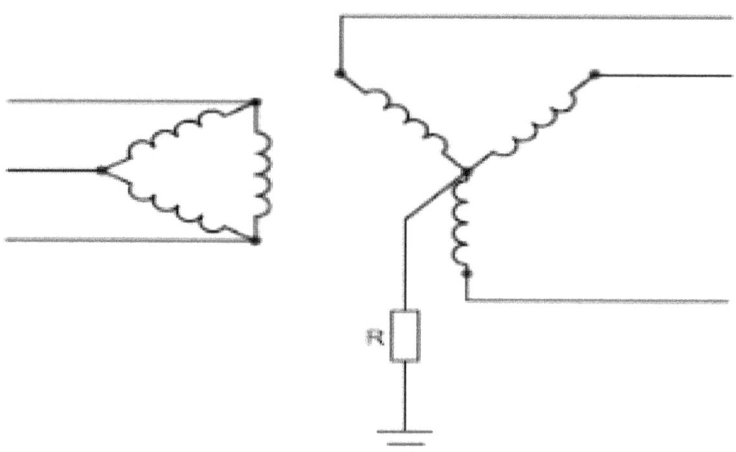

The reasons for limiting the current by resistance grounding include the following:

a) To reduce burning and melting effects in faulted electric equipment, such as switchgear, transformers, cables, and rotating machines.

b) To reduce mechanical stresses in circuits and apparatus carrying fault currents.
c) To reduce electric-shock hazards to personnel caused by stray ground-fault currents in the ground-return path.
d) To reduce the arc blast or flash hazard to personnel who may have accidentally caused or happen to be in close proximity to the ground fault.
e) To reduce the momentary line-voltage dip occasioned by the occurrence and clearing of a ground fault.
f) To secure control of transient over voltages while at the same time avoiding the shutdown of a faulted circuit on the occurrence of the first ground fault (high resistance grounding). Resistance grounding may be either of two classes, high resistance, or low resistance, distinguished by the magnitude of ground-fault current permitted to flow. Although there are no recognized standards for the levels of ground-fault current that define these two classes, in practice there is a clear difference.

Figure 18. Scheme for detecting a ground fault on a high-resistance grounded system

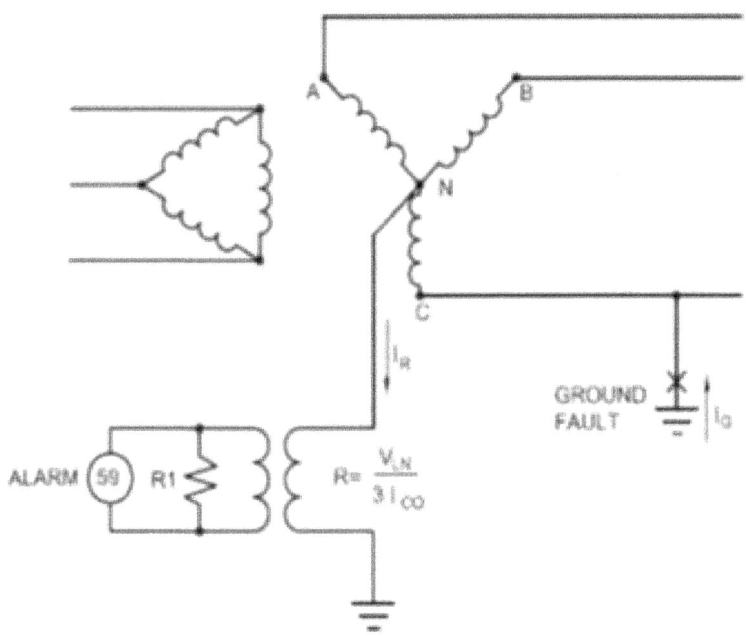

High-resistance grounding usually does not require immediate clearing of a ground fault since the fault current is limited to a very low level. The protective scheme associated with high-resistance grounding is usually detection and alarm rather than immediate trip out. A typical scheme for detecting a ground fault in a high-resistance grounded system is shown in (Figure 18). Under normal operation, the neutral point of the transformer is at zero potential. When a single line-to-ground fault occurs, the neutral point is raised to approximately line-to-neutral voltage. This rise in voltage is then detected using an over voltage relay, 59. A step-down transformer is typically used to reduce the line to neutral voltage of the system to a level (usually 120 V) acceptable to the relay. Since a ground fault may persist for an indefinite length of time, a continuous (rather than short term) rating should be imposed on the grounding resistor.

On the other side the low-resistance grounding is designed to limit ground-fault current to a range between 100 A, and 1000 A, with 400 A being typical. The neutral resistor, *R*, shown in figure***4.***16, is selected according to $R = V_{LN} / I_0$, where V_{LN}, is the system line to neutral voltage and I_0 is the desired ground-fault current. (Figure ***4.***15) illustrates the flow of currents for a single line-to-ground fault on a low-resistance grounded system. Since the combined effects of charging current and system source impedance will affect the ground-current value less than 0.5% in the typical range of utility supplied systems, it is permissible to ignore these effects in calculating the ground-fault resistance value. The general practice is to consider that the full system line-to-neutral voltage appears across the grounding resistor. Only in the case of systems supplied by small generators should departure from this general practice be considered.

Figure 19. Low-resistance grounded

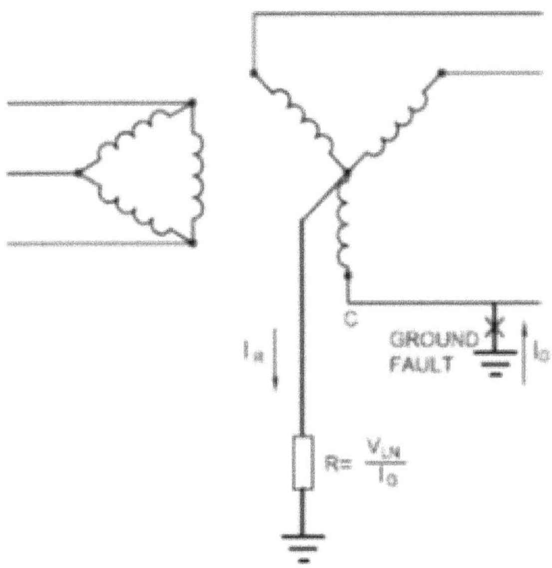

Low-resistance grounding has the advantage of facilitating the immediate and selective clearing of a grounded circuit. This requires that the minimum ground-fault current be large enough to positively actuate the applied ground-fault relay. One method of detecting the presence of a ground fault uses an over current relay, 51G. This method is presented in figure ***4.***19. When a ground fault occurs, the neutral potential is raised to approximately line-to-neutral voltage, resulting in current flow through the resistor. A typical turn's ratio for the current transformer is indicated. Upon indication that a ground fault has occurred, action would be initiated to disconnect the transformer from the secondary circuit.

Normal practice for automatic protection in low resistance grounding is to rate it for 10 s or 30 s, depending upon the degree of security appropriate for the application. In cases of faults that are not, or cannot be, disconnected by secondary breakers, the ability for prompt and automatic disconnection of the primary source is required. Suitable relaying and switching devices for this purpose are an integral part of the low-resistance system design as shown in (Figure 19).

Figure 20. Scheme for detecting a ground fault on a low-resistance grounded system

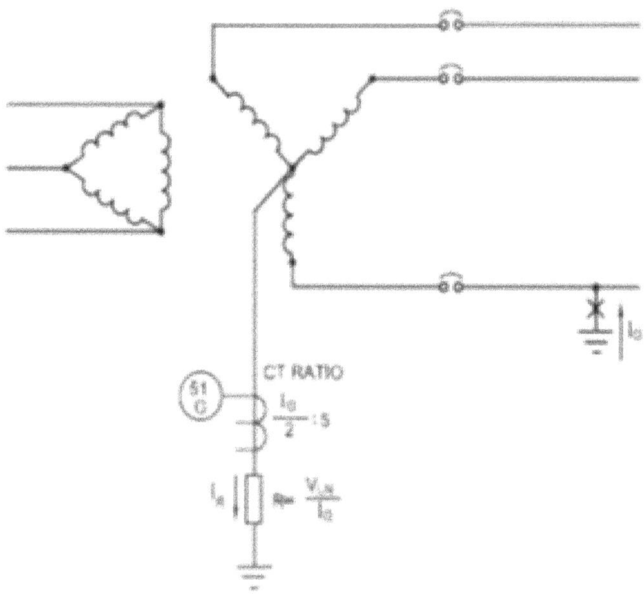

4.2.2.3 Reactance Grounding

The term reactance grounding describes the case in which a reactor is connected between the system neutral and ground, as shown in (Figure 18). Since the ground fault that may flow in a reactance-grounded system is a function of the neutral reactance, the magnitude of the ground-fault current is often used as a criterion for describing the degree of grounding. In a reactance-grounded system, the available ground-fault current should be at least 25% ($X0 = 10X1$) and preferably 60% ($X0 = 3X1$) of the three-phase fault current to prevent serious transient over voltages. The term $X0$, as used, is the sum of the source zero sequence reactance, $X0$, plus three times the grounding reactance, $3Xn$, ($X0 = X0$ source $+ 3Xn$). This is considerably higher than the level of fault current desirable in a resistance grounded system, and therefore reactance grounding is usually not considered an alternative to low-resistance grounding.

Reactance grounding is typically reserved for applications where there is a desire to limit the ground-fault duty to a magnitude that is relatively close to the magnitude of a three phase fault. Use of neutral grounding reactors to provide this fault limitation will often be found to be a less expensive application than use of grounding resistors if the desired current magnitude is several thousand amperes.

These circumstances may arise in one of two possible instances. One potential setting is where a large substation feeds a medium-voltage distribution system, and the total zero-sequence impedance of the step-down transformers in the station causes the single line- to-ground-fault current to greatly exceed the magnitude of a three-phase fault, and ground-fault limitation is desired to keep the total fault current within the reasonable limits. These conditions tend to occur most often in electric utility distribution practice.

In this system, instead of resistance, a reactance is connected between neutral and earth with ratio of reactance more than 3. The system is represented in the (Figure 21). Let the earth fault occur on phase B.

Figure 21. Reactance grounding system

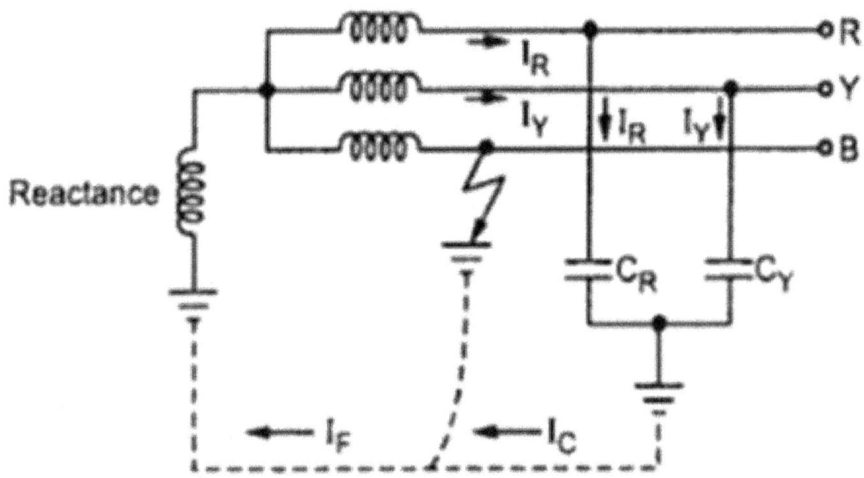

In addition to zero sequence impedance of the source and faulty phase up to point of fault, the fault current is dependent on reactance. By changing the value of reactance, the fault current can be varied. In practice this method is employed to give characteristics similar to solid earth system.

There is a drawback of this system. With increase in reactance, there is increase in transient voltages resulting from arcing. Hence it is not commonly employed though it ensures satisfactory relaying, partial grading of equipment insulation, less interference with neighboring circuits, and intermediate cost.

For reactance grounding, it is necessary that the magnitude of fault current should be at least 25% of the three-phase fault current. This is higher than the requirement on resistance grounding and thus it can be seen that the resistance grounding and the reactance grounding are not similar. The second instance is where there is a desire to serve single-line-to-neutral-connected load directly at the terminal voltage of generators, i.e., without an intervening generator isolation transformer. In this instance, a current will flow in the generator neutral as a result of unbalance between the loads on the three phases. A resistor in the neutral circuit of the generator will limit the flow of this unbalance, thereby limiting the ability of the system to carry unbalanced single-phase load. Medium-voltage generators are typically not designed to withstand the unbalanced mechanical forces associated with supplying ground-fault currents that exceed the magnitude of current that the machine would produce to a three-phase fault at its terminals, thereby making solid grounding of the neutral undesirable. Use of low-reactance grounding to limit the ground-fault magnitude to a level slightly lower than the three-phase level is a way to resolve these application constraints. The conditions that favor low-reactance grounding of generators are relatively rare, so this practice is somewhat obscure.

4.2.2.4. Resonant Grounding (Ground-Fault Neutralizer)

A ground-fault neutralizer is a reactor connected between the neutral of a system and ground. The reactor, X_{LN}, is specially selected, or *tuned*, to resonate with the distributed capacitance, V_{CN} of the system so that a resulting ground-fault current is resistive and low in magnitude. A resistance, *r*, is shown depicting reactor losses. The resulting ground-fault current is in phase with the line to neutral voltage so that current zero and voltage zero occur simultaneously. If the ground fault is in air, such as an insulator flashover, it may be self-extinguishing.

This method of grounding formerly was occasionally seen in high-voltage transmission practice. A key requirement is that because the resonant circuit must be retuned if the distributed parameters of the associated circuit are changed, the ideal application is one that does not involve frequent circuit switching or reconfiguration (Gouda Osama El-Sayed, (2017)).

4.3 ZIGZAG GROUNDING TRANSFORMER

One type of grounding transformer commonly used is a three-phase zigzag transformer with no secondary winding. The internal connection of the transformer is illustrated in (Figure 22).

Figure 22. Zigzag grounding transformer: (a) core windings, (b) system connection

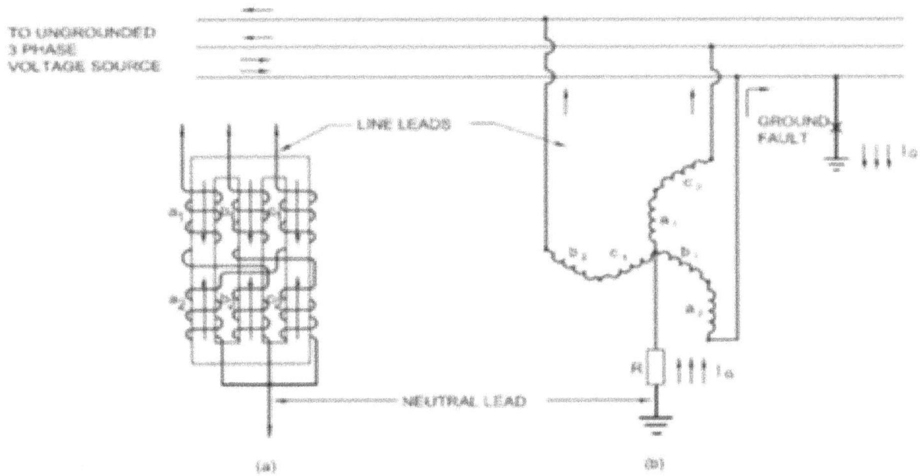

The impedance of the transformer to balance three-phase voltages is high so that when there is no fault on the system, only a small magnetizing current flows in the transformer winding. The transformer impedance to zero-sequence voltages, however, is low so that it allows high ground-fault currents to flow. The transformer divides the ground-fault current into three equal components; these currents are in phase with each other and flow in the three windings of the grounding transformer. The method of winding is seen from (Figure 23) to be such that when these three equal currents flow, the current in one section of the winding of each leg of the core is in a direction opposite to that in the other section of the winding on that leg. This tends to force the ground-fault current to have equal division in the three lines and accounts for the low impedance of the transformer-to-ground currents.

A zigzag transformer may be used for effective grounding, or impedance can be inserted between the derived neutral of the zigzag transformer and ground to obtain the desired method of grounding. This transformer is seldom employed for medium-voltage, high-resistance grounding. An example of low-resistance grounding is shown in Figure 23. The over current relay, 51G, is used to sense neutral current that only flows during a line-to-ground fault.

Figure 23. Low-resistance grounding of system through a zigzag grounding transformer with neutral sensing current relay

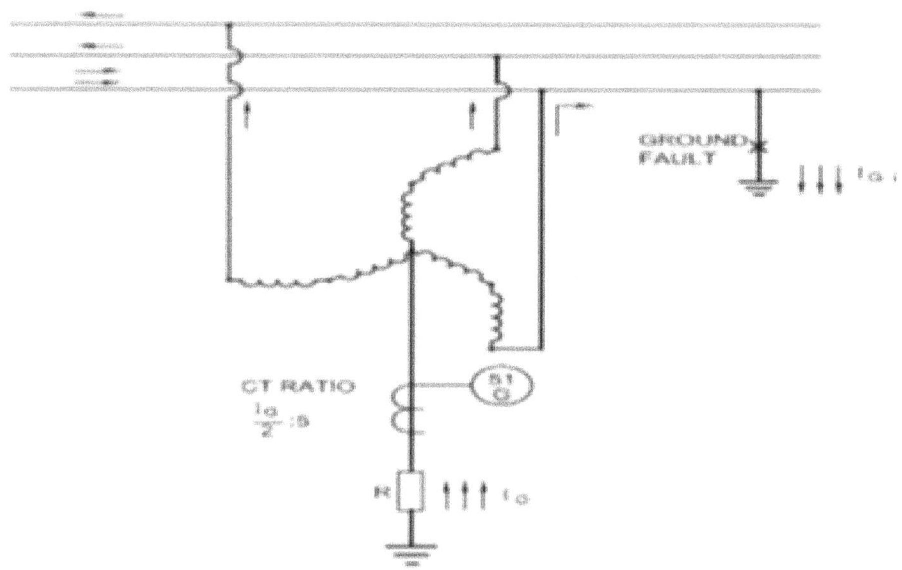

4.4 WYE-DELTA GROUNDING TRANSFORMER

A wye-delta connected three-phase transformer or transformer bank can also be utilized for system grounding, as shown in Figure 24 and Figure 25. As in the case of the zigzag transformer, it can be used for effective grounding or to accomplish resistance-type grounding of an existing ungrounded system. The delta connection must be closed to provide a path for the zero-sequence current, and the delta voltage rating is selected for any standard value. A resistor inserted between the primary neutral and ground, as shown in Figure 24 and Figure 25 provides a means for limiting ground-fault current to a level satisfying the criteria for resistance-grounded systems. For this arrangement, the voltage rating of the wye winding need not be greater than the normal line-to-neutral system voltage. A neutral sensing current relay, 51G, is shown for detection of a single line-to-ground fault. For high-resistance grounding it is sometimes more practical or economical, as illustrated in Figure 23, to apply the limiting resistor in the secondary delta connection. For this configuration the grounding bank must consist of three single-phase transformers with the primary wye neutral connected directly to ground. The secondary delta is closed through a resistor that effectively limits the primary ground-fault current to the desired low level. For this alternative application, the voltage rating of each of the transformer windings forming the wye primary should not be less than the system line-to-line voltage.

Figure 24. Wye-delta grounding transformer showing current flow

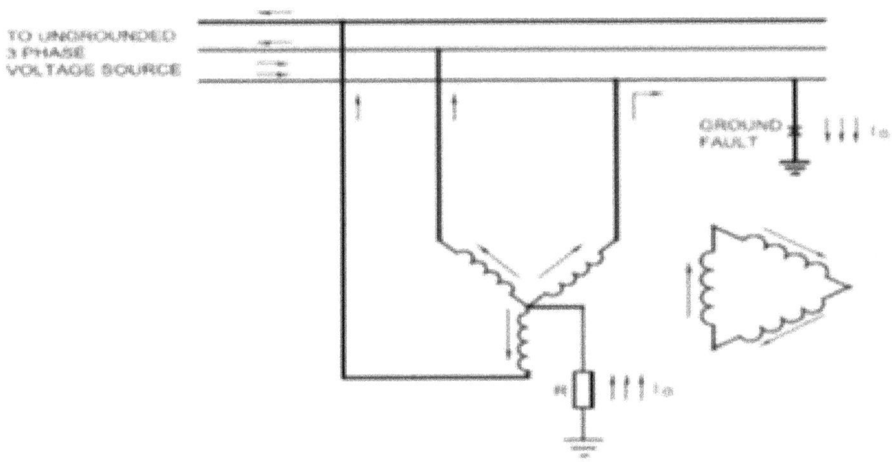

Figure 25. Low-resistance grounding of system through a wye-delta grounding transformer with neutral sensing current relay

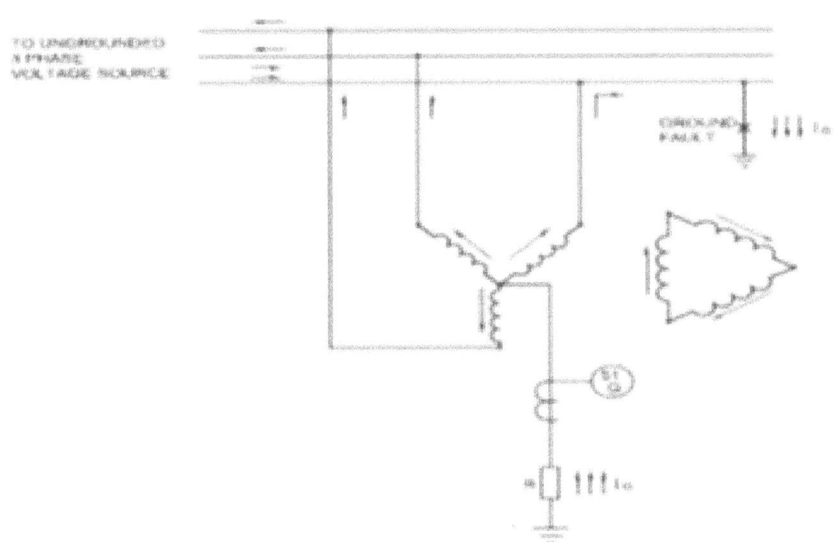

The rating of a three-phase grounding transformer or bank, in kilovolt ampere (kVA), is equal to the rated line-to-neutral voltage in kilovolts times the rated neutral current. Most grounding transformers are designed to carry their rated current for a limited time only, such as 10 s or 1 min. Consequently, they are much smaller in size than an ordinary three-phase continuously rated transformer with the same rating. It is generally desirable to connect a grounding transformer directly to the main bus of a power system, without intervening circuit breakers or fuses, to prevent the transformer from being inadvertently taken out of service by the operation of the intervening devices. (In this case the transformer is considered part of the bus and is protected by the relaying applied for bus protection.) Alternatively, the grounding transformer should be served by a dedicated feeder circuit breaker, as shown in Figure 25, or connected between the main transformer and the main switchgear, as illustrated in Figure 26. If the grounding transformer is connected as shown in Figure 22, there should be one grounding transformer for each delta-connected bank supplying power to the system, or enough grounding transformers to assure at least one grounding transformer on the system at all times. When the grounding transformer is connected, it is included in the protective system of the main transformer. Figure 26 shows high-resistance grounding of system through a wye-broken delta grounding transformer with neutral sensing voltage relay.

Figure 26. High-resistance grounding of system through a wye-broken delta grounding transformer with neutral sensing voltage relay

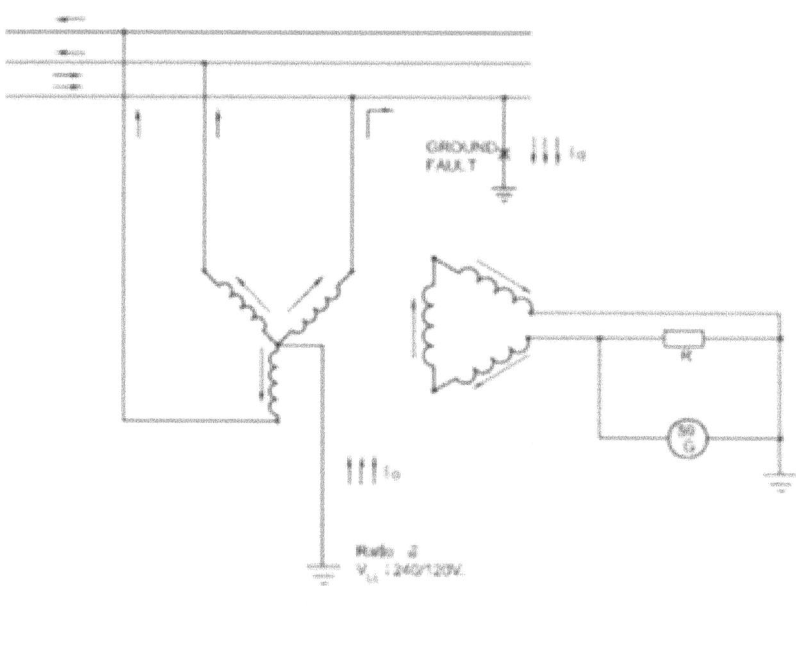

4.5 GROUNDING AT POINTS OTHER THAN SYSTEM NEUTRAL

In some cases, low-voltage systems (600 V and below) are grounded at some point other than the system neutral to obtain a grounded electrical system. This is done because delta transformer connections do not provide access to the three-phase system neutral. Two systems are in general use as follows:

1. Corner-of-the-delta grounded systems

This type of grounding is no longer popular and is not widely used in industrial systems.

2. One phase of a delta system grounded at midpoint

This provides neutral grounding for the single-phase 120/240 V and grounding for the 240 V three-phase system. It is not recommended for voltages over 240 V.

Connection of grounding transformers in delta connected or ungrounded power system to obtain neutral for system grounding: (a) circuit feeder breaker, (b) connected between main transformer and main switch gear.

4.6 LOCATION OF SYSTEM GROUNDING POINTS SELECTION

Each system as described in this chapter is characterized by its isolation from adjacent grounding systems. The isolation is provided by transformer primary and secondary windings. The NEC defines such a system as separately derived. A separately derived system is one "whose power is derived from a generator, transformer, or converter winding and that has no direct electrical connection, including a solidly connected grounded circuit conductor, to supply conductors originating in another system."

Therefore, the new system created by a transformer or generator requires the establishment of a new system ground if it is required or desired that this system to be grounded. The system ground point should always be at the power source as required or permitted by the NEC, including exceptions for multi-source systems.

4.7 DELTA WYE TRANSFORMER

In a delta-wye connected transformer, with the load-side neutral grounded, zero-sequence components of current can flow in the secondary wye-connected windings due to a ground fault. Zero-sequence current is then induced into the primary windings of the transformer and circulates in the delta connection. Positive and negative-sequence currents pass through the transformer combining to produce high current in two of the primary phase conductors. A ground fault on the secondary of the delta-wye connected transformer appears as a line-to-line fault on the primary. If the neutral of the wye-connected windings is not grounded, then zero-sequence current cannot flow and the system becomes ungrounded. Figure 27 shows Zero-sequence impedances of different transformers configurations.

Zero-sequence components of current can flow through a wye-wye connected transformer if a neutral path exists on both sides of the transformer. An example is shown in Figure 28 where a delta-wye connected transformer, T1, supplies power to a wye-wye connected transformer, T2. A fault on the load side of T2 produces zero-sequence current, which flows in the primary and secondary windings of that transformer. Zero-sequence current is permitted to flow in the primary of T2 because a path exists in the delta-wye connected transformer T1. Disconnecting any of transformer neutrals, on either T1 or T2, would prevent the flow of zero-sequence current in both transformers, except as allowed by magnetizing reactance.

Depending upon the connections to the transformer, the use of a wye-wye transformer can result in a single system, or its load side may be a separately derived system.

Figure 27. Zero-sequence impedances of different transformers configurations

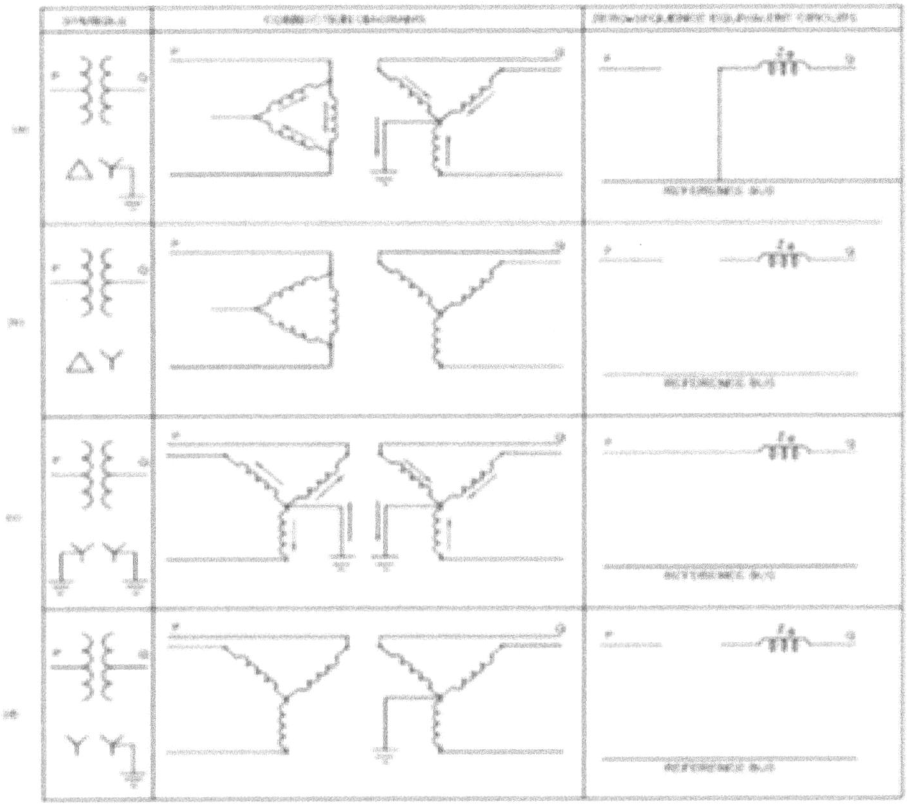

4.8 WYE-WYE TRANSFORMER

A wye-wye transformer, T2, is shown in (Figure 28) with the primary and secondary neutrals interconnected and grounded. This transformer configuration is used on solidly grounded utility distribution systems, particularly underground systems, to prevent Ferro resonance when the supply switches can be operated one pole at a time. The utilities ground the primary neutral point to minimize the neutral-to-earth voltage throughout the length of the distribution line and by default on underground systems using bare concentric neutral cables. They ground the secondary neutral to provide an effectively grounded low-voltage service.

Note that this multiple grounding of the primary at each transformer is not essential to prevent Ferro resonance or provide secondary grounding as long as the fourth conductor is brought to the primary neutral of the transformer. The neutral-to-transformer case and ground connection minimizes secondary neutral-to ground voltage during a fault between primary and transformer case. In an industrial distribution system, the physical length of the circuit will usually be short enough so that excessive neutral-to-ground voltages will not be present even if there is no ground at the wye-wye transformer common neutral terminals, as shown in (Figure 24). The NEC normally prohibits grounding of the neutral on the load side of the service disconnect, but allows multi grounding of the neutral of an outdoor overhead line or direct burial cable with bare neutral if the circuit voltage is over 1000 V.

Figure 28. Transformer connections illustrating the flow of zero-sequence current resulting from a line to ground fault

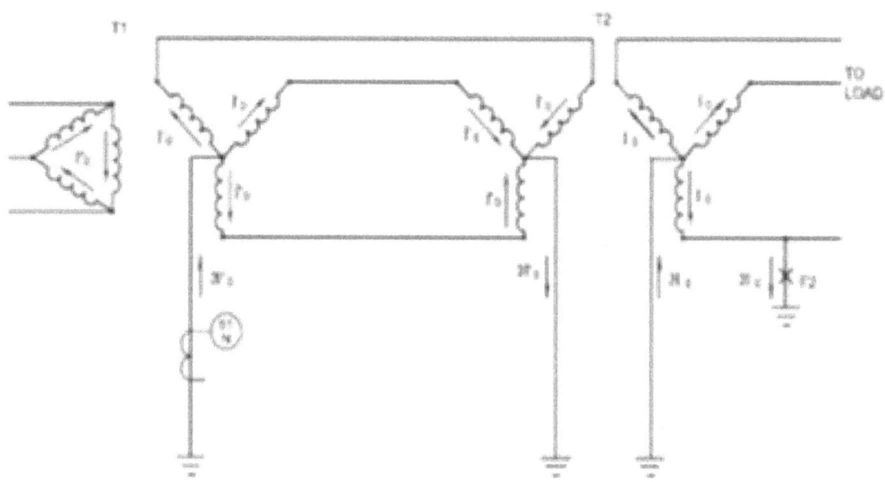

As shown in (Figure 24), with a continuous connection from the source neutral to the primary and secondary neutrals of the wye-wye transformer, the output of the transformer would not constitute a separately derived system as defined in the NEC. If the neutral is grounded at the source, T1, the output of the wye-wye transformer will be a continuation of the grounded system, though at the secondary voltage of the transformer. A fault, F2, on the load side of the wye-wye connected transformer, T2, will produce zero-sequence components of current in its primary windings. This zero-sequence current will flow back to the secondary neutral terminal of source transformer, T1. However, this current flowing through 51G cannot determine whether the fault is located before or after the wye-wye transformer, nor can residual or zero-sequence ground detection schemes on the output of T1. The main benefit of this transformer connection is to utilize the standard wye-wye transformer that contains an internal primary-to-secondary neutral connection suitable for utility practice as shown in (Figure 29). The circuit supplied by the wye-wye connected transformer shown in (Figure 30) can be considered a separately derived system, since there are no direct metallic connections between the primary and secondary of the transformer. Primary and secondary ground faults are separately measured and relayed. The secondary of the transformer will not be grounded unless a connection to earth is made. The secondary could be impedance grounded. Secondary neutral grounding will also require a connection from the neutral of the primary source to the primary neutral of the wye-wye transformer to supply zero sequence current. Unlike the delta-wye transformer, the wye-wye transformer itself is not a source of zero-sequence current. Grounding can be achieved without a primary neutral connection if a phase of the secondary rather than the neutral is grounded, since no zero-sequence current is involved. The effect is then identical to corner grounding of a delta-delta transformer.

Figure 29. Grounded wye-grounded wye transformer with multi-grounded common neutral

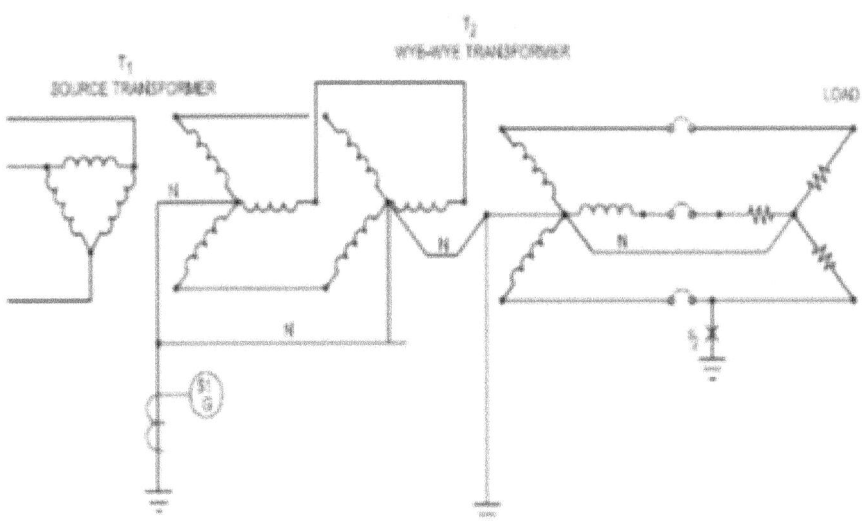

Figure 30. Grounded wye-grounded wye transformer with separately grounded neutrals

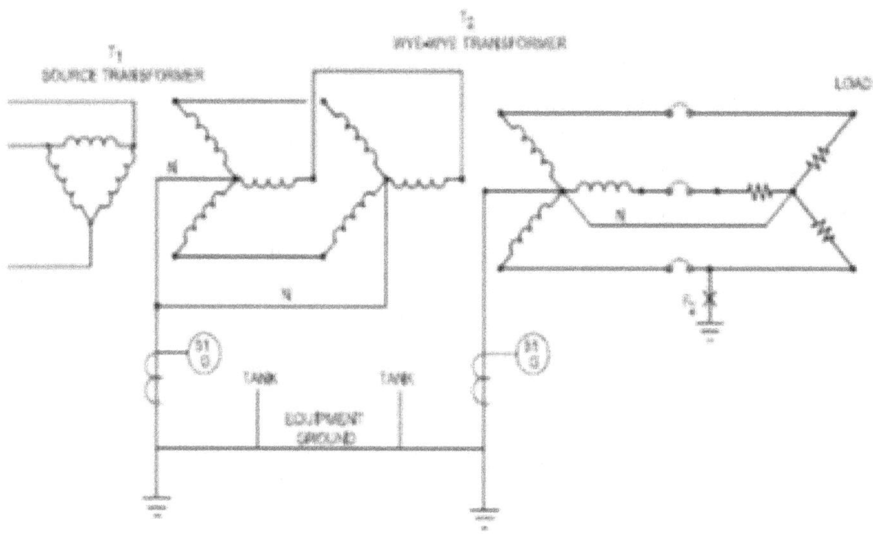

If a delta tertiary is added to a wye-wye transformer it will not be necessary to supply zero-sequence current from the primary source, since the tertiary will act as a source of zero-sequence current. Thus, the wye-wye transformer can be considered a part of a single multi-voltage system if the neutrals are interconnected or can be considered to create a separate system if they are not. The symmetry of the wye-wye allows it to provide grounding for its load-side system even though the source and load side may be interchanged at any time.

In a single power source when a system has only one source of power (gen. or transformer), grounding may be accomplished by connecting the source neutral to earth either directly or through a neutral impedance (Figure 31).

Provision of a switch or circuit breaker to open the neutral circuit is not recommended. It is not desirable to operate the system ungrounded by having the ground connection open while the generator or transformer is in service. In the event that some means of disconnecting the ground connection is required for measurement, testing, or repair, a disconnecting link should be used and only opened when the system is de-energized.

Figure 31. Grounding for systems with one source of power: (a) solidly grounded, (b) R or Z grounded

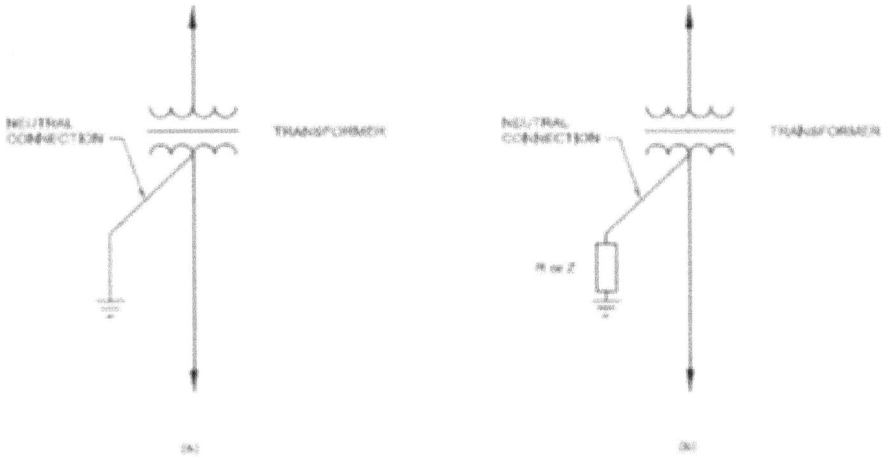

In Multiple power sources for installation of interconnected multiple power sources (i.e., generators or power transformers), operated in parallel, system grounding can be accomplished using one of the two following methods:

a) Each source grounded, with or without impedance (Figure 32).
b) Each source neutral connected to a common neutral bus, which is then grounded, with or without impedance (Figure 31).

For solidly grounded systems, with multiple sources, where all sources must be solidly grounded, it is acceptable to separately ground each power source as shown in part a) of (Figure 30) unless third harmonics are present or if it results in exceeding the fault capability of the generators. Levels of fault current in systems where generators are paralleled with transformer sources on a four-wire basis must be calculated using symmetrical component sequence values for the sources appropriately combined in the system.

Commercial computer programs are now available that will calculate branch currents for unbalanced faults in systems with both utility and generator sources. There can be a significant increase in the total system ground-fault current as compared to the sum of the fault current available from sources when not in a combined system, while the increase in generator currents can be proportionally even greater.

Where sources are in close proximity, or where the system is four wires, the common neutral or ground bus as shown in part a) of (Figure 31) should be used. In a four-wire system the sources would not be considered as separately derived. If the power sources are not in close proximity, common ground point is not recommended. The impedance in the neutral bus connection may become large enough to prevent effectively grounding the neutral of the source at the remote location. The interconnection may inadvertently become open, allowing the transformer to operate ungrounded. For impedance grounded systems it is acceptable to separately connect each neutral to ground through individual impedance (part b) of (Figure 30). Each impedance rating should allow sufficient current to satisfy the criteria for the grounding system being used. Individual neutral switching devices (automatic or manual) are not recommended, since incorrect operation may allow a power source to operate ungrounded. System relaying is more complex when there are multiple ground-fault sources.

The fault current sensed by the feeder is variable, depending on the number of ground-fault current sources that are connected at the time of the fault. When individual source impedances are used for low- or high-resistance grounding, circulation of third harmonic currents between paralleled generators is usually not a problem since the impedance limits the circulating current to tolerable values. When the total ground-fault current from several individual impedances exceeds about 1000 A, a common ground point and single impedance should be considered to provide a single acceptable value of ground-fault current of Figure 35. The advantage of this connection is that the maximum fault current is known, and selective relaying can be used to open tie breakers and selectively isolate the faulted bus.

Figure 32. Grounding for systems with multiple power sources (Method 1): Solidly grounded

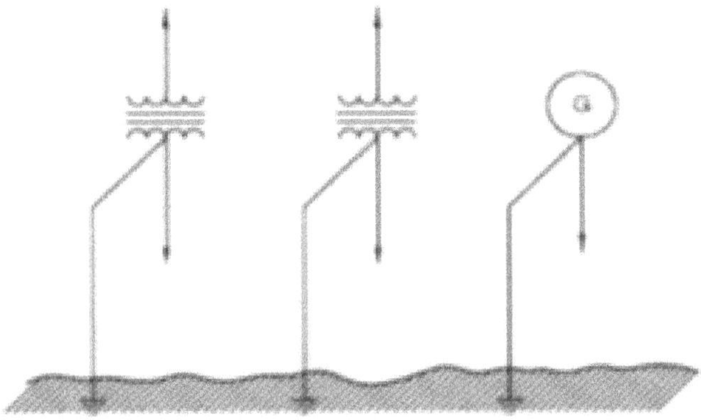

Figure 33. Grounding for systems with multiple power sources (Method 1): R or Z grounded

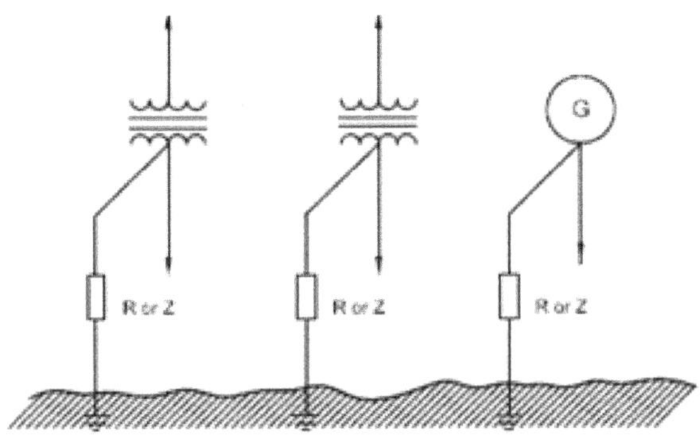

The primary purpose of neutral disconnecting devices in impedance grounded systems, as shown in Figure 35, is to isolate the generator or transformer neutral from the neutral bus when the source is taken out of service, because the neutral bus is energized during ground faults. A generator or transformer disconnected from the power bus, but with an unbroken connection of its neutral to the neutral bus, would have all of its terminals elevated with respect to ground during a ground fault. Disconnecting devices should be metal enclosed and interlocked in such a manner as to prevent their operation except when the transformer primary and secondary switches or generator main and field circuit breakers are open. On low-voltage systems the use of four-pole breakers may provide adequate interlocking. In this case line-to-neutral voltage should not be used for synchronizing.

Figure 34. Grounding for systems with multiple power sources (Method 2): Solidly grounded

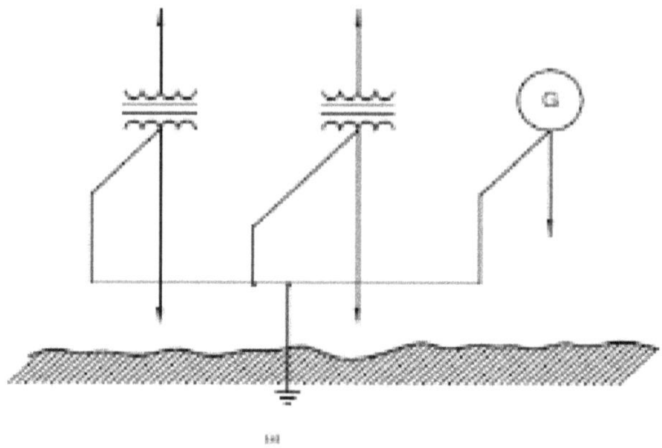

Figure 35. Grounding for systems with multiple power sources (Method 2): R or Z grounded

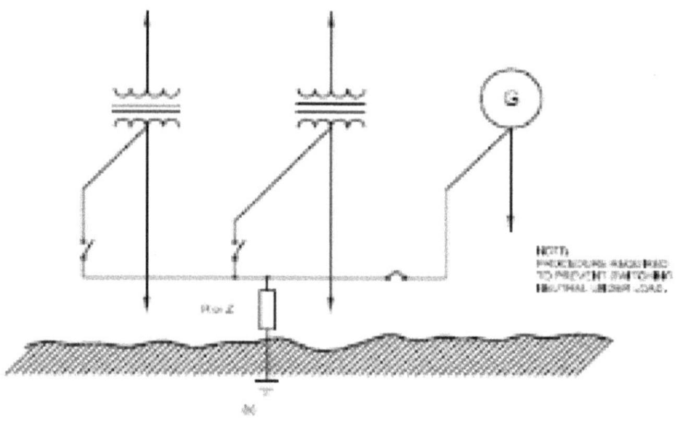

In the case of multiple transformers, all neutral isolating devices may be normally closed because the presence of delta-connected windings (which are nearly always present on at least one side of each transformer) minimizes the circulation of harmonic current between transformers. Generators that are designed to suppress zero-sequence harmonics, usually by the use of a two-thirds pitch winding, will have negligible circulating currents when operated in parallel; therefore, it is often found practical to operate these types of generators with the neutral disconnect device closed. This simplifies the operating procedure and increases assurance that the system will be grounded at all times, because interlocking methods can be used.

It is sometimes desirable to operate with only one generator neutral disconnecting device closed at a time to eliminate any circulating harmonic or zero-sequence currents.

In addition, this method provides control over the maximum ground-fault current and simplifies ground relaying. When the generator whose neutral generator is to be shut down, another generator is grounded by means of its neutral disconnecting device before the main and neutral disconnecting device of the first one are opened. This method has some inherent safety considerations that must be recognized and addressed in order to ensure continual safe operation. The procedures required to permit only one disconnecting device to be closed with multiple sources generally do not permit the use of conventional interlocking methods to ensure that at least one neutral disconnecting device will be closed. Therefore, this method should only be used where strict supervision of operating procedures is assured.

4.9 GROUNDING LOCATIONS SPECIFIED BY THE NEC

The following locations for ground connections are required, or permitted, by the NEC for the most common types of power system grounding. This is not intended to be a complete listing of code requirements, and the current edition of the NEC should be consulted for details or recent changes as well as to determine whether grounding is required or prohibited. The purpose is to call attention to location requirements and not to interpret the requirements. On service-supplied systems of 50 V to 1000 V, system grounding, when required or elected, should be made at the service entrance, between the load ends of the service drop or lateral and the neutral land point. If the service is supplied from a transformer external to the building one additional grounding point is required external to the building. If a grounded conductor extends past the service entrance switch, it should have no further grounds on this extension except as noted by the various exceptions in the NEC to this requirement, as follows (Gouda Osama El-Sayed, (2017)).

Where dual services feed a double-ended bus, a single ground at the center point of the neutral bus is allowed to replace those previously listed. If more than one building is fed from a single service there should be a system grounding connection made at the entrance to each building. However, if an EGC is run with the load conductors, this ground connection can be eliminated so as to avoid elevating non-current carrying enclosures above ground potential due to load drop in the neutral conductor. Grounding connections should not be located or connected so as to cause objectionable currents in grounding conductors or grounding paths.

Separately derived circuits, if required or elected to have a system ground, should be grounded between the source and the first disconnecting device. System grounding connections downstream of the disconnecting device have the same rules as for service supplied circuits. The point of grounding for systems shall be the neutral or common conductor where one exists; otherwise the point shall be a phase conductor.

On systems over 1000 V, a transformer-derived neutral may also be used as the attachment point for a system ground. This method is not mentioned for effective grounding of low-voltage systems. High-voltage and medium-voltage systems may also have multiple neutral grounds where the conductors are overhead outdoors or where they are directly buried with a bare neutral conductor.

4.10 GROUNDING OF INDUSTRIAL AND COMMERCIAL GENERATORS

Generators produce slightly no sinusoidal voltages because of saturation and imperfect winding and flux distribution. Industrial generators therefore produce odd harmonic voltages, with the third harmonic voltage being as much as 10%. These harmonic voltages can cause heating from circulating currents in a closed loop. This is one reason why most industrial generators have their internal windings connected in wye rather than delta. The third harmonic voltage produced in the generator's windings are in phase and additive. This would cause third harmonic current to circulate within the delta connected windings, as shown in (Figure 4.32).

The circulating current would create additional heating within the generator thereby reducing some of its thermal capacity. Generators that operate in a delta connection allow for this in their design. If the generator windings are designed with a two-thirds pitch, the third harmonic voltage can be suppressed, but the zero-sequence impedance will be lowered increasing the ground-fault current.

A grounded generator connected to a delta-wye transformer is shown in (Figure 37). Any third harmonic voltage, *V3*, produced by the generator would be impressed on the primary of the transformer. Since the third harmonic voltages are in phase, the voltage difference across each winding of the transformer's delta will equal zero and no third harmonic, or multiples of the third harmonic, current can be expected to flow.

Figure 36. Circulation of third harmonic current in a delta-connected generator

Any current flowing as a result of a line-to-ground fault on the secondary side of the transformer will appear, as shown in (Figure 34), as a line-to-line fault at the generator output. This type of fault is the most damaging to the generator because of its negative sequence content. There will be no zero-sequence current flow in the generator even though the generator is grounded. Zero-sequence current will circulate in the delta winding of this transformer. The physical limitations imposed by generator construction result in less available insulation thickness, with a resulting reduction in voltage-impulse withstand as compared to no rotating electrical equipment. Thus, special attention should be given to limiting voltage to ground by the grounding of generator neutrals.

Figure 37. Analysis of third harmonic current (no zero-sequence loop)

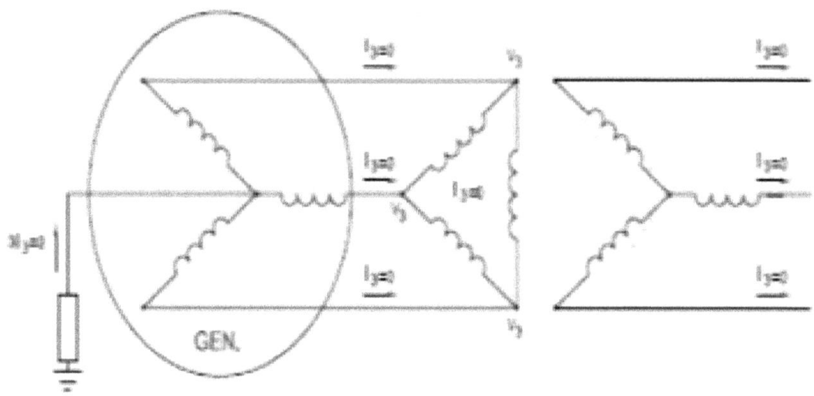

Figure 38. Zero-sequence currents during a wye side fault

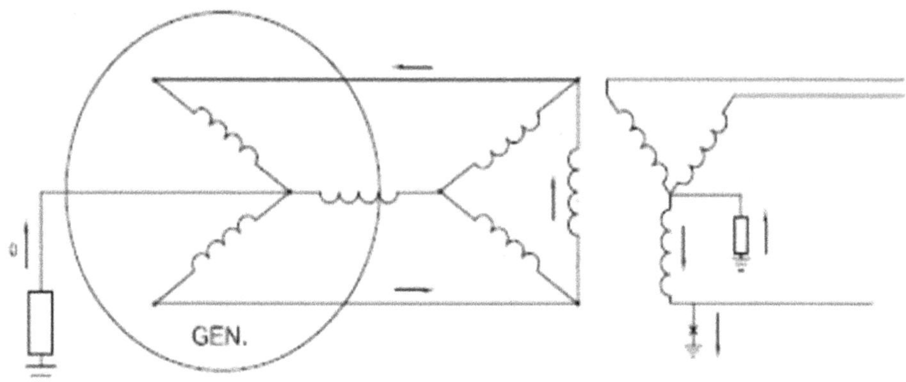

Internal ground faults in solidly grounded generators can produce large fault currents. These currents can damage the laminated core, adding significantly to the time and cost of repair. Such currents persist until the generator voltage decays; since they are not capable of being interrupted by the generator circuit breaker. Both magnitude and duration of these currents should be limited whenever possible.

REFERENCES

ANSI/IEEE C37.5-1979. (n.d.). IEEE Standards, https://standards.ieee.org/ieee/C37.5/6581/

Characterestics of TT, TN and IT systems. (n.d.). Wiki Electrical installation, Presented by Schneider (2016), http://www.electrical-installation.org/enwiki/Characteristics_of_TT,_TN_and_IT_systems

Easy power. (n.d.). Grounding. https://www.easypower.com/resources/article/grounding-analysis-ground-fault-current

El-Sayed, G. O. (2017). *Design Parameters of Electrical Network Grounding Systems, book*. Advances in Computer and Electrical Engineering. IGI Global.

Chapter 5
Grounding Electrodes Performance Under Lightning Strikes Including the Influence of Soil Ionization With Frequency and Soil Resistivity Variations

Osama El-Sayed Gouda
Faculty of Engineering, Cairo University, Egypt

ABSTRACT

In this chapter, ATP transmission line technique and FEM method are used to study the characteristics of the grounding electrodes that are installed in uniform and non-uniform soil including the effect of soil ionization with frequency, soil resistivity, and permittivity variations. The critical breakdown field strength of two layer soil with different values of the reflection factor is presented. The transient probabilistic grounding potential rise and the transient impedances of horizontal and vertical electrodes are calculated in non-uniform soils including the effect of soil ionization with frequency and soil resistivity variations. Finally, the conductivity and permittivity variations on the soil ionization electric field are considered in the used model.

DOI: 10.4018/979-8-3693-2969-6.ch005

5.1 INTRODUCTION

The function of the grounding system is to protect the electrical networks, devices and people from any fault occurring. The behavior of grounding electrodes against high-frequency lightning is completely different from that of the steady state or power frequency faults. The lightning strokes usually have large amount of current that changes the electrical performance of the grounding electrodes surrounding by the soil and causes the soil ionization.

One of the most severe faults that occurs in the electrical power network is a lightning current stroke. This lightning stroke fault has heavy current with high frequency content that changes the achievement of the grounding electrodes. A typical grounding system contains vertical or horizontal electrodes or in medium and high voltages substations it can be designed as a grid buried in the soil. When the high impulse current discharges into the earth through the grounding system the soil breakdown phenomenon occurs (IEEE Guide for safety in AC substation, (2000)).

It has been found in the literatures that the injection of high impulse current in the soil determines the degree of soil resistance non-linearity and reduction done by soil ionization and consequently the transient ground potential rise of the ground surface decreases. Therefore, it is noticed that the soil breakdown improves the efficiency of the grounding systems.

Two main processes have been advanced to explain the increase of soil conduction during high impulse current discharges, firstly the thermal heating process and secondly the soil ionization process. In the thermal heating process, the discharged current increases the temperature of the existing water filling among the soil grains. Due to the heating process, the resistivity of the heated water decreases, which in turn causes the resistivity of the bulk soil and consequently reduces the grounding electrode resistance.

In the soil ionization process, due to lightning current with high frequency content, the performance of the soil electrical parameters such as the conductivity and permittivity causes the change in grounding system performance (Cavka D., Mora N., Rachidi F., Poljak D., (2013)). In the opinion of the authors, there is another factor affecting the soil ionization process, which is the variation in the solubility and ionisability of the electrolytes contained in the site natural soil. This suggested factor has to be investigated in the future. Damir Cavka et al. (Cavka D., Mora N., Rachidi F., Poljak D., (2013)), (Cavka D., Mora N., Rachidi F., (2014)) Idir Djamel et al. (Djamel I., Slaoui F.H., Georges, S., (2016)), Leonid Grcev (Grcev L., (2005)), (L. Grcev et al. Grcev L., Grceva S., (2009)), Rafael Alipio et al. (Alipio R., Visacor S., (2012)), (Adriana G. Pedrosa et al. Pedrosa A.G., Shroeder, M.A.O., Afonso, M.M., Alípio, R., Assis, S.C., Oliveira, T.A.S., Braga, A.R., (2010)), and S. Visacro (Visacro, S., (2007)) studied the performance of the grounding electrodes in an uniform soil when lightning current took place in the network. They considered the effect of frequency on the surrounding soil in their research. Many methods based on different theories in the frequency domain with an appropriate fast Fourier Transform have been developed. Such these methods are circuit theory (Ramamoorty M., Narayanan M.M.B., (1989)), (Mokhtari M., Malek Z.A., and Salam Z., (2015)), transmission line model (Andre M., Mattos F., (2005)), electromagnetic field theory (Grcev, L., Haznadar, Z., (1988)), (Grcev, L., (1992)), hybrid methods (Papalexopoulos, A.D., Meliopoulos, AP., (1987)) and the generalized modified mesh current method (Baba, Y., Nagaoka, N., Ametani, A., (2005)) . CIGRE grounding electrode resistance model is widely used to determine the minimum grounding electrode resistance obtained at current peak (CIGRE, (1991)).

This chapter studied the critical electric field intensity as a function of the soil electrical conductivity, soil permittivity and the frequency content of lightning impulse. The article investigated the effective radius of soil ionization when lightning impulse frequency content and soil resistivity variations were considered. The reflection factor effect on the transient grounding potential rise (TGPR), transient earth surface potential (TESP) and transient impedance of the grounding electrode (TIGE) have been investigated and the soil permittivity and conductivity variations with time were included. It can be concluded that the reflection factor has significant impact on the equivalent radius of the electrode. It is observed that the transient voltage peak value decreases sharply when the soil ionization phenomenon is considered and the impulse impedance of the soil that has negative reflection factor is lower than that in case of positive reflection factor.

5.2 SOIL IONIZATION MODELING UNDER LIGHTNING STROKES

In the soil ionization process, the electric field enhancement in air voids enclosure among the soil grains, causing the soil breakdown occurrence. Since the resistance of the ionized air is much smaller than the resistance of the soil grains, the equivalent soil resistance decreases. Hence, it is concluded that soil ionization is mostly accepted as the main factor in the soil breakdown phenomenon. The performance of grounding system when soil ionization phenomenon occurs in uniform soil is investigated in (Cavka D., Mora N., Rachidi F., (2014)) , and in non-uniform soil in (Omar, K., et al, (2017)). CIGRE Working Group proposed an empirical model to calculate the grounding resistance as a function of high impulse current when the soil ionization influence is considered. This Group suggested equation to compute the grounding electrode resistance with soil ionization consideration as the following (CIGRE, (1991))

$$R_i = \frac{R}{\sqrt{1 + i(t)/I_g}} \quad (5.1)$$

Where R is the low current grounding electrode resistance in [Ω], $i\ (t)$ is the peak impulse current in [kA] and I_g is the limit current in [kA] at which the soil ionization occurs and is given in equation (5. 2)

$$I_g = \frac{E_c \rho}{2\,\pi R^2} \quad (5.\ 2)$$

Where ρ is the soil resistivity in [Ω.m] and E_c is the soil critical electric field intensity in [kV/m], which is recommended by CIGRE to be 400 kV/m (ClGRE (1991)).

One of the authors proposed a model taking into account the geometry of the ionized zone as the new geometry of the grounding electrode. This happened because the arc resistance is considered to be zero due to the dissipation of large currents through the earth. Discharge channels near to electrode will be formed when the electric field exceeds its critical value and formed the enlargement of the dissipation area.

5.3 SOIL CHARACTERISTICS UNDER LIGHTNING STROKES

5.3.1 Apparent Resistivity of Non-Uniform Soil

In two soil horizontal layers, the apparent soil resistivity ρ_a can be calculated according to IEEE 80 by the following formulas (IEEE Guide for safety in AC substation, (2000));

$$\rho_a = \frac{\rho_1}{\left[1 + \left[\left(\frac{\rho_1}{\rho_2}\right) - 1\right]\left[1 - e^{\frac{1}{k(z+2h)}}\right]\right]} \text{ For } \rho_2 < \rho_1 \quad (5.\ 3)$$

Where h *is* the burial depth of grounding grid, k is the reflection factor $k = [(\rho_2 - \rho_1)/(\rho_1 + \rho_2)]\ (5.\ 4)$, z is the top layer thickness, ρ_1 is the upper soil resistivity and ρ_2 is the lower soil resistivity (Nahman, J., Paunovic, L., (1979)).

5.3.2 Critical Breakdown Field Strength in Uniform and Non-Uniform Soils

The critical breakdown field strength (E_c) is commonly defined as the value at which soil breakdown happens. Many authors investigated the soil breakdown phenomena when subjected to high-frequency lightning (Omar, K., et al, (2017)). According to Scholars' (ClGRE, (1991)) experiments and measurements of the (E_c) values are considered to be from tens to thousands kV/m depending on the soil composition. The critical breakdown field strength (E_c) as constant value is recommended by CIGRE (ClGRE, (1991)) or calculated by E.E. Oettle equation (Oettle, E. E., (1988)), which proposed the relation between the critical breakdown field strength (E_c) and the soil resistivity as the following

$$E_c = 241\rho^{0.215} \quad (5.\ 5)$$

The critical electric field intensity as a function of the soil resistivity is considered by (Cavka D., Mora N., Rachidi F., (2014)) , unfortunately the influence of the frequency variation on the soil ionization phenomenon is ignored by some investigators (Djamel I., Slaoui F.H., Georges, S., (2016)).

Recently, some researchers included the effect of frequency content on the performance of the grounding system when soil ionization phenomenon occurs in uniform and non- uniform soils (Omar, K., et al, (2017)). Manna and Chowdhuri (Manna, T.K., Chowdhury, P., (2007)) proposed relation between the critical breakdown field strength (E_c) and the soil dielectric constant and its conductivity as given in equation (5. 6)

$$E_c = 8.6083\,\varepsilon_g^{-0.0103}\,\sigma_g^{-0.15264} \quad (5.\ 6)$$

Where σ_g is the soil conductivity (in mS/cm), and ε_g is the soil relative permittivity.

Fgure 1 presents the critical breakdown field strength using Scott expression, Messier expression and Portola expression. In this figure the following values are considered, $\rho_1/\rho_2 = 300/100$ when k = -1/2, $\rho_1/\rho_2 = 500/100$ when k = -2/3, $\rho_1/\rho_2 = 100/300$ when k = 1/2, $\rho_1/\rho_2 = 100/500$ when k = 2/3 and finally when k = 0 the soil is uniform and its resistivity is taken as 100 Ω.m. Figure 1 shows the critical breakdown field strength of two layers of soil with different values of the reflection factor, when the grounding electrode is subjected to 30 kA lightning high frequency content. The critical breakdown field strength is calculated also by applying expression of Scott *et al.*, in which the soil dielectric constant ε_r and the soil conductivity σ are functions in the lightning strokes frequency (Scott, J.H., Carroll, R.D., Cunningham, D.R., (August 1964)) as the following:

$$\varepsilon_r(f) = 10^D \ (5.\ 7)$$

$$where\ D = 5.491 + 0.946\,log_{10}(\sigma_{100Hz}) - 1.097\,log_{10}(f) + 0.069\,log_{10}^2(\sigma_{100Hz})$$

$$-0.1141.097\,log_{10}(f)\,log_{10}(\sigma_{100Hz}) + 0.067\,log_{10}^2(f) \ (5.\ 8)$$

$$\sigma(f) = 10^n \ [\text{MS/m}] \ (5.\ 9)$$

$$where\ n = 0.028 + 1.098\,log_{10}(\sigma_{100Hz}) - 0.068\,log_{10}(f) + 0.036\,log_{10}^2(\sigma_{100Hz})$$

$$-\ 0.046\,log_{10}(f)\,log_{10}(\sigma_{100Hz}) + 0.0180.067\,log_{10}^2(f) \ (5.\ 10)$$

Where: f is the supply frequency in [Hz] and σ_{100Hz} is the soil conductivity at 100 Hz in [M S/m].

$$\varepsilon_r(f) = \frac{\varepsilon_\infty}{\varepsilon_0}\left(1 + \sqrt{\frac{\sigma_{DC}}{\pi f \varepsilon_\infty}}\right) \ (5.\ 11)$$

$$\sigma\left(f\right) = \sigma_{DC}\left(1 + \sqrt{\frac{4\pi f \varepsilon_\infty}{\sigma_{DC}}}\right) \ [\text{S/m}] \ (5.\ 12)$$

Figure 2 shows the critical breakdown field strength of two layers soil with different values of the reflection factor using Messier formula (Messier, M., (1985)) for calculating the soil dielectric constant and its conductivity and Manna and Chowdhuri model for the calculations of the critical breakdown field strength. Based on the laboratory measurements of tests done on many soil samples using supply frequency ranges from 40 Hz to 2 MHz Visacor and Portela developed empirical equations for the soil permittivity and conductivity calculations as a function of supply frequency as the following

$$\varepsilon_r(f) = 2.34 \times 10^6 \left(\frac{1}{\sigma_{100Hz}}\right)^{-0.535} .f^{-0.597} \ (5.\ 13)$$

$$\sigma(f) = \sigma_{100Hz}\left(\frac{f}{100}\right)^{0.072} \quad (5.14)$$

Calculations of the critical breakdown field strength of two layers soil with different values of the reflection factor as given in Figure 3 are done by using Visacor *et al* formulas [42] for calculating the soil dielectric constant and the soil conductivity and Manna and Chowdhuri model for the calculations of the critical breakdown field strength. Figures 1 through 3 proved that regardless of the lightning high frequency content, the value of the critical breakdown field of the homogeneous soil having 100Ω.m is less than its value in the two-layer soil, whether the reflection coefficient is positive or negative. By comparing the critical breakdown field strength results in Figure 1 and Figure 2, it is noticed that there are marked differences in the critical breakdown field values at low frequency, while at high frequency less noticeable changes are noticed with the reflection factor variation.

In Figure 3 using Visacor et al. formula (Visacro, S., (2007)) for the calculations of both the dielectric constant and the conductivity of the soil and Manna and Chowdhuri for the calculations of the critical breakdown field strength, a sharp decrease in the critical breakdown field strength value with the increase of the lightning high frequency is observed. This is probably due to the dependence of Scott's and Messier formulas on the dc conductivity σ_{DC} to calculate the soil dielectric constant and its conductivity while on the other hand Visacor et al. formula considered the change in the soil dielectric constant and its conductivity as a function of conductivity σ_{100Hz} at 100Hz. The values of σ_{DC} and Hz σ_{100Hz} are considered in the calculations as 0.01 S/m and 0.098 S/m respectively (Markus Loewer, et al., May 23 - 27, (2016)).

Figure 1. Critical breakdown field strength using Scott expression

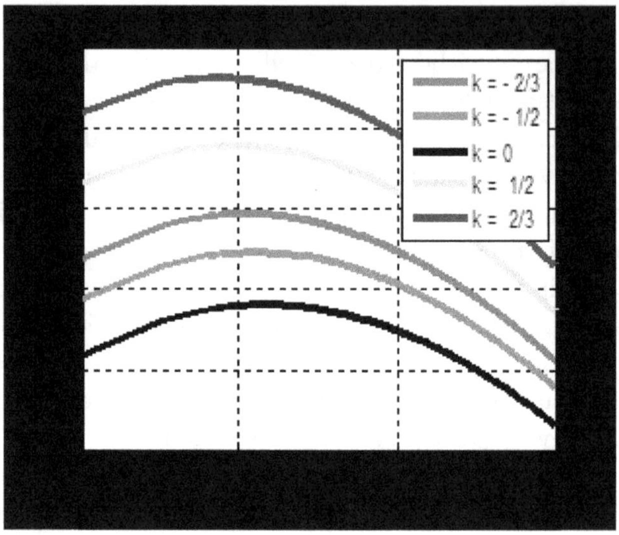

Figure 2. Critical breakdown field strength using Messier expression

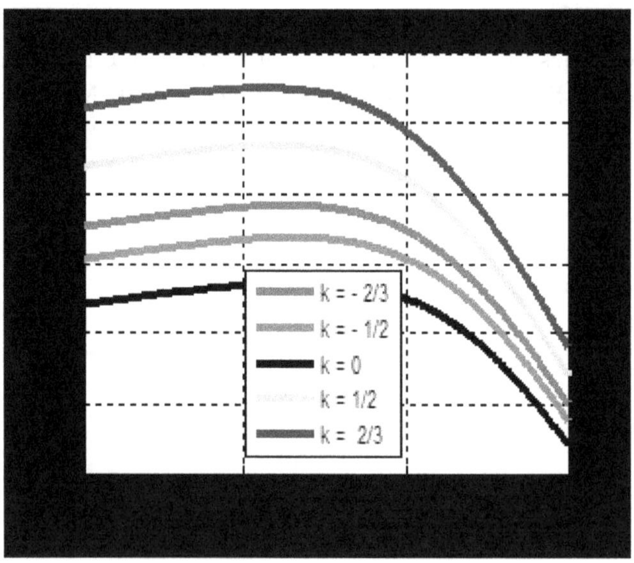

Figure 3. Critical breakdown field strength using Portola expression

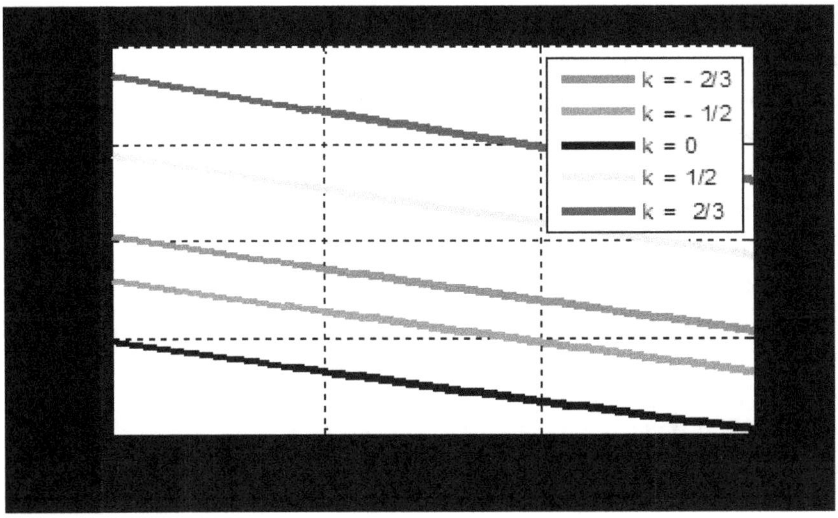

5.3.3 Lightning Current Impulse Model

In this chapter, two lightning current waveforms corresponding to the first and subsequent lightning strikes are used. The Heidler's lightning current function (HF) is chosen to represent the current waveform (Grcev, L., (2009)), (Heidler, F., Cvetic, J., (2002)).

$$i(t) = \frac{I_0}{\eta}\frac{\left(\frac{t}{\tau_1}\right)^n}{1+\left(\frac{t}{\tau_1}\right)^n}e^{(-t/\tau_2)} \quad (5.\ 15)$$

$$\eta = e^{-(\tau_1/\tau_2)(n(\tau_2/\tau_1))^{1/n}} \quad (5.\ 16)$$

Where t is the time in second, I_0 is the amplitude of the current pulse, τ_1 is the front time constant, τ_2 is the decay time constant, n is exponent having values between 2 to 10 and η is the amplitude correction factor. The first stroke waveform is reproduced by one and the subsequent stroke is reproduced by the sum of two Heidler's functions (Heidler, F., Cvetic, J., (2002)).

5.4 EFFECTIVE ELECTRODE RADIUS INCLUDING SOIL IONIZATION EFFECT

Due to lightning current the soil ionization occurs when the leaked current in the earth exceeds its critical value. The occurrence of the soil ionization by lightning current pulses leads to an increase of the grounding electrode radius subsequently, decrease in the ground resistance, transient voltage, and transient impedance are happened. The new effective radius of grounding electrode $\boldsymbol{r_i}$ in meter is proposed by equation (5. 17) which is a general equation for both vertical and horizontal rods (Liu, Y.Q., et al, (2004)), (Mokhtari, M., et al (2016)). In this equation $\boldsymbol{\rho}$ is the soil resistivity in Ω.m, $\boldsymbol{I_m}$ is the maximum value of the stroke current in kA, $\boldsymbol{E_c}$ is electric critical field in kV/m and $\boldsymbol{l}$ is the electrode length in m.

$$r_i = \frac{\rho I_m}{2 \pi E_c l} \quad (5.\ 17)$$

Figure 4 shows the effective radius of the soil ionization around the electrode with different reflection factors, calculated by various soil models when subjected to first and subsequent return current stroke. According to National Electric Code (NEC), 250.52(A) (5) in 2017 (National Electric Code (NEC), (2017)) and the field experience, the electrode length is considered to be 3 m and the laying depth of horizontal grounding electrode is 0.5 m. The actual electrode diameter was 0.014 m.

Table 1 gives the effective electrode radius including soil ionization effect at first and second lightning strokes versus the reflection factor. The variation of the soil resistivity as a result of the soil ionization around a vertical and horizontal rod due to the effect of the first stoke, at instance when the impulse current reached its peak value is given in Figures 6 and 7 by using FEM and COMSOL multiphysics program. It is noticed that the resistivities of the soil layers are decreased within the area around the rod. The reduction in the soil resistivity reaches about 38% of the steady state value of the soil resistivity and it is uniformly distributed around both the vertical and the horizontal rods.

Figure 4. Equivalent radius of horizontal electrode at soil ionization phenomena, the electrode has 3 m length with different reflection factors using Messier model at first lightning stroke

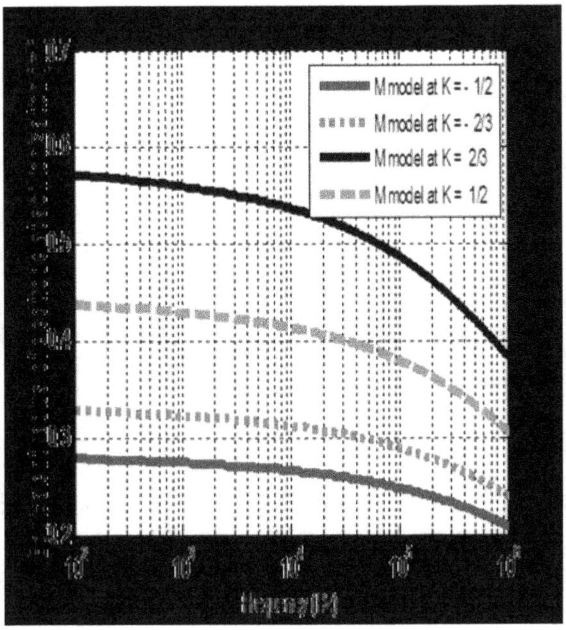

Figure 5. Equivalent radius of horizontal electrode at soil ionization phenomena, the electrode has 3 m length with different reflection factors using Messier model at subsequent lightning stroke

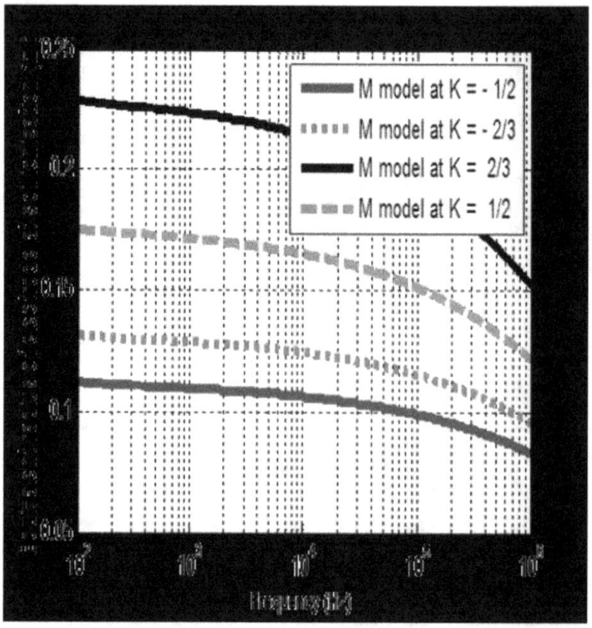

Figure 6. Peak value of impulse current using FEM

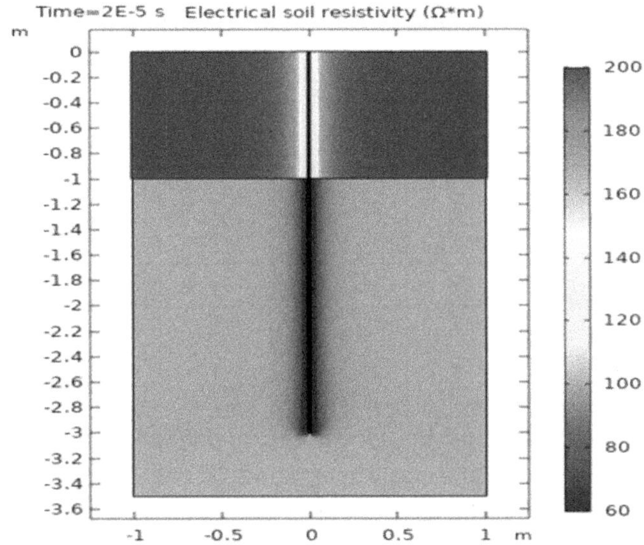

Figure 7. Peak value of impulse current using COMSOL

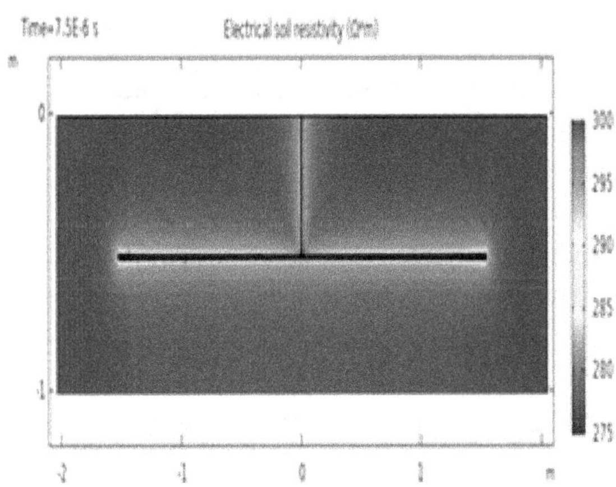

Table 1. Effective electrode radius including soil ionization effect at first and second lightning strokes verses reflection factor, (1) Effective electrode radius in [meter] without frequency effect, (2) Effective electrode radius [meter] including the soil

ionization effect using Messier expression, (3) Effective electrode radius [meter] including the soil ionization effect using Scott expression, (4) Effective electrode radius [meter] including the soil ionization effect using Visacor et al expression, K is the reflection factor

Lightning stroke	K	(1)	(2)	(3)	(4)
First stroke	2/3	0.5396	0.2935	0.5107	0.3908
	1/2	0.4123	0.3971	0.3980	0.2992
	0	0.2063	0.2038	0.2070	0.1504
	-2/3	0.3084	0.3005	0.3033	0.2242
	-1/2	0.2618	0.2566	0.2598	0.1905
Subsequent stroke	2/3	0.2183	0.1780	0.1718	0.1354
	1/2	0.1668	0.1402	0.1355	0.1037
	0	0.0834	0.0747	0.0727	0.0521
	-2/3	0.1247	0.1079	0.1046	0.0777
	-1/2	0.1059	0.0930	0.0903	0.0660

From Table 1 and Figures 4 to 7, it can be concluded that the reflection factor has a significant impact on the equivalent radius of the electrode. Increasing the reflection factor leads to influenced effect in the equivalent radius increase. Conversely, the increase in the lightning stroke frequency leads to a decrease in the equivalent radius of electrode.

5.5 TRANSIENT PERFORMANCE OF GROUND ELECTRODES

In this section, the performance of grounding electrodes (horizontal or vertical) in non-uniform soil including the effect of the soil ionization with frequency and soil resistivity variations is investigated. The investigated grounding electrode has 3 m length and 7 mm radius, the laying depth of the horizontal grounding electrode is considered as 0.5 m, and the thickness of first layer is assumed to be 1m. The soil relative permittivity is considered to be 10 for first layer and 8 for the second layer. Finally, the reflection factors are taken as -1/2, -2/3, 1/2, 2/3, and k = 0 with uniform soil resistivity 100 Ω.m.

5.5.1 Modeling of Grounding Electrode Under Lightning

In this chapter IEEE-80 standard and transmission line model (TLM) in ATP are used to simulate the grounding electrodes. In (TLM) method any grounding electrode is divided into N segments each one contains grounding resistance, capacitance and inductance elements. The formulas of R_g, C_g and L_g calculations are as given in (Leonid, G., (2009)). The resistance, grounding capacitance and electrode inductance of the vertical electrode can be calculated by using equations 5.18, 5.19 and

(5. 20) respectively (Leonid, G., (2009)). Figure 8 shows the electrodes segments and their equivalent circuit.

Figure 8. Electrodes segments, (a) horizontal electrode, (b) vertical electrode, (c) equivalent circuit

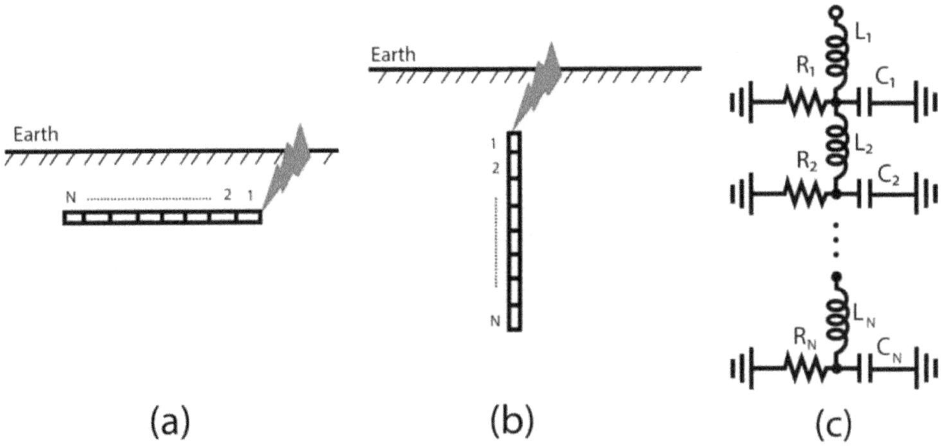

$$R_g = \frac{\rho}{2\pi}\ln(\frac{4l}{\mathbf{a}} - 1) \quad (5.18)$$

$$C_g = \frac{\rho\varepsilon}{R_g} \quad (5.19)$$

$$L_g = \frac{\mu}{2\pi}\left[\ln\frac{2l}{\mathbf{a}} - 1\right] \quad (5.20)$$

Where, μ is the soil permeability (H/m), a is the radius of grounding electrode which is replaced by r_i when the ionization process is started, ρ is the soil resistivity, l is the length of grounding electrode, and d is the laying depth of grounding grid.

The resistance, inductance and capacitance of the horizontal electrode can be calculated by using the relations given in 5.21, 5.22 and 5.23 (Leonid, G., (2009)) where the resistance R_i of the ith segment with a length of l_i can be calculated by equation 5.21, where ρ is the soil resistivity and h is the burial depth of the grounding electrode and a is the radius of grounding electrode.

$$R_i = \frac{\rho}{2\pi l_i}\left[\frac{2h+a}{l_i} + ln\frac{l_i + \sqrt{l_i^2 + a^2}}{a} - \sqrt{1+\left(\frac{a}{l_i}\right)^2} + ln\frac{l_i + \sqrt{l_i^2 + 4h^2}}{2h} - \sqrt{1+\left(\frac{2h}{l_i}\right)^2}\right] \tag{5.21}$$

In single horizontal rod the inner self-inductance of a grounding conductor under high frequency can be ignored comparing with its external self-inductance because of the strong skin effect, so the self- inductance L_i of the ith segment can be calculated by

$$L_i \approx \frac{\mu_o l_i}{2\pi}\left(ln\frac{2l_i}{a} - 1\right) \tag{5.22}$$

Where l_i is the length of the ith segment of grounding electrode and μ_o is the permeability of free space. The shunt capacitance C_i of the ith segment with a length of l_i and an ionization zone radius of a_i in an infinite medium is

$$C_i(a_i) = \frac{2\pi\varepsilon l_i}{\frac{a_i}{l_i} + ln\frac{l_i + \sqrt{l_i^2 + a_i^2}}{a_i} - \sqrt{1+\left(\frac{a_i}{l_i}\right)^2}} \tag{5. 23}$$

In the above equation ε is the soil permittivity. When the electrode is buried in a depth of h, the capacitance is obtained based on image theory.

The transient behavior of the grounding electrode is investigated by using ATP-EMTP (ATP rule book, (1995)) and the electrode parameters. The number of the segments required for an accurate simulation is dependent on the highest frequency of the injected impulse, the higher the frequency, the larger the number of sections is required. However, the difficulty of incorporating the frequency dependent soil properties limits the application of the approach in transient analysis, since the lightning current has frequency components. To investigate the effect of lightning current's front time T_f, on the soil resistivity and permittivity, the equivalent frequency f_{eq} of the lightning current is determined using (Mokhtari, M., Abdul-Malek, Z., & Wooi, C. L., (2016)).

$$f_{eq} = \frac{1}{4T_f} \tag{5. 24}$$

In the above equation f_{eq} is used in frequency dependent soil properties and T_f is the lightning current's front time. The equivalent frequencies for the first and the subsequent return stroke currents are considered 31.25 kHz and 312.5 kHz, respectively. Table 6.2 shows the effect of front time on the performance of horizontal grounding electrode in non-uniform soil for reflection factor = -2/3 at 30 kA lightning stroke. From Table 2 it is noticed that the front time has noticeable effect on the (TGPR) and transient impedance

Table 2. Front time effect on TGPR and transient impedance

Front time T_f	**Frequency (kHz)**	**(TGPR) (kV)**	**Transient Impedance (?)**
80	31.25	666.7	22.22
0.8	312.5	321.3	26.7

5.5.2 Transient Grounding Voltages in Double Layer Soil

5.5.2.1 Transient Ground Potential Rise (TGPR)

Figure 8 shows the transient ground potential rise when the horizontal and vertical rods are subjected to 30 kA first lightning strokes using Messier expression, Scott expression and Visacor and Portela formulas at constant $\rho_1/\rho_2 = 300/100$ as given in Fig. 5.4(a). The rod length is considered to be 3 m with 0.014 m diameter. From this figure it is observed that the transient voltage peak value decreases sharply when the soil ionization phenomenon is considered regardless of the use of any method in the calculation. Figures 9 through 12 give the calculated transient ground potential rise voltages using Messier expression and Scott *et al.* expression to include the effect of frequency and soil resistivity variations on the soil ionization electric field. Figure 10 contains the calculations using Messier expression at different reflection factors for horizontal electrode. In Figure 11, calculations are done using Messier expression at different reflection factors for vertical electrode and in Figure 12, the calculations are carried out using Scott expression at different reflection factors for horizontal electrode.

It is noticed that Scott *et al.* and Messier (Messier, M., (1985)) expressions calculations are very close in magnitudes and shapes. Marked difference is noticed when Visacor *et al. formula* (Markus Loewer, et al., May 23 - 27, (2016)) is used as given in Figures 13 through 16 It is noticed also that the reflection factor has an influence effects on the grounding potential rise; the minimum value is obtained in uniform soil case when the soil resistivity was 100 Ω.m. Finally, it is observed that horizontal electrodes gave lower peak values comparing with the vertical electrodes.

Figures 9 through 12 give similar relations of TGPR and reflection factors in case of subsequent lightning strokes strike and horizontal and vertical grounding electrodes. The results indicate that the transient grounding potential rise peak values are reduced compared with the TGPR of first lightning stroke due to decreasing in the effective radius of grounding electrode. As given in Figures 17 through 20 increasing the reflection factor increases the transient grounding potential rise TGPR. The time interval of subsequent lightning stroke TGPR and the time required to reach to the stroke peak value are shorter compared with the first lightning stroke grounding potential rise.

Figure 9. First lightning stroke TGPR of horizontal and vertical electrodes (a) at k= -1/2, horizontal electrode using Messier expression, Scott expression and Visacor and Portela formulas with soil ionization phenomena

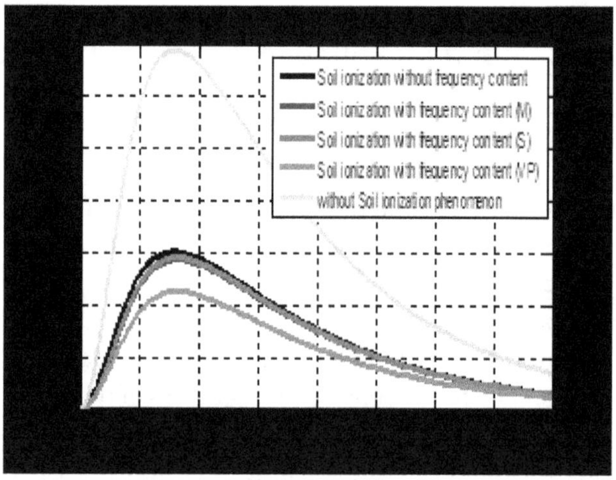

Figure 10. First lightning stroke TGPR of horizontal and vertical (b) using Messier expression at different reflection factors for horizontal electrode

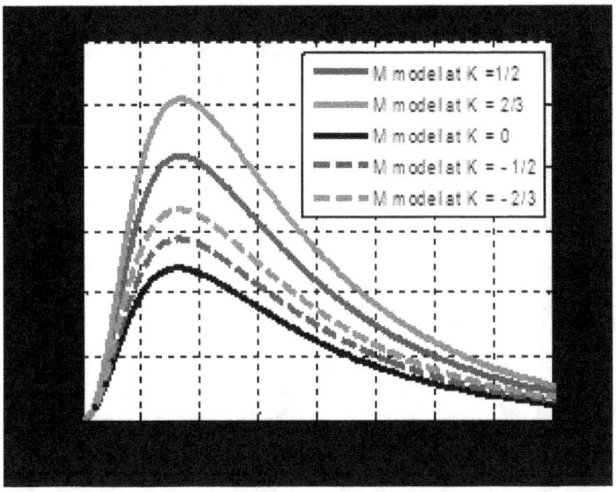

Figure 11. First lightning stroke TGPR of horizontal and vertical using Messier expression at different reflection factors for vertical electrode

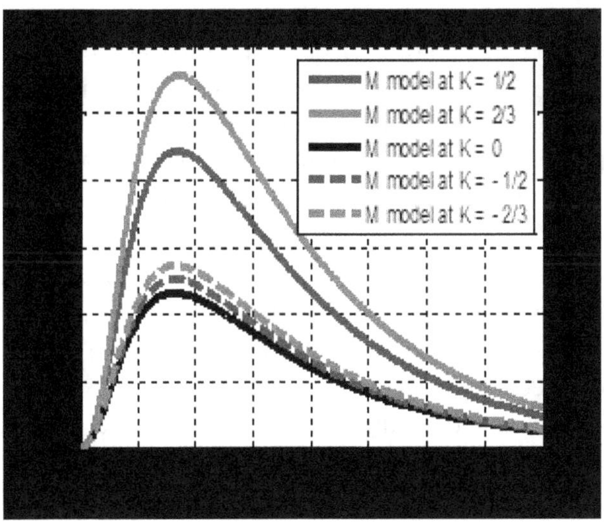

Figure 12. First lightning stroke TGPR of horizontal and vertical using Messier expression at different reflection factors for vertical electrode using Scott expression at different reflection factors for horizontal electrode

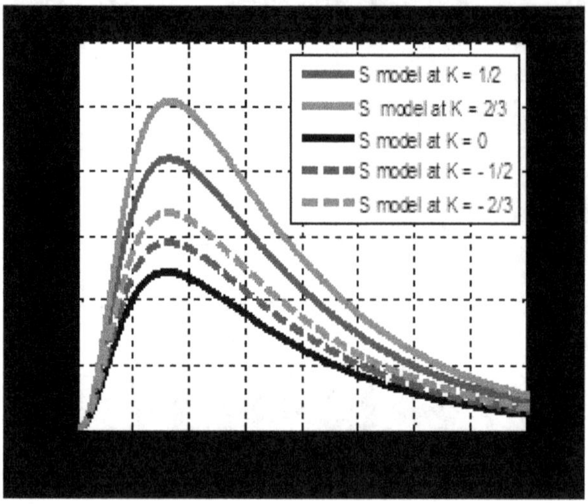

5.5.2.2 Transient Earth Surface Potential (TESP)

The earth surface potential of horizontal and vertical grounding electrodes including the effect of soil ionization for 3 m electrode length and different values of reflection factors is given in Figures 21 through 24 for first lightning stroke. The grounding electrode is simulated by ATP-EMTP (ATP rule book, (1995)). Messier expression and Scott expression are used to include the effect of frequency and soil resistivity variations on the soil ionization process. Figure 17 gives the TEPS in case of horizontal electrode using Messier expression, Figure 18 gives the TEPS in case of vertical electrode using Messier expression, Figure 19 shows the TEPS in case of vertical electrode using Scott expression and, Figure 20 gives the TEPS in case of horizontal electrode using Scott expression.

From this figure it is noticed that the TEPS is higher for the non-uniform soil (different values of reflection factors) than TEPS in case of 100 Ω.m uniform soil. It is observed also that the peak value of TEPS in the case of horizontal electrode is much lower than that in case of vertical electrode. As the results form TEPS curves the touch voltages of horizontal electrode are lower than the touch voltages of vertical electrode. Similar calculations are done in case of subsequent lightning stroke as given in Figures 21 through 24. A remarkable reduction in TEPS is noticed comparing with the values obtained in first lightning stroke calculations given in Figures 17 through 20.

Figure 13. Subsequent lightning stroke GPR of horizontal using Messier expression with soil ionization phenomena

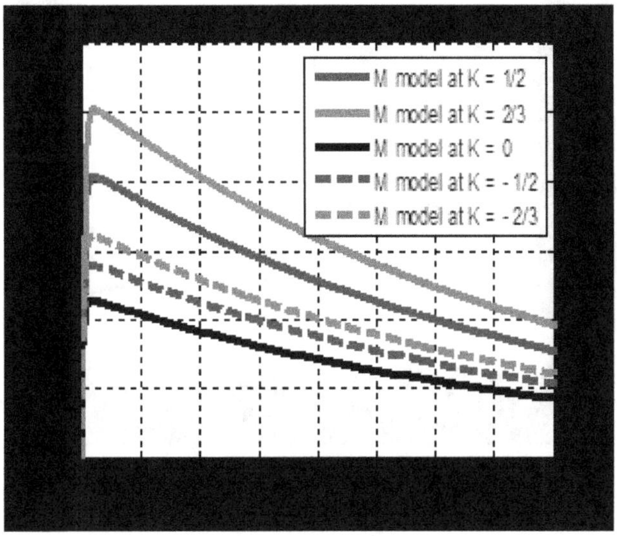

Figure 14. Subsequent lightning stroke GPR of vertical using Messier expression with soil ionization phenomena

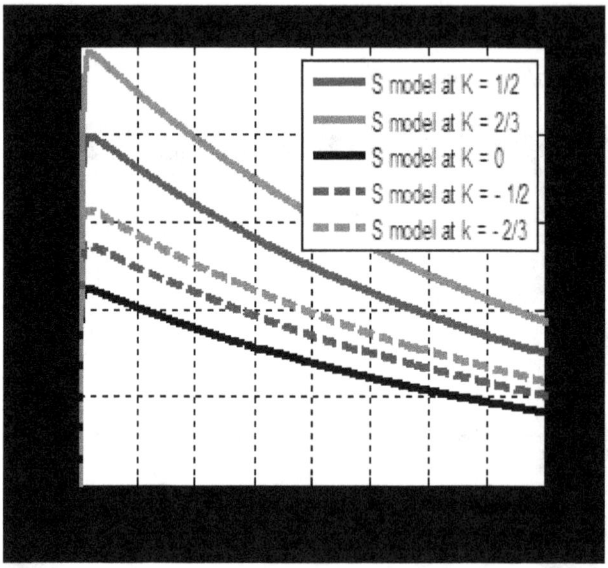

Figure 15. Subsequent lightning stroke GPR of horizontal electrode using Scott expression with soil ionization phenomena

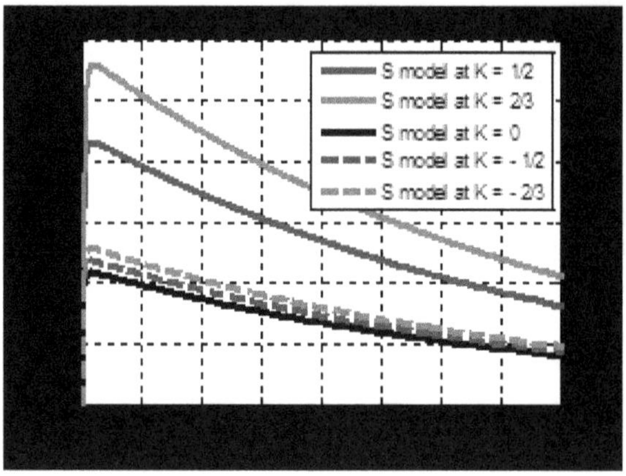

Figure 16. Subsequent lightning stroke GPR of vertical electrode using Scott expression with soil ionization phenomena

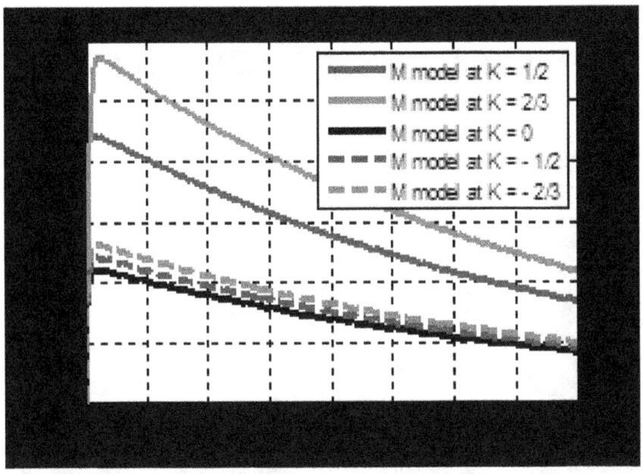

Figure 17. Earth surface potential (ESP) at first lightning stroke soil respecting ionization phenomena at time 5 μs horizontal using Messier expression with soil ionization phenomena

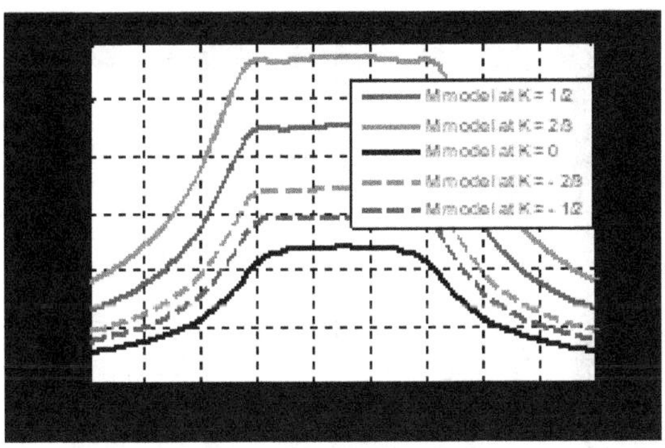

Figure 18. Earth surface potential (ESP) at first lightning stroke soil respecting ionization phenomena at time 5 μs vertical using Messier expression with soil ionization phenomena

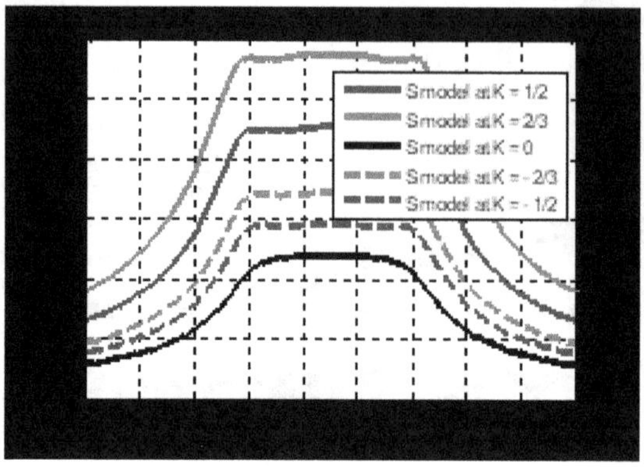

Figure 19. Earth surface potential (ESP) at first lightning stroke soil respecting ionization phenomena at time 5 μs vertical electrode using Scott expression with soil ionization phenomena

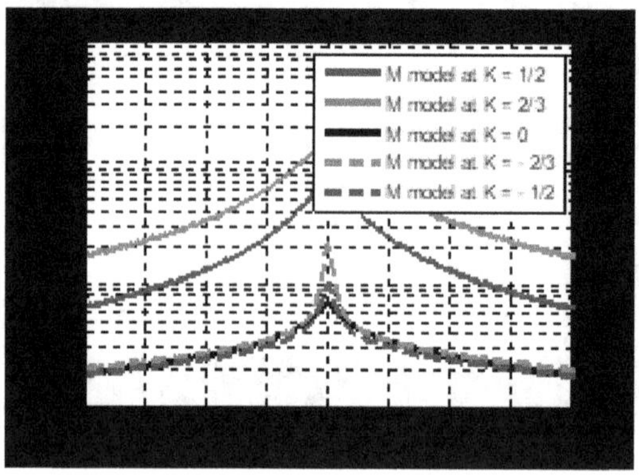

Figure 20. Earth surface potential (ESP) at first lightning stroke soil respecting ionization phenomena at time 5 μs horizontal and vertical electrode using Scott expression with soil ionization phenomena

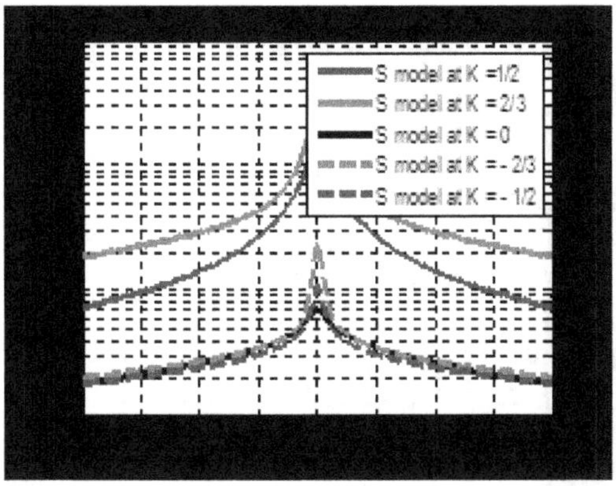

Figure 21. Earth surface potential (ESP) at subsequent lightning stroke respecting ionization phenomena at time 5 μs horizontal using Messier expression with soil ionization phenomena

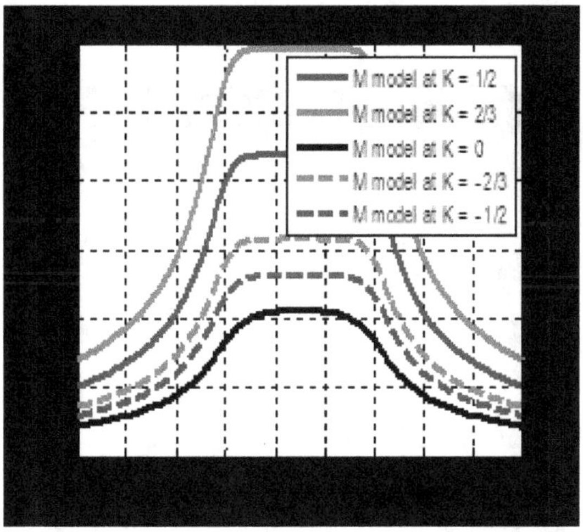

Figure 22. Earth surface potential (ESP) at subsequent lightning stroke respecting ionization phenomena at time 5 μs vertical using Messier expression with soil ionization phenomena

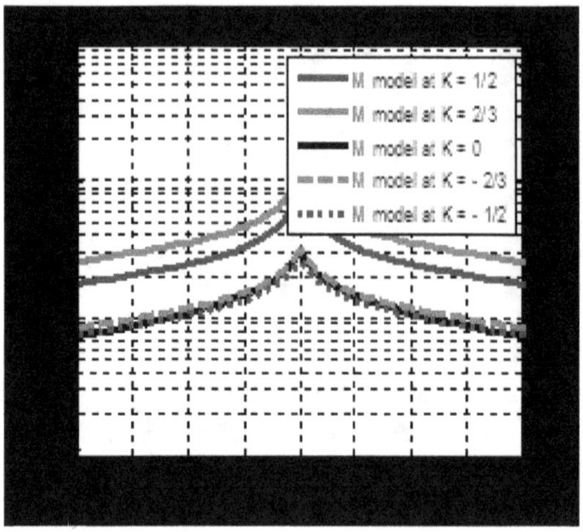

Figure 23. Earth surface potential (ESP) at subsequent lightning stroke respecting ionization phenomena at time 5 μs horizontal electrode using Scott expression with soil ionization phenomena

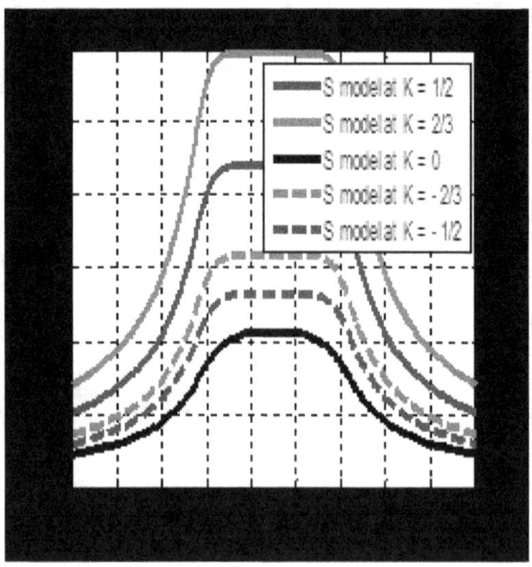

Figure 24. Earth surface potential (ESP) at subsequent lightning stroke respecting ionization phenomena at time 5 μs vertical electrode using Scott expression with soil ionization phenomena

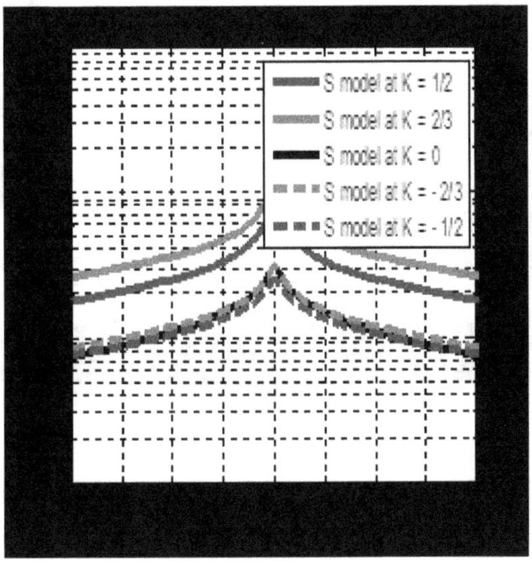

5.5.2.3 Transient Impedance of the Grounding Electrode

Figures 25 through 28 show the transient impedance of horizontal and vertical grounding electrodes including the soil ionization phenomenon and frequency content effects. In this figure the electrode is subjected to 30 kA first and subsequent lightning stroke at different values of reflection factors and at 100 Ω.m uniform soil resistivity.

Figure 25. Transient impedance at first lightning stroke respecting ionization phenomena horizontal using Messier expression with soil ionization phenomena

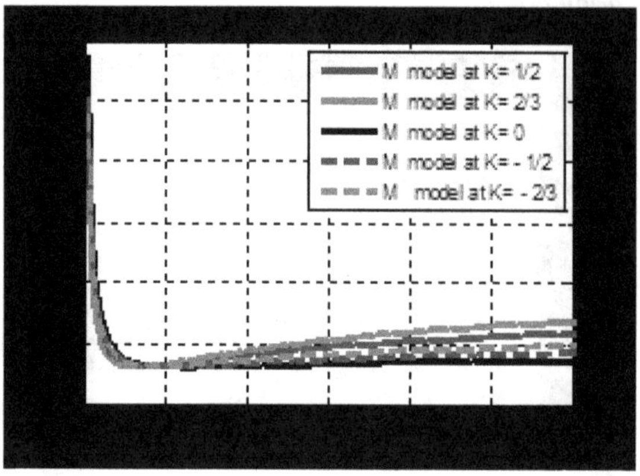

Figure 26. Transient impedance at first lightning stroke respecting ionization phenomena vertical using Messier expression with soil ionization phenomena

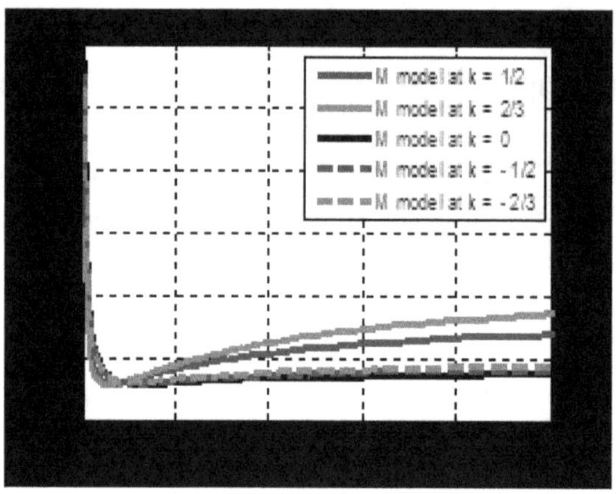

Figure 27. Transient impedance at first lightning stroke respecting ionization phenomena horizontal electrode using Scott expression with soil ionization phenomena

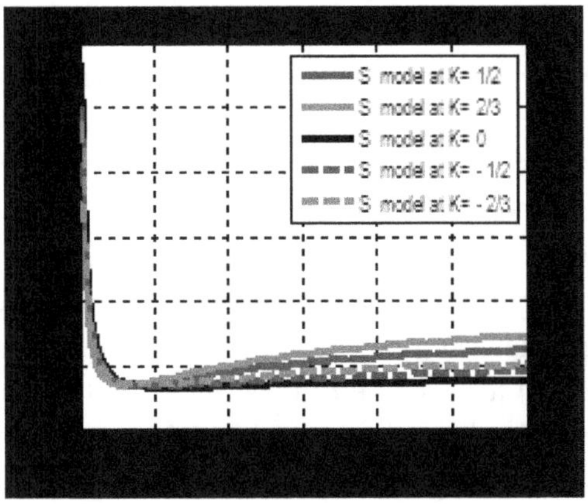

Figure 28. Transient impedance at first lightning stroke respecting ionization phenomena vertical electrode using Scott expression with soil ionization phenomena

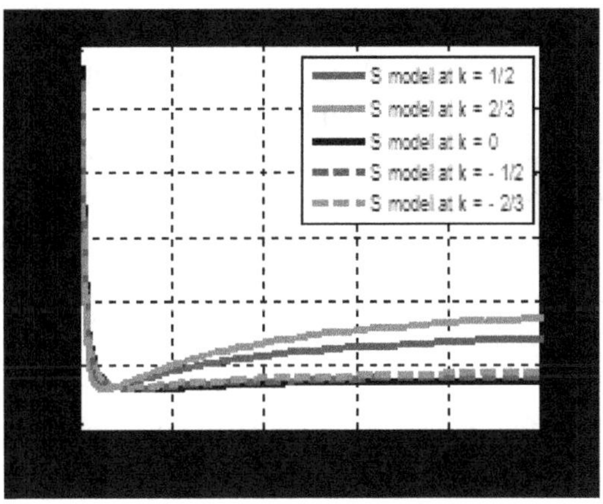

In Figures 25 through 28, Messier and Scott expressions are used to investigate the influences of the frequency and soil resistivity variations on the soil ionization in the calculations of the transient impedance. From this figure, it is noticed that impulse impedance of the soil that has negative reflection factor is lower than the impulse impedance in case of positive reflection factor. Also, it is seen that the computed impulse impedance is low for all reflection factors than impulse impedance of the grounding system in 100 Ω.m uniform soil. It is observed that after some micro seconds the impulse impedance returns to its resistance value at power frequency. It is noticed also that the reflection factor has no effect on the transient impedance of grounding electrode until reaching to about 0.5 μs in case of both horizontal and vertical rods using Messier and Scott expressions.

Figures 29 and 30 give the impulse impedances of horizontal and vertical grounding electrodes respectively with different lengths at different reflection factors using Messier model for first lightning stroke. Figure 31 shows the impulse impedance for the subsequent lightning stroke in horizontal rod and Figure 32 shows the simulation and experimental results of transient impedance under soil ionization models according to similarity approach given in (Koehler, Fabian, Swingler, J., (2018)), (Chisholm, W.A., Janischewskyj, W., (1989)), Nixon et al. (Nixon, K.J., (2006)) and CIGRE (CIGRE, (1991)) for a single rod and by using transmission line method. Figure 29 shows the relation between the horizontal grounding electrode length and the impulse impedance for different reflection factors including the effect of soil ionization with frequency and soil resistivity variations. From this figure it is noticed that increasing the horizontal electrode length reduces the grounding impedance when subjected to first lightning stroke until reaching to 20 m as given in the figure. After that length, a slight increase in the grounding impedance is noticed may be due to the increase in the electrode inductance. As shown in the same figure it is noticed that the reflection factor has remarkable decrease of the grounding impulse impedance with the increase of the electrode length until reaching to constant value. The electrode length at this value is called the effective length of grounding electrode. Also, it is observed that in the first lightning stroke case the impulse impedance reaches to effective length before subsequent lightning stroke.

On the contrary of that it is observed from Figure 29 that the increase of the horizontal electrode length decreases the grounding impulse impedance, and the reflection factor has inconsiderable effect on the grounding impedance. Similar calculations are done using vertical grounding electrode and there is no noticeable difference with previous characteristics as given in Figure 30. Figure 31 shows the subsequent lightning stroke impedance in horizontal rod.

Figure 29. Impulse impedances of grounding electrodes

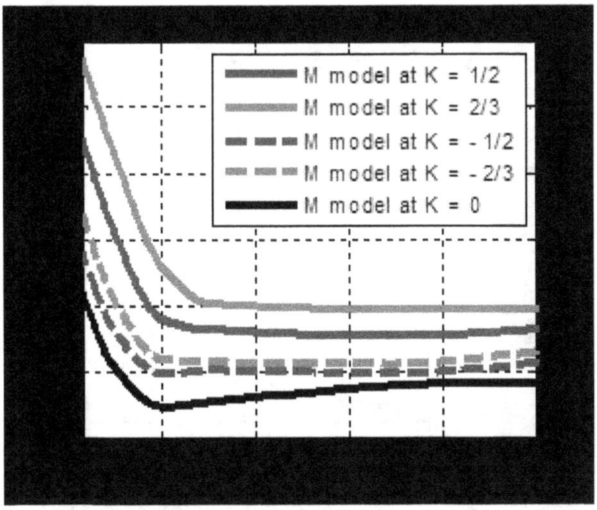

Figure 30. Impulse impedances of horizontal and vertical grounding electrodes respectively with different lengths at different reflection factors using Messier model for first lightning stroke

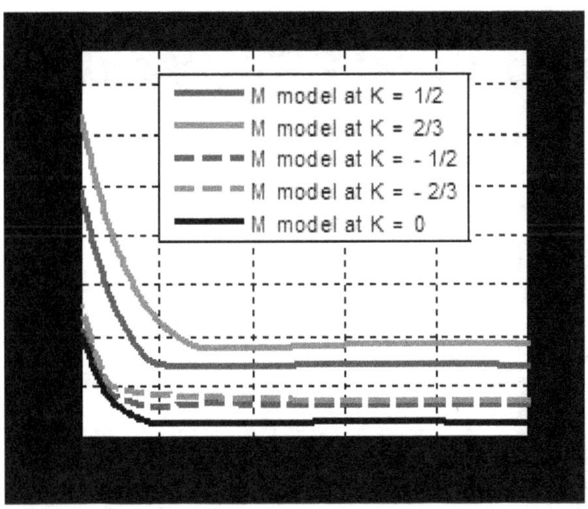

Figure 31. Impulse impedances of grounding electrodes for the subsequent lightning stroke in horizontal rod

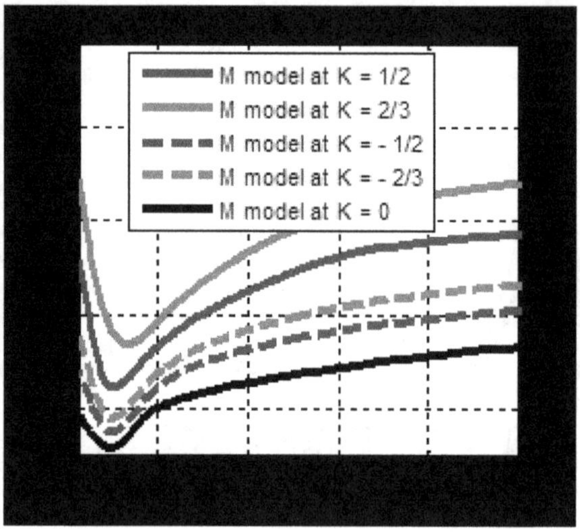

Figure 32. Simulation and experimental results of transient impedance under soil ionization models according to similarity approach (Nixon et al.) and CIGRE for a single rod by using transmission line technique

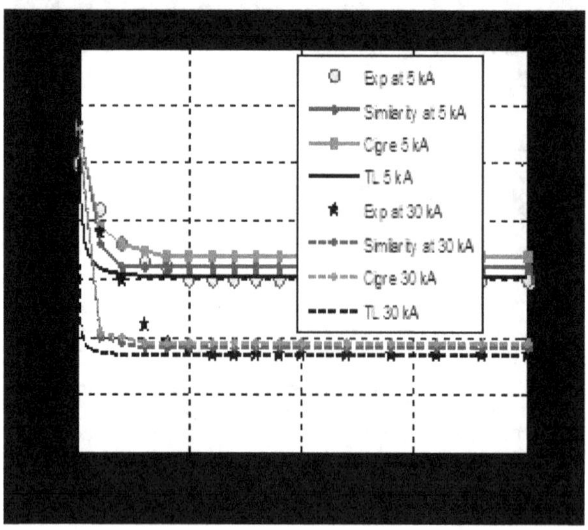

For the purpose of the verifications, the calculated values of impedances are compared with that obtained by similarity approach (Koehler, Fabian, Swingler, J., (2018)), (Chisholm, W.A., Janischewskyj, W., (1989)), experimental measurements done by Nixon et al. (Nixon, K.J., (2006)), and also compared with CIGRE model calculations (CIGRE, (1991)). The data used in this case are: $\rho_1/\rho_2 = 4.48$, the upper layer thickness is taken as 1.9 m; the lower layer thickness is considered 2.4 m until reaching to the water table. The vertical electrode length is taken as 3 m. The lightning wave shapes in laboratory tests done by Nixon, K.J. (Nixon, K.J., (2006)) and calculations are 5 kA and 30 kA with front/tail times 4/35 μs. In the case under study the critical breakdown field strength is considered to be 300 kV/m. As it seen in Fig. 5.8 (d) the calculated values done by using the present method are close to the experimental measurements more than that obtained by W.A. Chisholm, W. Janischewskyj (Chisholm, W.A., Janischewskyj, W., (1989)) and CIGRE model (CIGRE, (1991)) .

Table 3. Comparison between experimental measurement of TGPR by Nixon et al. (Nixon, K.J., (2006)), by Liu, Y.Q (Liu, Y.Q., et al, (2004)) and calculated by the present study

Calculated Max (TGPR) by the present model using the data given in Ref. (Nixon, K.J., (2006))	483.2 (kV)
Measured Max TGPR by Nixon et al (Nixon, K.J., (2006))	492 (kV)
Calculated Max (TGPR) by the present model using the data given in Ref.] (Liu, Y.Q., et al, (2004))	185 (kV)
Measured Max (TGPR) by Liu, Y.Q (Liu, Y.Q., et al, (2004))	175 (kV)

This may be because the current method takes into account the effect of soil ionization with the variations in soil conductivity and permittivity with the stroke frequency variations. The authors believe that the little difference between the measured and calculated values may be due to the change in solubility and ionisability of the soil electrolytes under the stroke electric field which is not considered in the calculations.

For further confirmation of the results obtained in this article, the TGPR calculated results are compared with the experimental and results obtained by references (Nixon, K.J., (2006)), (Liu, Y.Q., et al, (2004)) as given in Table 3. From the tabulated results it can be noticed that the difference is within +1.7% to - 5.7% which proves the used method in this article has high credibility.

5.5.2.4 The Influence of Upper Soil Layer Thickness on the Grounding Electrode Performance

It is very interesting to investigate the effect of the upper layer thickness on the ground electrode performance. Table 4 gives the effect of thickness of the upper layer soil on maximum TESP, TGPR and transient impedance for horizontal grounding electrode in non-uniform soil for different reflection factors at 30 kA first lightning stroke at time = 5μs.

Table 4. Maximum TESP, TGPR and transient impedance verses the upper layer thickness

Reflection factor (k)	**Thickness of upper layer soil (m)**	**Max. (TESP) (kV)**	**Max. TGPR) (kV)**	**Max. Transient Impedance (Ù)**
2/3	0.8 (m) 1 (m) 1.5 (m)	1224 1154 1085	1045 1016 954.49	34.833 33.86 31.816
-2/3	0.8 (m) 1 (m) 1.5 (m)	680 690 774	645.73 666.7 714.62	21.52 22.22 23.82

From Table 4, it is noticed that with the increase of upper layer thickness the maximum values of the Transient Ground Potential Rise (TGPR), Transient Earth surface potential (TESP) and Transient Impedance decrease when the reflection factor is positive, i.e. ($\rho_2 > \rho_{1)}$. The opposite is happened when the reflection factor was negative, i.e. ($\rho_1 > \rho_{2)}$. This can be understood in both cases as a result of changing the value of the upper layer soil resistivity. Table 5 gives the influence of laying depth of the horizontal grounding electrode on the TGPR and the Transient Impedance.

From Table 5, it is noticed that increasing the burial depth of the electrode reduces TGPR and the Transient Impedance regardless of whether the reflection factor is positive or negative.

Table 5. Effect of burial depth of horizontal grounding electrode in non-uniform soil

Reflection Factor (k)	**Burial depth of grounding electrode (m)**	***(TGPR)* (kV)**	**Transient Impedance (Ù)**
-2/3	0.5 (m) 0.8 (m)	666.7 596.8	22.22 19.89
2/3	0.5 (m) 0.8 (m)	1016.03 754.87	33.86 25.16

5.6 DISCUSSION OF THE OBTAINED RESULTS

In this chapter, the grounding electrodes' performance under lightning strokes in uniform and non-uniform soil including the influence of soil ionization with frequency and soil resistivity variations is investigated. Constant values of E_c for all soil types is proposed for use as medium of grounding system by Oettle (Oettle EE., (1988)), Musa (Mousa AM., (1994)) and CIGRE (ClGRE, (1991)) are 1000, 300, and 400 kV/m, respectively. These articles considered that E_c is independent of the soil properties. E.E. Oettle equation Oettle, E. E., (1988) proposed relation to calculate the E_c value as a function in the soil resistivity. In this study the formula given in equation 5.6 suggested by Manna and Chowdhuri (Manna, T.K., Chowdhury, P., (2007), to correlate the soil resistivity and soil permittivity with the critical electric field value E_c (Oettle EE., (1988)), (Manna TK, Chowdhuri P., (2007)). Based on extensive frequency variations used in the calculating of the soil critical electric field values (in kV/cm), it is concludes that the used equation is applicable for any soil types. BY comparing the critical breakdown field strength results in Figure 1, and in Figure 2, it is noticed that marked difference in the critical breakdown field values at low frequency is noticed, while at high frequency less noticeable changes are observed with the reflection factor variation. Sharp decrease in the critical breakdown field strength value with the increase of the lightning high frequency is observed as given in Figure 3.

The ionization radius around each electrode has its individual value and varies dynamically with the changes of each soil parameters, reflection factor and the grounding electrode system such as the electrode length, burial depth, soil resistivity, soil permittivity, critical electric field and the maximum value of the stroke current and its frequency. From Table 1 and Figure 4 it can be concluded that the reflection factor has significant impact on the equivalent radius of electrode. Increasing the reflection factor leads to influenced effect in the equivalent radius increase.

The changes in magnitudes of transient grounding voltages and transient impedances of the horizontal and vertical rods when subjected to the first and subsequent lightning strokes in double layer soil have been occur because of the distribution of the electric field around the electrode changes by the variations in the current distribution along the electrode, the variation in the soil surrounding the electrode resistivity done by ionization process.

This chapter is characterized by providing a model for calculating the effect of the soil ionization phenomenon on the transient probabilistic ground potential rise, the transient impedances and transient earth surface potential taking into consideration the variations in soil conductivity and permittivity with the stroke frequency, such this model was not covered in previous articles.

IEEE-80 standard, transmission line technique in ATP and the charge simulation method have been used in this study. The critical breakdown field strength of two soil layers with different reflection factors is presented. The transient probabilistic ground potential rise (TGPR) in uniform and in two layers of soil is studied. The transient impedances of horizontal and vertical electrodes are calculated considering the effect of soil ionization including the influence of soil ionization with frequency and soil resistivity variations. It is concluded that the reflection factor has significant impact on the equivalent radius of electrode. It is observed that the transient voltage peak value decreases sharply when the soil ionization phenomenon is considered and the impulse impedance of the soil that has negative reflection factor is lower than that in case of positive reflection factor. For the verifications of the present study the calculated impulse impedances and TGPR are compared with experimental results obtained by Nixon et al. and Liu, Y.Q, where a good agreement is noticed. Also, it is noticed that with the increase of upper layer thickness the maximum values of the transient ground potential rise, transient earth surface potential and transient impedance decrease when the reflection factor is positive, i.e. ($\rho_2 > \rho_{1)}$. The opposite is happened when the reflection factor was negative, i.e. ($\rho_1 > \rho_{2)}$. Finally, the paper pointed out that the change in solubility and ionisability of the soil electrolytes under the stroke electric field which is not considered in the calculations and may need more investigation in the future. This may lead to the difference between the measured and calculated values.

5.7 THE PENETRATION OF THE VERTICAL ELECTRODE INTO THE LOWER SOIL LAYER

The Finite Element Method (Liu, Y.Q., et al, (2004)) is used to simulate the penetration of the vertical electrode into the lower soil layer. The simulation is based on the following equations:

$$\nabla . J = 0 \quad (5.25)$$

$$J = \sigma E + \frac{\partial D}{\partial t} + J_e \quad (5.26)$$

$$E = -\nabla V \quad (5.27)$$

In the above equations, J is the current density vector in (A/m[3]); σ is the conductivity of the soil surrounding the ground electrode in(S/m); E is the soil electric field in (V/m); and J_e is the density of the current in (A/m[3]); V is the soil potential in (Volt).The proposed model consists of an electrode buried in two layer soils, the upper boundary condition is considered as an electrical insulation boundary, which is characterized by:

$$n.J = 0 \quad (5.28)$$

In the above equation, n is the outward normal direction to the surface of that boundary side. The other three boundaries are taken as floating potential. In this study, the soil ionization was taken when the electric field exceeds its critical value under the influence of the lightning current, and the soil resistivity value in this case is according to what was stated in Equation (5.29) (Mokhtari, M., et al (2016))

$$\rho = \rho_0 e^{(-t/\tau)} \quad (5.29)$$

In equation (5.39) ρ_0 is the soil resistivity at steady-state, t is the time equals zero at the instant of *local field = the critical field* and τ is the soil ionization constant. The decrease in the soil resistivity around the electrode occurs as a result of the ionization process according to Equation (30). Through the de-ionisation procedure, the soil resistivity ρ surrounding the electrode will be changed according to equation (5.30) (Mokhtari, M., et al (2016))

$$\rho = \rho_i + (\rho_0 - \rho_i)(1 - e^{-t/\tau_1})(1 - E/E_c)^2 \quad (5.30)$$

In the above equation ρ_i is the minimum value reached by the soil resistivity around the ground electrode as a result of the ionization procedure, τ_1 is the de-ionisation time constant of the soil surrounding the electrode and finally E is the value of the electric field. The increase in soil resistivity from ρ_i to ρ_0 results from the removal of ionization (de-ionisation) from the soil.

Figure 33. Penetration of the vertical electrode into the lower soil layer, electrode length is 6m, upper layer thickness is 5m ρ_1=2000 Ω.m/ρ_2=500 Ω.m

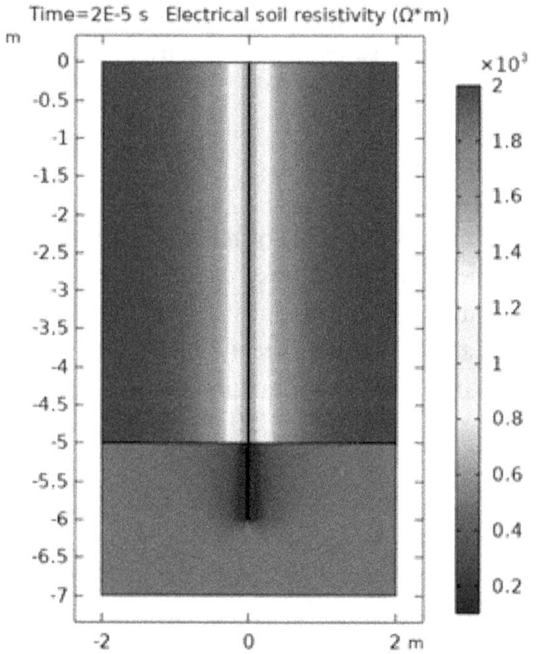

Figure 34. Penetration of the vertical electrode into the lower soil layer, electrode length is 6m, upper layer thickness is ρ_1=500 Ω.m/ρ_2=2000 Ω.m

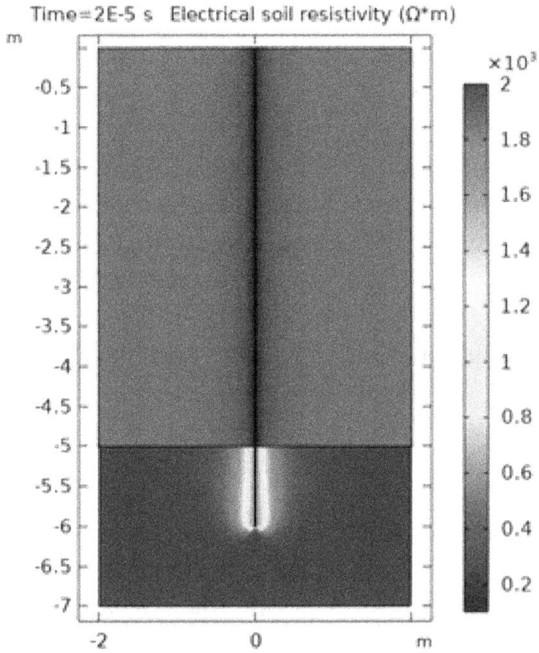

The soil critical field E_c is selected to be 110 kV/m, furthermore, the time constants are chosen the same values as those reported by Liu and Darveniza (Liu, Y.Q., et al, (2004)), (τ = 2 µs, τ_1 = 3ìs). Figure 4 illustrates the penetration of the vertical electrode into the lower soil layer, the electrode length is 6m, and upper layer thickness is 5m. In Figure 33, ρ_1=2000 Ω.m/ρ_2=500 Ω.m and in Figure 34 ρ_1=500 Ω.m/ρ_2=2000 Ω.m. Comparing the two figures it is noticed that the soil layer that has a greater resistivity leads to an increase in the diameter of the ionized soil layer, and thus the effective diameter of the electrode.

5.8 PARAMETERS AFFECTING THE GROUNDING ELECTRODE UNDER LIGHTNING

5.8.1 Soil Resistivity and Lightning Frequency Impacts on the Radius of the Ground Electrode During Soil Ionization

Due to the occurrence of lightning strikes, the value of the electric field increases in the area surrounding the ground electrode, causing an ionization event in the soil. When the electric field exceeds its critical value, ionization of the soil surrounding the ground electrode leads to an increase in its effective radius., which necessarily leads to a decrease in the step and touch voltages, in the ground potential rise around the electrode, and in the transient impedance of the electrode. The effective radius r_i of the electrode can be obtained by the use of equation (5.31).

$$r_i = \frac{\rho I_m}{2\pi E_c l} \tag{5.31}$$

In equation (5.31) ρ is the soil resistivity in Ω.m, I_m can be defined as the peak value of the lightning currentin kA, the critical fieldE_c is in kV/m and l is the electrode length in meter.The calculations of the critical field are carried out using Chowdhuri- Manna formula as illustrated in equation (5.31). Messier formula (Messier, M. A., (1985)), (Pedrosa, A. G., et al., (2010)) is used to calculate both soil permittivity and conductivity as given in equations (9) and (10). Chowdhuri- Manna and Messier formulas are selected in the computations because they have proven their accuracy compared to other algorithms, the steps used in the calculations is illustrated in Figure 35.

Figure 35. Steps of calculations flowchart

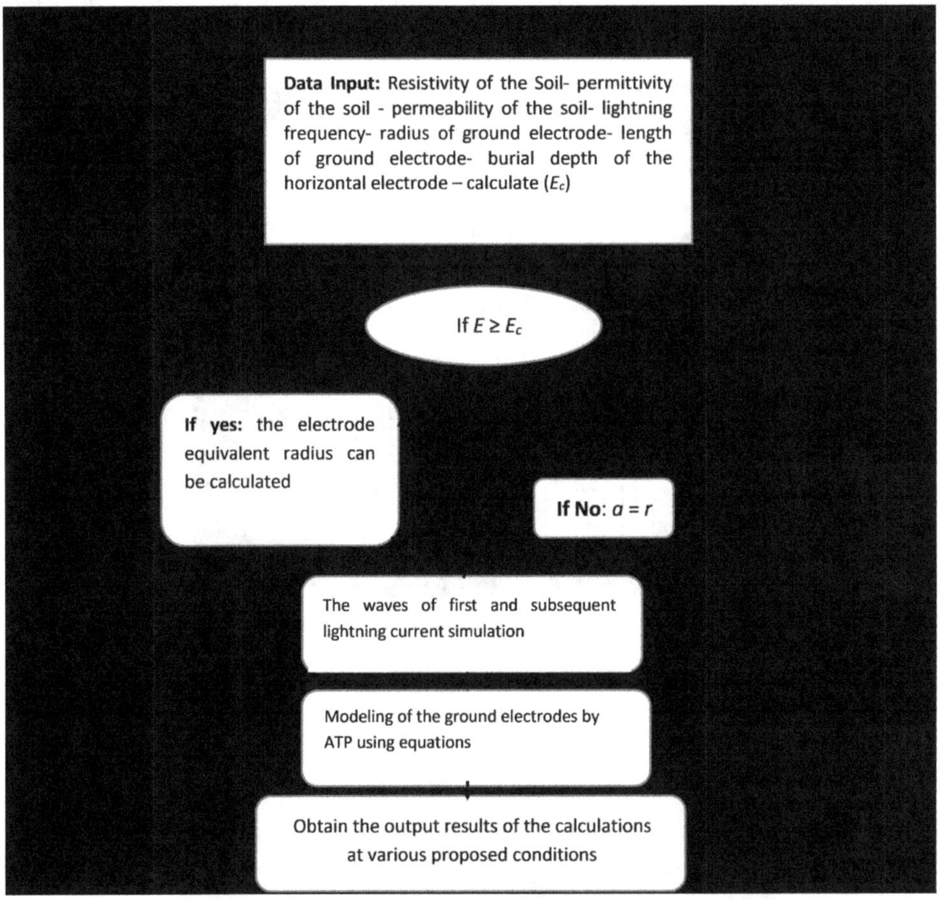

Figure 36 illustrates the effective radius of the ground electrodes, vertical and horizontal, as a function of the lightning frequency when soil ionization occurs; the electrode length was 6 meters and 7 mm radius. The computations is done at ionization of homogeneous soil under the influence of first lightning wave ionization of homogenous soil under the influence of subsequent wave at two-layer soil ionization under the influence of first lightning wave at two-layer soil ionization under the influence of subsequent wave.

Figure 36. The effective radius of the ground electrodes as a function of the lightning frequency when soil ionization occurs at ionization of homogeneous soil under the influence of first lightning wave

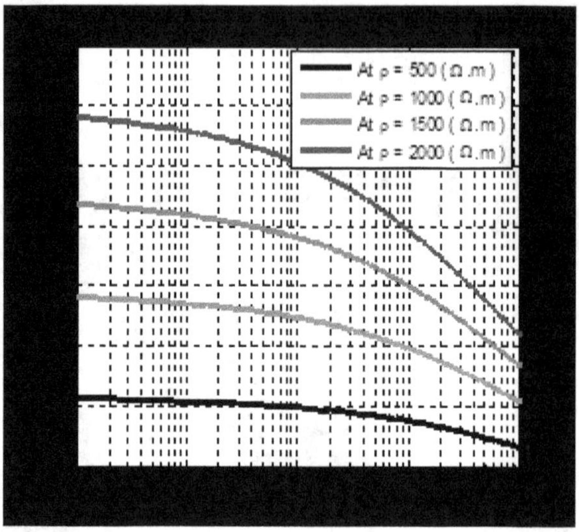

Figure 37. The effective radius of the ground electrodes as a function of the lightning frequency when soil ionization occurs at ionization of homogenous soil under the influence of subsequent wave

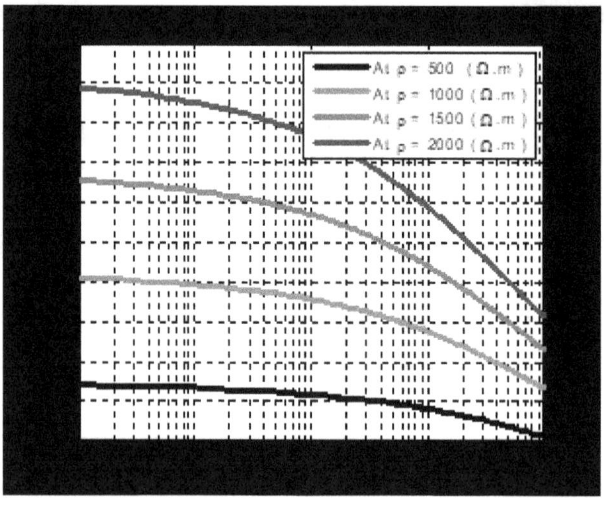

Figure 38. The effective radius of the ground electrodes as a function of the lightning frequency when soil ionization occurs at two-layer soil ionization under the influence of first lightning wave

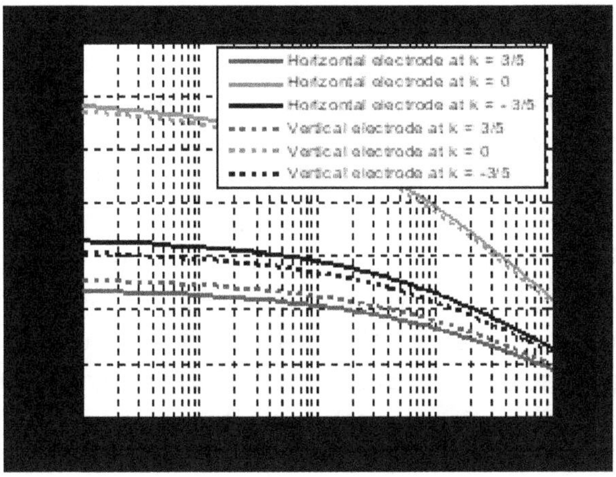

Figure 39. The effective radius of the ground electrodes as a function of the lightning frequency when soil ionization occurs at two-layer soil ionization under the influence of subsequent wave

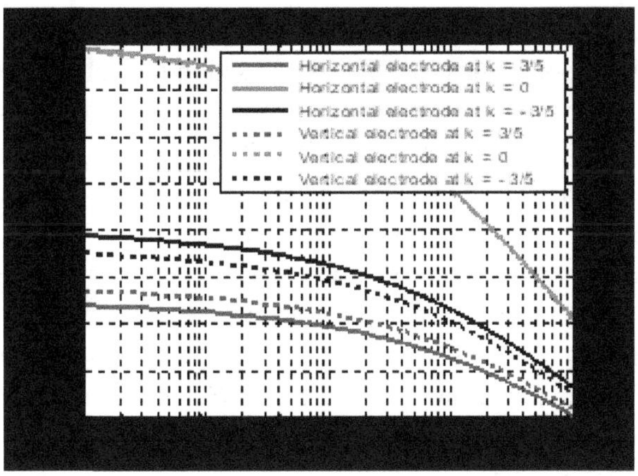

The peak values of the lightning current are chosen to be: 30 kA for first wave and 12 kA for subsequent wave. The effective radius is a function of the frequency range between 1000 kHz and 0.1 kHz.

The soil resistivity is changed between 500 and 2000 Ω.m in homogenous soil and the reflection factors are taken as -3/5, 3/5 in heterogeneous soils, at k = 0 the soil resistivity is considered 2000 Ω.m, ρ_1/ρ_2 =2000/500 when k = -3/5 and ρ_1/ρ_2 =500/2000 when k = 3/5. As it is illustrated in Figures 35 and 36, in homogeneous soil, with a decrease in the frequency of the lightning wave and at the same time an increase in the soil resistivity, the effective radius of the ground electrode increases if it is struck by first and subsequent waves. It is also noted that the effective radius of the ground electrode decreases if the opposite occurs. The same phenomenon is occurred in the case of two-layer soil as shown in Figures 37 and 38 for horizontal and vertical ground electrodes, but a slight difference was observed between the values of the effective radius in the cases of horizontal and vertical electrodes.

This difference may reach 6% when the frequency of the lightning wave equal 0.1 kHz, while the reflection factor is -3/5 and + 3/5. This difference can be neglected when the lightning wave frequency reaches 1 MHz at the same reflection coefficient, while the difference reaches zero in the case of the reflection factor equal to zero as illustrated in Figures 38 and 39.

5.8.2 The Influences of the Peak Values of the Lightning Strike on the Ground Potential Rise (GPR) Around the Ground Electrode

Figure 12 illustrates the maximum values of the ground potential rise verses the peaks of lightning when horizontal and vertical electrodes were struck by lightning waves. The lightning peak values have been changed from 40 to 160 kA, taking into account the effect of soil ionization. The data used in the calculations of (GPR) by employing Messier model are: the length of the electrode is 6 m, its diameter is 14 mm, ρ1/ρ2 =2000/500 (k = -3/5), ρ_1/ρ_2 =500/2000 (k = 3/5) and at ρ_1=$\rho_{2=}$ 2000 – 500 Ω.m, (k = 0) in case of homogenous soil.

From Figures 40 to 43 it is noticed that (GPR) values in case of horizontal electrodes are higher than their values in case of using vertical electrodes of the grounding system, whatever the lightning current values. It has also been noted that the positive reflection coefficient, which means that the resistivity of the lower soil layer is higher than the upper one, gives a higher (GPR) in the soil surrounding the electrode than in the case of a negative reflection coefficient, in which the resistivity of the lower soil is less than the upper one. These findings can be confirmed by reviewing Figures 39 and 40.

As it is noticed in Figures 42 and 43, the values of the ground resistivity have powerful influence in (GPR) values. It can be said with certainty that increasing the resistivity of the soil around the electrode increases the (GPR) values, regardless of the electrode configuration.

Figure 40. Relation between the maximum values of the ground potential rise and the peaks of lightning vertical ground electrode, k = 3/5, k = 0and k = -3/5

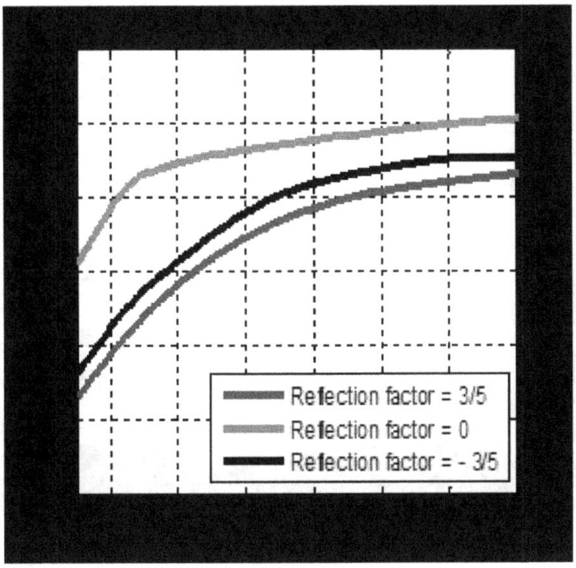

Figure 41. Relation between the maximum values of the ground potential rise and the peaks of lightning horizontal ground electrode, k = 3/5, k = 0and k = -3/5

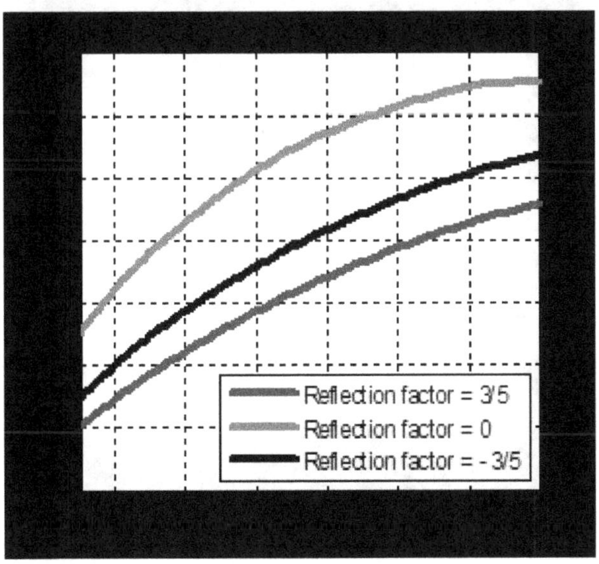

Figure 42. Relation between the maximum values of the ground potential rise and the peaks of lightning vertical ground electrode, $\rho_1=\rho_{2=}$ *2000 – 500 Ω.m*

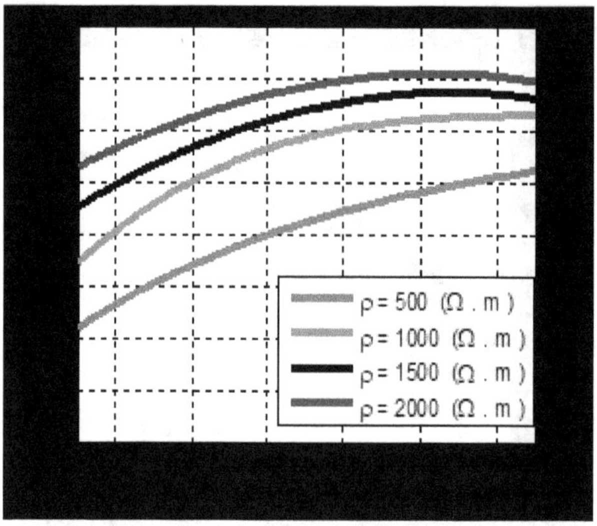

Figure 43. Relation between the maximum values of the ground potential rise and the peaks of lightning horizontal ground electrode $\rho_1=\rho_{2=}$ *2000 – 500 Ω.m*

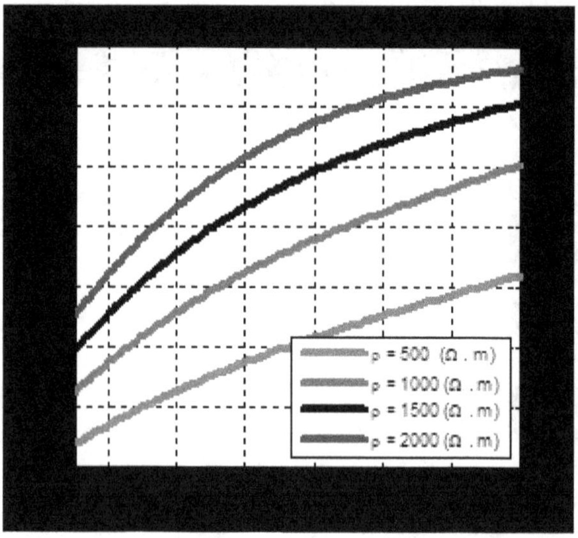

5.8.3 The soil resistivity influence on the impulse impedance the ground electrodes

The impulse impedance is defined as the ratio between the peak values of the transient (GPR) and the lightning current, as given in equation (5.32).

$$Z(impulse) = \frac{Peak\ value\ of\ v(t)}{Peak\ value\ of\ i(t)} \quad (\Omega) \qquad (5.32)$$

Figures 44 and 45 illustrate the relation between the impulse impedance of the ground electrode and the lightning current peaks. The calculations are done when ρ1/ρ2 =2000/500 (k = -3/5), ρ_1/ρ_2 =500/2000 (k = 3/5) and at $\rho_1=\rho_2=$ 2000 – 500 Ω.m, (k = 0) in case of homogenous soil. From this figure, it is clear that there is a difference in the impulse impedance between the positive and negative reflection coefficient cases, which may reach 10.9% when the value of the lightning current was 30 kA. This difference decreases as the value of the peak lightning current increases when using vertical electrode. The differences in impulse impedance magnitude increase with using the horizontal electrodes, reaching 22.3% at a lightning strike of 30 kA. This confirms the effect of the upper layer of soil on the impulse impedance of the electrode.

The difference in the impulse impedance of homogeneous and two-layer soils with negative reflection factors reaches 48% and 36% in the cases of horizontal and vertical electrodes, respectively. It is also shown that increasing the resistivity of homogeneous soil leads to an increase in the impulse impedance, regardless of the electrode configuration and the value of the lightning wave current. This is evident from Figures 46 and 47.

Figure 44. Relation between the impulse impedance and the peaks of lightning vertical ground electrode, k = 3/5, k = 0and k = -3/5

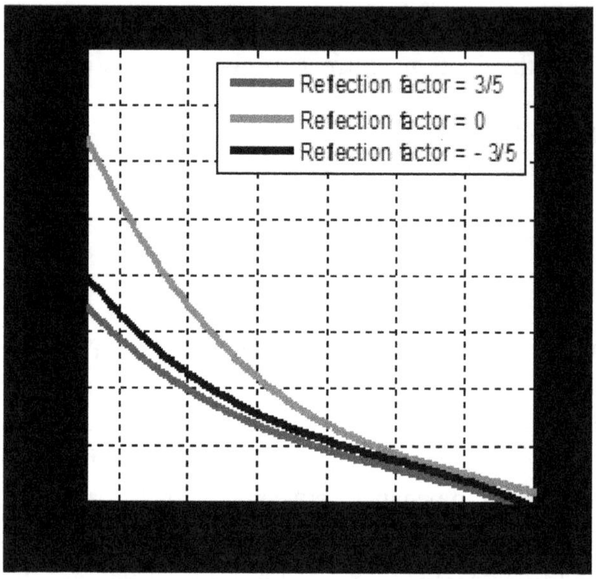

Figure 45. Relation between the impulse impedance and the peaks of lightning horizontal ground electrode, k = 3/5, k = 0and k = -3/5

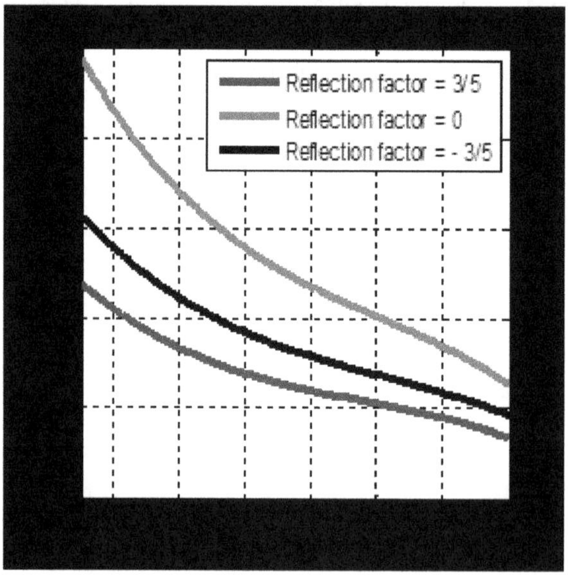

Figure 46. Relation between the impulse impedance and the peaks of lightning vertical ground electrode, $\rho_1 = \rho_{2=}$ 2000 – 500 Ω.m

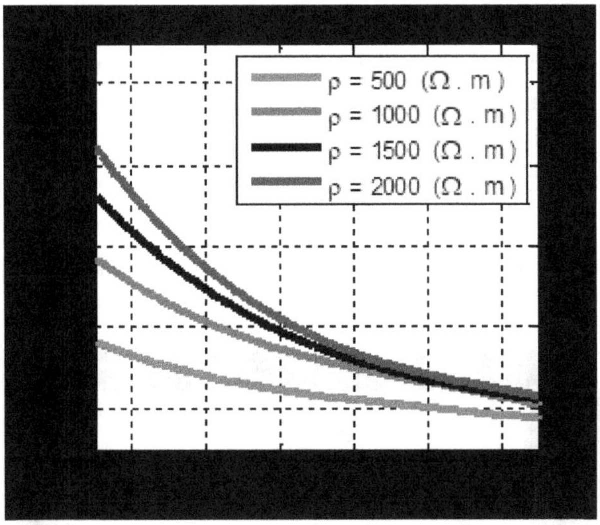

Figure 47. Relation between the impulse impedance and the peaks of lightning horizontal ground electrode, $\rho_1 = \rho_{2=}$ 2000 – 500 Ω.m

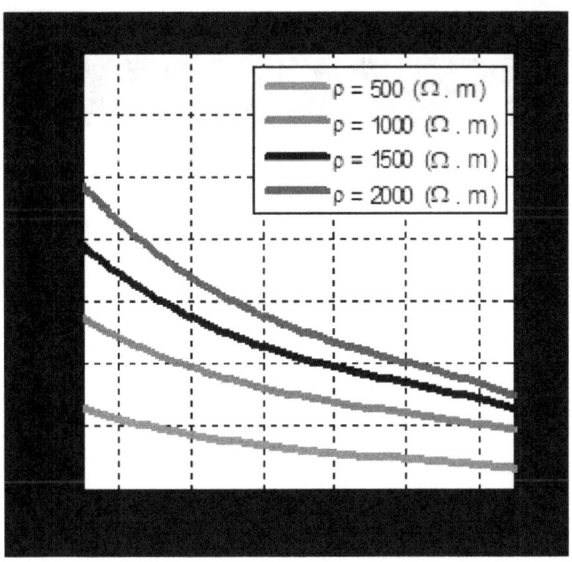

5.8.4 Front Time Influence on the Maximum (GPR) in Case of the Soil Ionization

To study the influences of the front time on the ground potential rise (GPR), a lightning wave with a peak current equal to 100 kA is used. Messier's model has been employed for this. Figures 48 and 49 show the relationship between the GPR and wave the front time in cases of k = -3/5, k = 0 and k = 3/5, using horizontal and vertical electrodes.

Figure 48. Relation between the (GPR) and the front time, the lightning current peak is 100 kA vertical ground electrode, k = 3/5, k = 0and k = -3/5

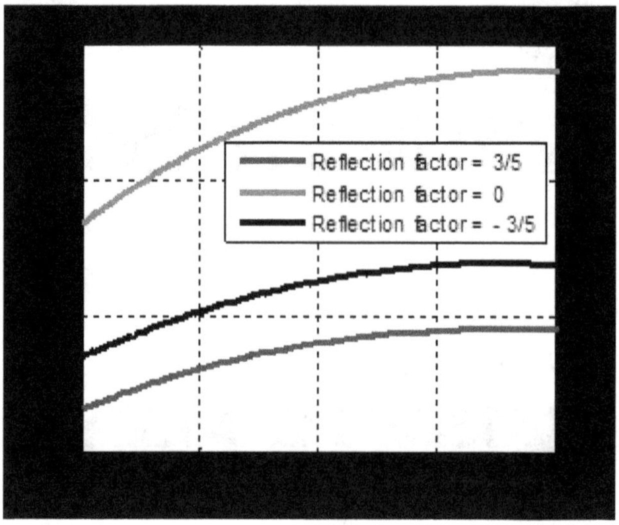

Figure 49. Relation between the (GPR) and the front time, the lightning current peak is 100 kA horizontal ground electrode, k = 3/5, k = 0and k = -3/5

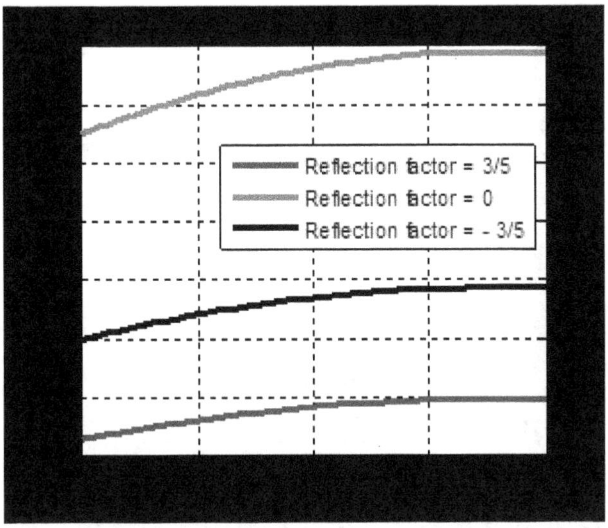

Figure 50. Relation between the (GPR) and the front time, the lightning current peak is 100 kA vertical ground electrode, $\rho_1 = \rho_{2=}$ 2000 – 500 Ω.m

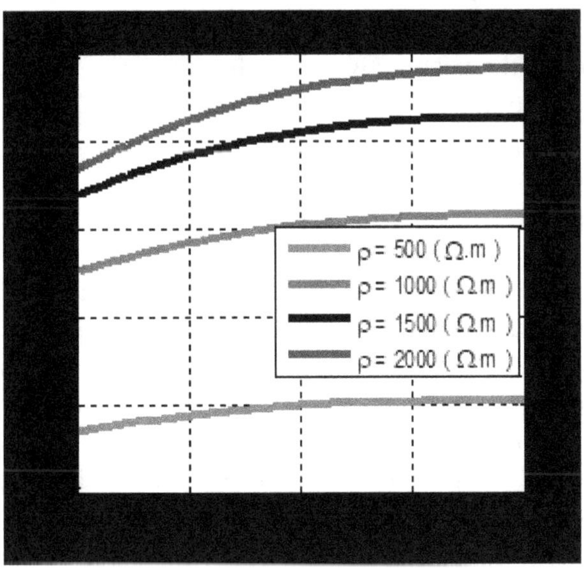

Figure 51. Relation between the (GPR) and the front time, the lightning current peak is 100 kA horizontal ground electrode, $\rho_1=\rho_{2=}$ 2000 – 500 Ω.m

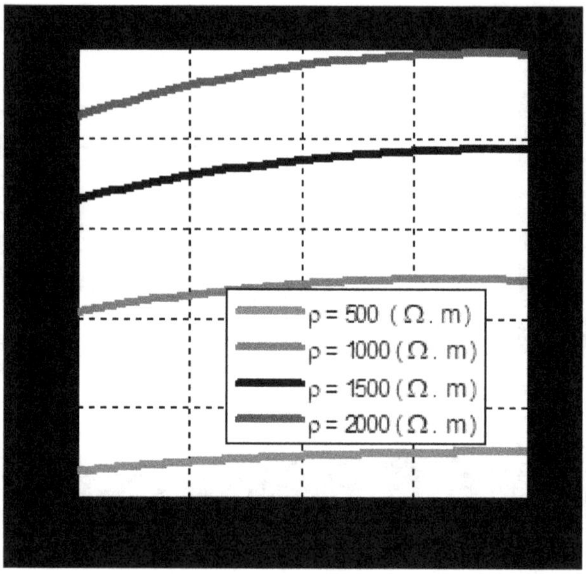

The effect of front time on (GPR) was examined in the case of homogeneous soil, as illustrated in Figures 50 and 51, it was found, in general, that increasing the front time leads to an increase in (GPR), regardless of the value of the soil's resistivity or the reflection coefficient in the two-layer soil, but the rate of increase of (GPR) depends on the soil resistivity as well as the reflection factors in addition to the wave the front time

5.8.5. Front Time Influence on the Maximum (GPR) in Case of the Soil Ionization

Similar computations have been carried out to study the effect of wave front time on the impulse impedance. The results are as illustrated in Figures 52 and 53, for two-layer soils with $k = -3/5$, $k = 0$ and $k = 3/5$ for vertical and horizontal ground electrodes. The computations of homogenous soil are illustrated in Figures 54 and 55. In general, it has been confirmed from the previous figures it is observed that increasing the wave front time leads to an increase in the impulse impedance of ground electrodes, but there are factors influence the rate of increase, such as the reflection factors of the two layer soils, the lightning front time, and the resistivity of homogenous soil

5.8.6 Impact of the Electrode Length on the Impulse Impedances

In this part of the chapter, a study is conducted on the effect of electrode length on its impulse impedance in the case of two-layer, k = 3/5 and k = -3/5, and homogeneous soils, k = 0, with the use of Messier model. The calculations are performed in the case of ionization and non-ionization of the soil. The first lightning stroke with 30 kA is applied. Figures 56 and 57 illustrate the results obtained for different reflection factors.

Figure 52. Relation between the impulse impedance and the front time, the lightning current peak is 100 kA vertical ground electrode, k = 3/5, k = 0and k = -3/5

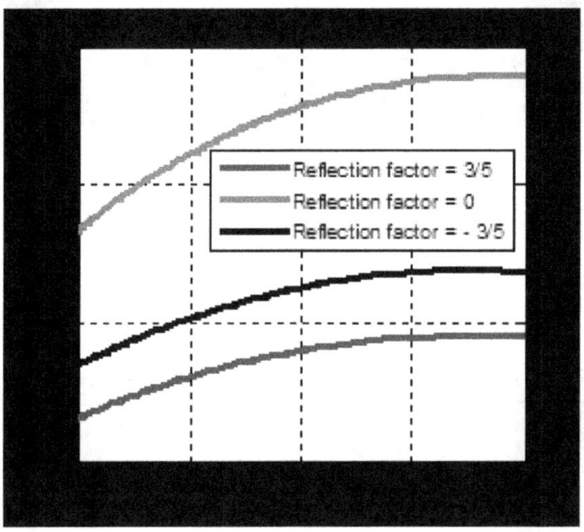

Figure 53. Relation between the impulse impedance and the front time, the lightning current peak is 100 kA horizontal ground electrode, k = 3/5, k = 0and k = -3/5

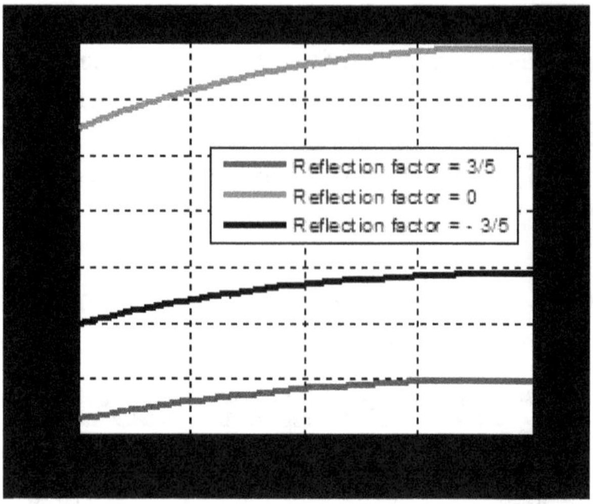

Figure 54. Relation between the impulse impedance and the front time, the lightning current peak is 100 kA vertical ground electrode, ρ_1=$\rho_{2=}$ 2000 – 500 Ω.m

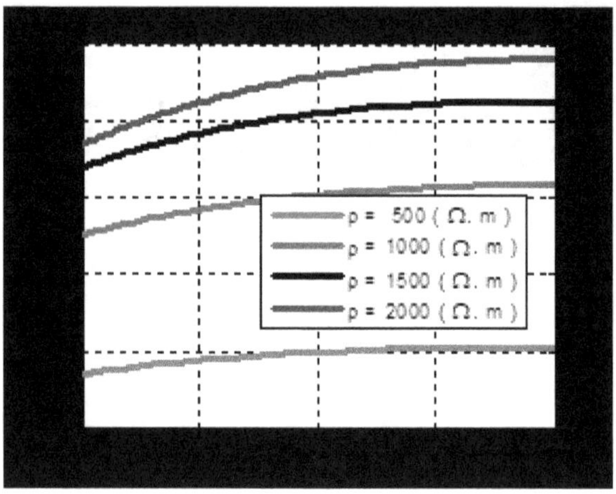

Figure 55. Relation between the impulse impedance and the front time, the lightning current peak is 100 kA horizontal ground electrode, ρ_1=$\rho_{2=}$ 2000 – 500 Ω.m

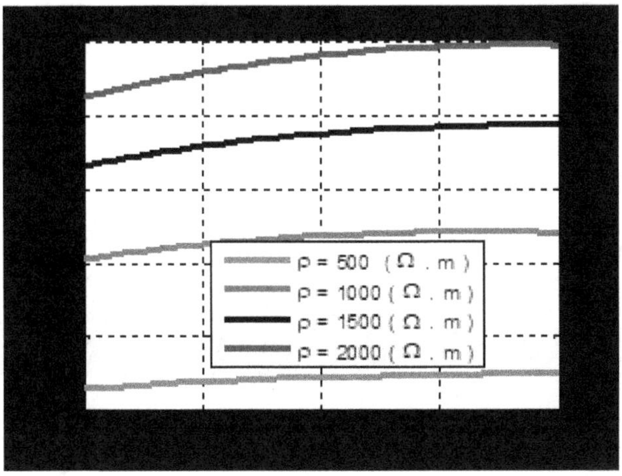

From this figure it is noticed that with the increase in the electrode length, the impulse impedance decreases, regardless of the ground electrode configuration. In addition, a further decrease in the impulse impedance occurs when the soil is ionized until reaching to a certain length, which can be defined as the critical length, or the effective length, at that length (also defined as the saturated length), the ground electrode impedance, reaches a steady state and therefore does not change with increasing the electrode length.

From Figures 56 and 57 it is also noted that the critical value of the vertical electrode is constant regardless of the change in the reflection factor, whether positive or negative. However, in the case of the horizontal rod, the critical length in the case of the positive reflection coefficient is lesser than its value in the case of the negative reflection factor. It was also noted that the critical value of the vertical electrode is always shorter than the horizontal one. It is worth noting here that with the soil ionization, the impedance of the short electrodes is approximately the same value, regardless of the values of the reflection coefficients. From this it is clear that the ionization process clearly affects the critical length of the grounding rods.

Finally it has been observed that the reflection coefficient, the electrode configuration, and soil ionization or de-ionization affect the rate of decrease in the electrode impedances until the critical value of the electrode length is reached, at which point the impedance values are mostly equal.

Figure 56. The influences of the upper layer thickness of two soil layer on the grounding electrodes characteristics

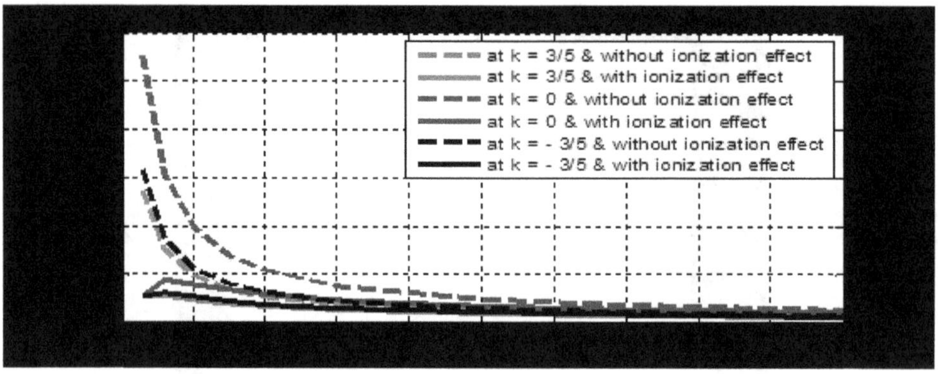

Figure 57. The influences of the upper layer thickness of two soil layer on the grounding electrodes characteristics

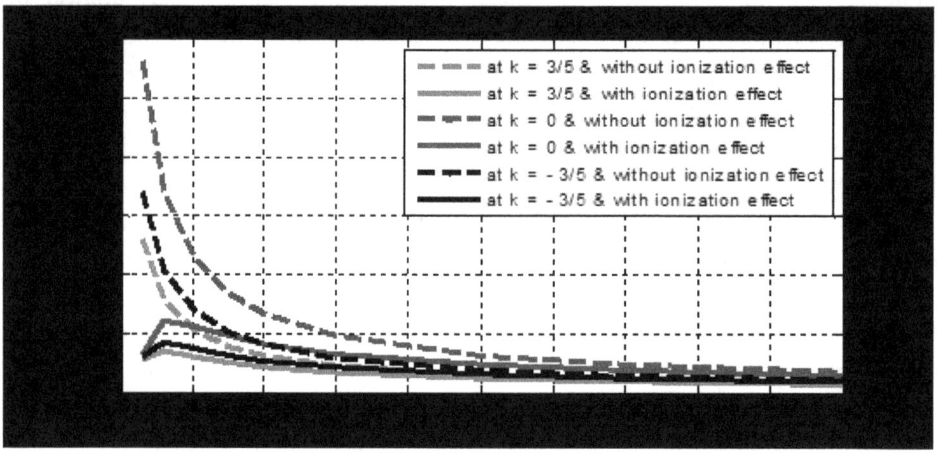

5.8.7 The Influences of the Upper Layer Thickness of Two Soil Layer on the Grounding Electrodes Characteristics

Because most grounding electrodes rarely extend practically through the upper layer of the two layer soils, it is essential to investigate the influence of the upper layer thickness on the behavior of grounding electrodes that are struck by lightning.

The equivalent radius of the electrode is calculated by using Messier model while the peak current of lightning was 100 kA. Figure 58 gives the electrode equivalent radius verses the upper layer thickness of two-layer soils with k = 3/5 and k = -3/5 for vertical and horizontal electrodes. The computations are carried out on electrode has 6 meters length and 7mm radius.

Figure 58. Influences of the upper layer thickness of two soil layer on the grounding electrodes characteristics

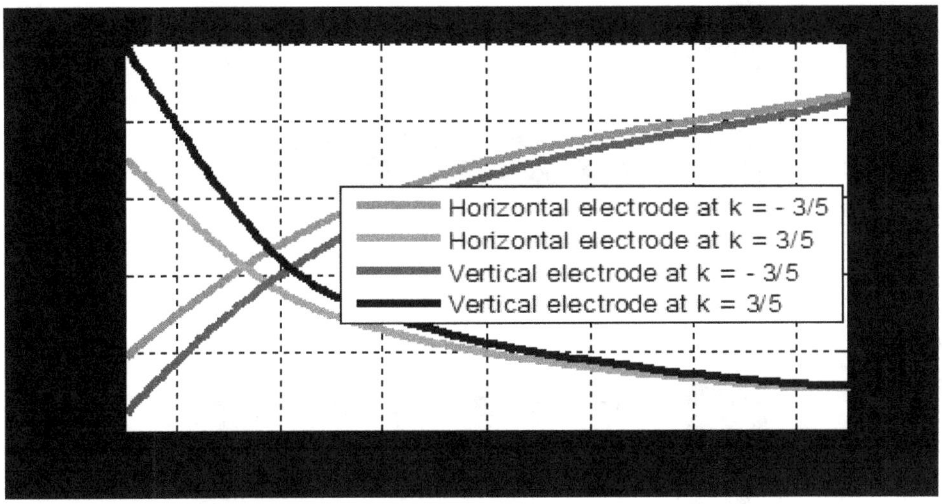

It is clear from Figure 58 that with an increase in the thickness of the first layer in two-layer soil, the equivalent electrode radius increases in the case of k = -3/5, whether the electrode is horizontal or vertical, while in the case of k = 3/5 the opposite occurs. It is clear from Figure 6.18 that as the thickness of the first layer increases, the ground potential rise around the electrode increases at k = -3/5, while the opposite occurs at k = 3/5. This phenomenon occurs whether the electrode is vertical or horizontal. The peak current of lightning used in this figure is 100 kA.

Figure 59. Impact of the upper layer thickness on the (GPR) vertical electrode

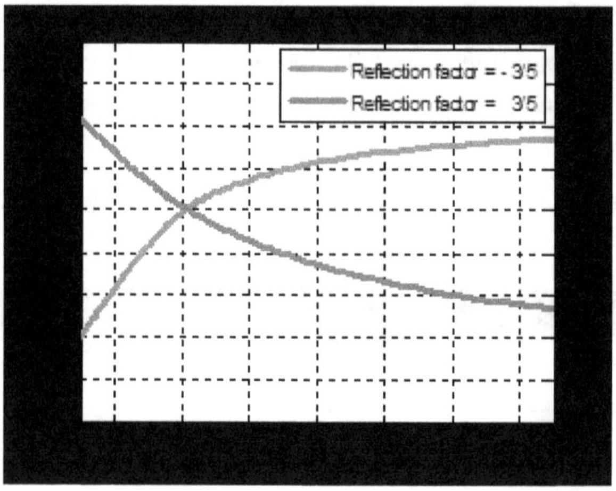

Figure 60. Impact of the upper layer thickness on the (GPR) horizontal electrode

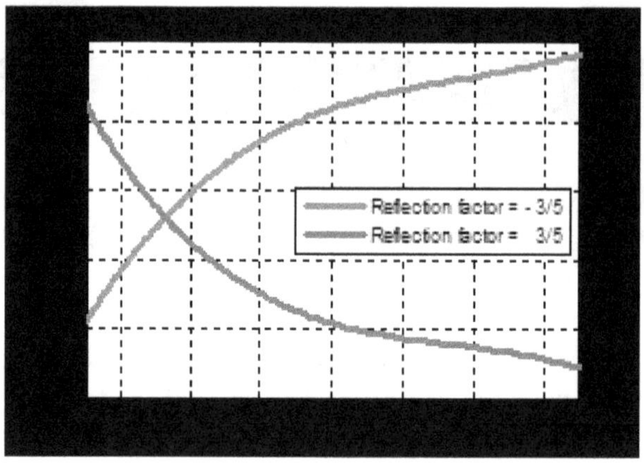

The relationships between the impulse impedance and the thickness of the upper layer in ionized soil, which consists of two layers, in the case of the horizontal and vertical poles are given in Figures 61 and 62. It is observed that an increase in impedance occurs with the increase in the thickness of the upper layer when k = -3/5, while the opposite occurs when k = 3/5. The peak current of lightning during the computation was 100 kA.

Figure 61. Impact of the upper layer thickness on the impulse impedance vertical electrode

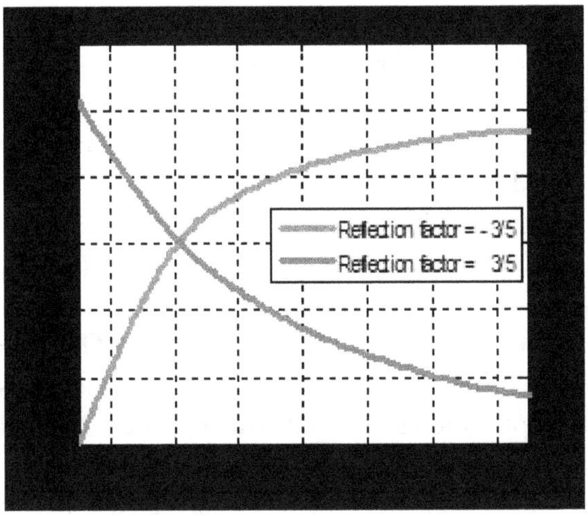

Figure 62. Impact of the upper layer thickness on the impulse impedance horizontal electrode

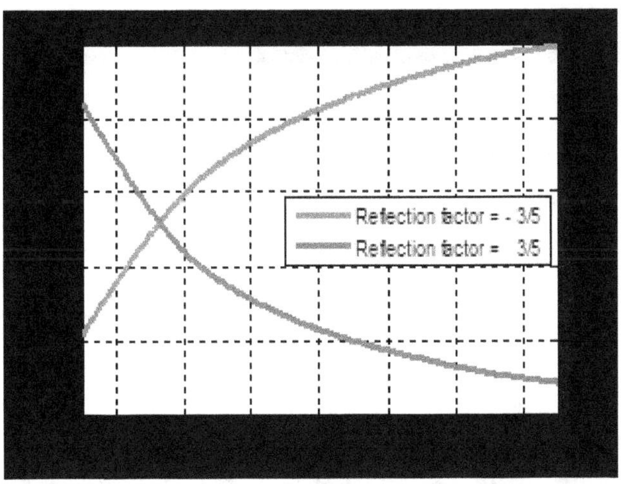

5.9 RESULTS VERIFICATIONS

Due to the lack of the field measurements and laboratory capabilities that represent reality, two approaches are used to verify the results obtained. The first one is to compare the proposed method results with the calculated and measured results done by the others. The second is conducting some experimental measurements on a laboratory model, taking into account the scaling factor. To verify the results obtained, the calculated impulse impedances are compared with those measured in the laboratory by reference (Alipio R., Visacor S., (2012)).

The same data used by R.Alipio (Alipio R., Visacor S., (2012)) are employed in the present calculations, which are 10 m length, 3 mm radius and laying depth of 30 cm horizontal electrode. The soil resistivity is $\rho = 1600$ (Ω.m) and the same lightning current wave with different values of different front time is used in the present calculations. The soil ionization is ignored. The results are illustrated in Figure 20..

From this figure it is noticed that the obtained results by the use of the present algorithm agree to some extent with the laboratory measurements reported by R. Alipio (Alipio R., Visacor S., (2012)). The difference is noticeable with the results of the short current front time.

Figure 63. Comparison between calculated results and results measured (Alipio R., Visacor S., (2012))

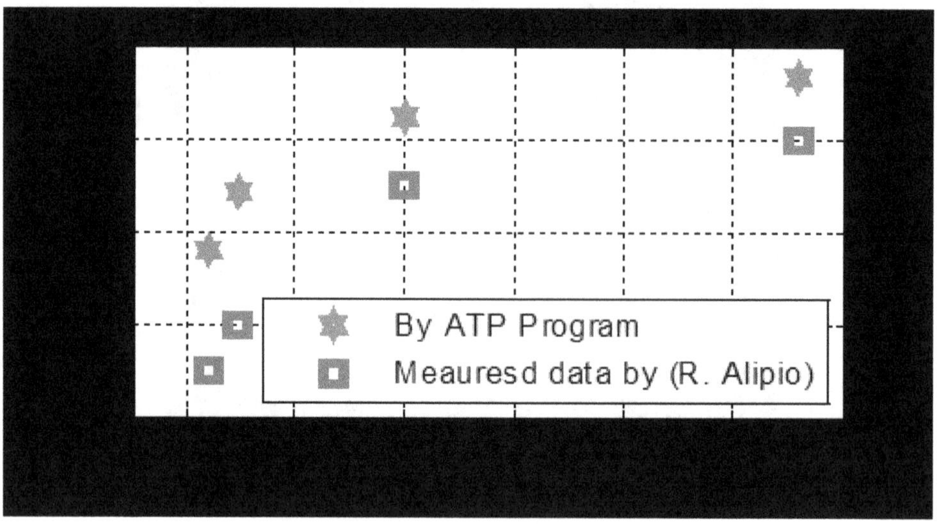

Figures 64 and 65 illustrate the relation between the impulse impedances of the vertical and horizontal grounding electrodes and the reflection coefficients by using the present algorithm, Messier model, and by using Scott model equations.

Comparing the results shown in Figures 64 and 65, it turns out that there is a match between the two methods.

Figure 64. The impulse impedances verses the reflection factors by using Messier model "M'' (Messier, M.A., (August 1980)), and Scott model "S'' (Scott, J.H., Carroll, R.D., Cunningham, D.R., (August 1964)), vertical electrode

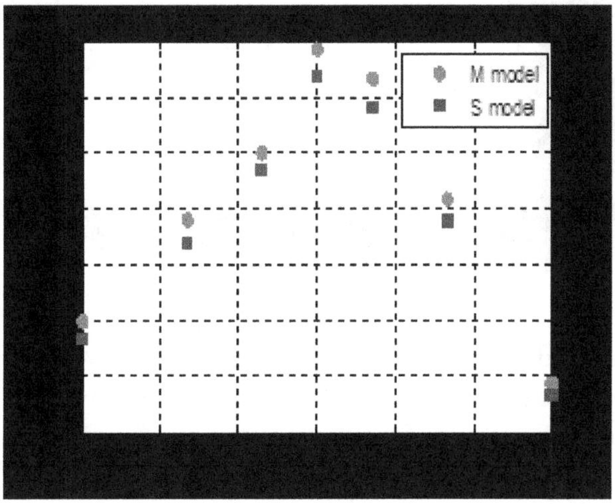

Figure 65. The impulse impedances verses the reflection factors by using Messier model "M'' (Messier, M.A., (August 1980)), and Scott model "S'' (Scott, J.H., Carroll, R.D., Cunningham, D.R., (August 1964)), horizontal electrode

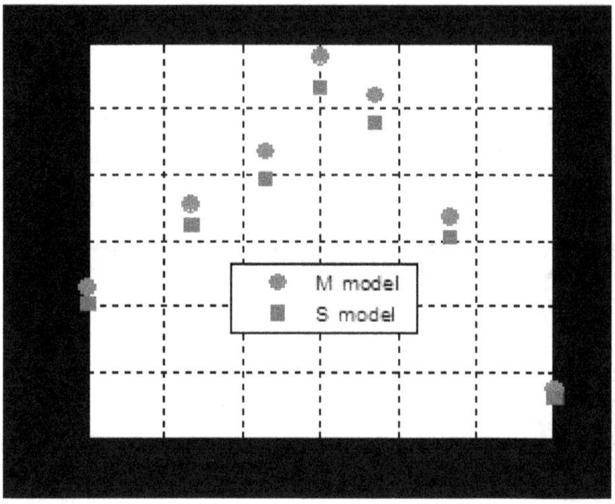

The percentage differences between the results of the proposed algorithm and measured values illustrated in Figures 64 and 65 are given in Table 6. The table contains also the percentage differences between the use of Messier model "M'' and Scott model "S''

Table 6. The percentage difference between the simulation results, experimental measurements (Alipio R., Visacor S., (2012)) and other models

Front time in (Figure 12)	**% Difference in impulse impedance**	Reflection Factors (Figure 6.21(a))	**% Difference in impulse impedance**	Reflection Factor (Figure6.21(b))	**% Difference in impulse impedance**
1.2	11.6	-0.6	3.3	-0.6	2.4
1.5	13.04	-0.35\	2.8	-0.35\	3.2
3	4.9	-0.15	1.4	-0.15	3.8
6.8	3.921	0	2.4	0	5.08
		+0.15	2.46	+0.15	3.6
		+0.35	2.8	+0.35	3.2
		+).6	1.85	+).6	1.47

In addition, the results of transient impedance obtained by the present model are compared with the laboratory measurements carried out by Nixon et al., as well as those reported by CIGRE algorithm (ClGRE, (1991)), and results obtained by G. Ala et al. *et al.* (G. Ala, F. Viola, P. L. Buccheri, and P. Romano, (May 2008)), *Nixon et al.* (Nixon, K.J., (2006)). The data used in the calculations are the same as those used in the mentioned references: the upper layer thickness is 1.9 m, $\rho_1/\rho_2 = 4.48$ and the electrode is vertical with a length of 3 m. The lightning wave shape has peak current 30 kA with 4/35 μs front/tail times.

Figure 66. Measured values of the transient impedance compared with the present calculations under soil ionization and G. Ala, et al. (G. Ala, F. Viola, P. L. Buccheri, and P. Romano, (May 2008)), Nixon et al. (Nixon, K.J., (2006)), and CIGRE (CIGRE, (1991))

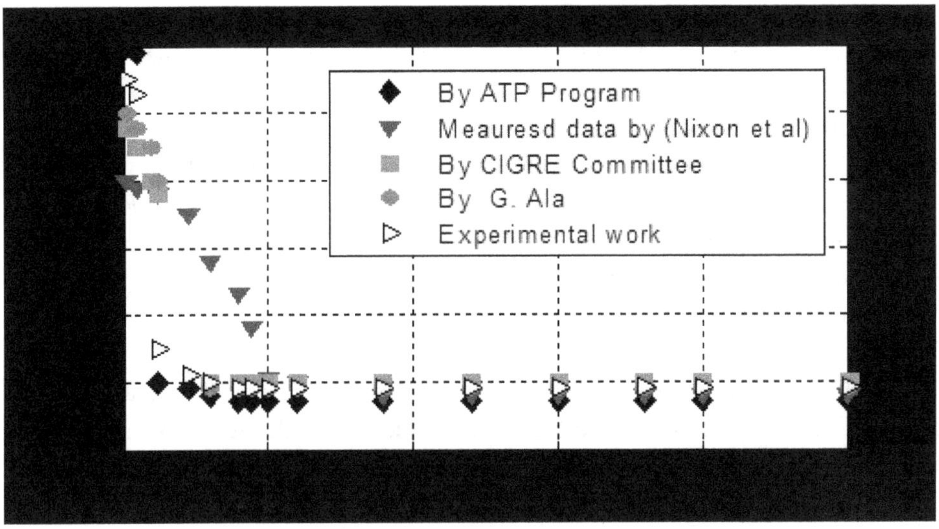

Figure 66 illustrates the mentioned comparisons, as it is observed in this figure, the calculated transient impedance using the proposed method are more consistent with laboratory measurements compared to those calculated by CIGRE method (CIGRE, (1991)). This can be understood because the present method considered the impact of the soil ionization phenomenon while others ignored its influence. In this area, the authors recommend taking into account the effect of the solubility and invisibility of the soil electrolyte resulting from the high electric field, which causes differences between the calculated results and those measured. To further confirm the reliability of the calculations done in the article, the GPR calculated is compared with the measurements carried out by others (Nixon, K.J., (2006)) and (Liu, Y.Q., et al, (2004)) as illustrated in Table 7. As it is noticed the differences are between +1.69% and - 5.866%, which necessarily confirms the accuracy of the method used.

Table 7. Comparison between calculations of the present method and those measured by others

cases under investigation by the proposed method	Measured values
Computed GPR by using the proposed method and the same data of Ref. (Nixon, K.J., (2006)) = 483.2 (kV)	Measured GPR by using Nixon et al. algorithm [33] (Nixon, K.J., (2006)) = 492 (kV)
Computed GPR by using the proposed method and the same data of Ref. (Liu, Y.Q., et al, (2004)) = 185 (kV)	Measured GPR by using Liu, Y.Q algorithm [48] (Liu, Y.Q., et al, (2004)) = 175 (kV)

To validate the accuracy of the calculations, a laboratory model has also been used. It contains two identical tanks; each one has dimensions of 1m × 1m × 0.5 m, filled with tap water. One is installed above the other, the top ends of the lower tank is coated with plastic to obtain electrical isolation between the metallic bodies of the two tanks.

Only the electrical contact between the two mediums (representing the two layer soil) occurs through the metallic base of the upper tank. The lower tank is connected to earth. Figure 67(a) shows the components of the experimental setup of uniform soil simulation and two layers in Figure 67(b), the electrolytic resistivity was controlled by changing the salinity of the tap water by adding an amount of Sodium Chloride salt. The water model temperature of the two layers was about 26 °C during the measurements.

The lightning stroke is simulated with scaling factor 1: 2000 by the use of an automatic surge generator, type SG61000-5 and meets the requirements of IEC 61000-4-5 standards. Scaling factor 1: 100 is used for the grounding rod and the upper layer thickness. The peak value of the transient potential is measured using a probe connected with an oscilloscope as shown. The obtained results in yellow color have been illustrated. Equation (5.32) is used in the calculation of the transient impedance. Some differences with the results that are reported by others, the reason is due to the boundary effects of the tank walls. The chapter presented a study on the effect of homogeneous soil as well as two layers on the behavior of ground cathodes, whether horizontal or vertical, under the strikes of lightning waves, taking into account the phenomenon of soil ionization resulting from the impulse wave of the lightning current.

REFERENCES

Ala, G., Viola, F., Buccheri, P. L., & Romano, P. (2008, May). Finite difference time domain simulation of Earth electrodes soil ionization under lightning surge condition, IET Sci. *Measurement Techniques*, 2(3), 134–145.

Alipio, F., & Visacor, S. (2013). Frequency Dependence of Soil Parameters: Effect on the Lightning Response of Grounding Electrodes'. *IEEE Transactions on Electromagnetic Compatibility*, 55(1), 132–139. 10.1109/TEMC.2012.2210227

Alipio, R., & Visacor, S. (2012), How the Frequency Dependence of Soil Parameters Affect the Lightning Response of Grounding Electrodes, *International Conference on Lightning Protection*, 1-4. 10.1109/ICLP.2012.6344354

Andre, M., & Mattos, F. (2005). Grounding grids transient simulation. *IEEE Transactions on Power Delivery*, 20(2), 1370–1379. 10.1109/TPWRD.2004.839177

ATP rule book, (1995), Can/Am EMTP User Group.

Baba, Y., Nagaoka, N., & Ametani, A. (2005). Modeling of thin wires in a lossy medium for FDTD simulations. *IEEE Transactions on Electromagnetic Compatibility*, 47(1), 54–61. 10.1109/TEMC.2004.842115

Cavka, D., Mora, N., & Rachidi, F. (2014). A Comparison of Frequency-Dependent Soil Models: Application to the Analysis of Grounding Systems'. *IEEE Transactions on Electromagnetic Compatibility*, 56(1), 177–187. 10.1109/TEMC.2013.2271913

Chisholm, W. A., & Janischewskyj, W. (1989). Lightning surge response of ground electrodes'. *IEEE Transactions on Power Delivery*, 4(2), 1329–1337. 10.1109/61.25620

CIGRE, (1991), Guide to procedures for estimating the lightning performance of transmission lines,' WG 01 (Lightning), Study Committee 33

Djamel I., Slaoui F.H., & Georges, S. (2016), Transient Response of Grounding Systems under Impulse Lightning Current. Conf. Electric Power Quality and Supply Reliability (PQ), 71-75.

Grcev, L. (1992), Computation of transient voltages near complex grounding systems caused by lightning currents. *Proc. IEEE Int. Symp. Electromagnetic Compatibility*, 393-400. 10.1109/ISEMC.1992.626123

Grcev, L. (2009). Impulse efficiency of ground electrodes,'. *IEEE Transactions on Power Delivery*, 24(1), 441–450. 10.1109/TPWRD.2008.923396

Grcev, L., & Grceva, S. (2009). on high-frequency circuit of horizontal grounding electrodes,'. *IEEE Transactions on Electromagnetic Compatibility*, 51(3), 873–875. 10.1109/TEMC.2009.2023330

Grcev, L., & Haznadar, Z. (1988), A novel technique of numerical modeling of impulse current distribution in grounding systems. *Proc. Int. Conf. on Lightning Protection*, 165-169.

Grcev, L., & Popov, M. (2005). on high-frequency circuit equivalents of a vertical ground rod. *IEEE Transactions on Power Delivery*, 20(2), 1598–1603. 10.1109/TPWRD.2004.838460

Heidler, F., & Cvetic´, J. (2002). A Class of Analytical Functions to Study the Lightning Effects Associated with the Current Front'. *European Transactions on Electrical Power*, 12(2), 141–150. 10.1002/etep.4450120209

IEEE Guide for safety in AC substation, (2000), IEEE Std 80 -2000.

Koehler, F., & Swingler, J. (2018), Practical model for tower earthing systems in lightning simulations, Electric Power Systems Research, 158, 306-314.

Leonid, G. (2009). Modeling of grounding electrodes under lightning currents. *IEEE Transactions on Electromagnetic Compatibility*, 51(3), 559–571. 10.1109/TEMC.2009.2025771

Leonid, G. (2009). Time and Frequency – Dependent Lightning Surge Characteristics of Grounding Electrodes'. *IEEE Transactions on Power Delivery*, 24(4), 2186–2196. 10.1109/TPWRD.2009.2027511

Liu, Y. Q. (2004). An improved model for soil ionization around grounding system and its application to stratified soil. *Journal of Electrostatics*, 60(2–4), 203–209. 10.1016/j.elstat.2004.01.012

Loewer, Igel, & Norman. (2016), Electrical and Dielectric Properties of Soils in the mHz to GHz Frequency Range, Conference: 11th International Conference on Electromagnetic Wave Interaction with Water and Moist Substances (ISEMA), 449-461.

Manna, T. K., & Chowdhuri, P. (2007). Generalized equation of soil critical electric field EC based on impulse tests and measured soil electrical parameters. *IET Generation, Transmission & Distribution*, 2007(1), 811–817. 10.1049/iet-gtd:20060559

Manna, T. K., & Chowdhury, P. (2007). Generalized equation of soil critical electric field Ec based on impulse tests and measured soil electrical parameters. *IET Generation, Transmission & Distribution*, 1(5), 811–817. 10.1049/iet-gtd:20060559

Messier, M. A. (1980), The propagation of an electromagnetic impulse through soil: Influence of frequency dependent parameters', Mission Res. Corp., Santa Barbara, CA, USA, Tech. Rep. MRC-1980.

Messier, M. A. (1985). *Another soil conductivity model, internal rep.* JAYCOR.

Mokhtari, M., Malek, Z. A., & Salam, Z. (2015). An improved circuit based model of a grounding electrode by considering the current rate of rise and soil ionization factors'. *IEEE Transactions on Power Delivery*, 30(1), 211–219. 10.1109/TPWRD.2014.2347283

Mokhtari, M., Abdul-Malek, Z., & Wooi, C. L. (2016), Integration of frequency dependent soil electrical properties in grounding electrode circuit model. International Journal of Electrical and Computer Engineering, 6(2), 792.

Mousa, A. M. (1994). *The soil ionization gradient associated with discharge of high currents into concentrated electrodes. Power Delivery*. IEEE Transactions.

Nahman, J., & Paunovic, L. (1979). Resistance to earth of earthing grids buried in multi-layer soil'. *Electrical Engineering*, 88(4), 281–287. 10.1007/s00202-004-0282-y

National Electric Code (NEC). (2017), 250.52(A) (5), https://www.electricallicenserenewal.com/Electrical-Continuing-Education-Courses/NEC-Content.php?sectionID=259.1

Nixon, K. J. (2006), The Lightning Transient Behavior of a Driven Rod Earth Electrode in Multi-Layer Soil, PhD Thesis, University of the Witwatersrand, Johannesburg.

Oettle, E. E. (1988). A new general estimation curve for predicting the impulse impedance of concentrated earth electrodes'. *IEEE Transactions on Power Delivery*, 3(4), 2020–2029. 10.1109/61.194013

Oettle, E. E. (1988). *The characteristics of electrical breakdown and ionization processes in soil.* Trans. SAIEE.

Omar, K., Sofiane, C., Madjid, T., Madjid, T., Abdelouahab, M., & Noureddine, H. (2017). time-domain modeling of grounding systems impulse response incorporating nonlinear. *IEEE Transactions on Electromagnetic Compatibility*, 60(4), 1–9.

Papalexopoulos, A. D., & Meliopoulos, A. P. (1987). Frequency dependent characteristics of grounding systems. *IEEE Transactions on Power Delivery*, 2(4), 1073–1081. 10.1109/TPWRD.1987.4308223

Pedrosa, A. G., Alípio, R. S., Schroeder, M. A. O., & Afonso, M. M. (2010), Simulation of behavior electrical grounding Comparative analysis for performance in frequency-domain. *Proc. Int. Workshop Appl. Model. Simul.*, 1, 89–92.

Pedrosa, A. G., Shroeder, M. A. O., Afonso, M. M., Alípio, R., Assis, S. C., Oliveira, T. A. S., & Braga, A. R. (2010). Transient response of grounding electrodes for the frequency-dependence of soil parameters, IEEE/PES Transmission and Distribution Conference and Exposition. *Latinskaia Amerika = América Latina*, 839–845.

Ramamoorty, M., Narayanan, M. M. B., Parameswaran, S., & Mukhedkar, D. (1989). Parameswaran S., Transient performance of grounding grids. *IEEE Transactions on Power Delivery*, 4(4), 2053–2060. 10.1109/61.35630

Scott, J.H., Carroll, R.D., & Cunningham, D.R., (1964), Dielectric Constant and Electrical Conductivity of Moist Rock from Laboratory Measurements, US Dept. of interior geological survey technical letter, Special projects-12.

Visacor, S. (2007). A Comprehensive Approach to the Grounding Response to Lightning Currents. *IEEE Transactions on Power Delivery*, 22(1), 381–386. 10.1109/TPWRD.2006.876707

Chapter 6

Studying Direct Lightning Strike Impact on Human Safety Near HVTL Towers Considering Two Layer Soils and Ionization Influence

Osama El-Sayed Gouda
Faculty of Engineering, Cairo University, Egypt

ABSTRACT

This chapter contains the following points: design procedure of grounding system according to IEEE 80, methods for calculating the grounding grid resistance (Dwight's formula, Laurent and Niemann, Sverak's equation, Schwarz's formula, Dawalibi, Mukhedkar's formula, Chow and Salama's formula, Nahman's formula, and Heppe's method). It contains also the design of charts of grid earthing system and application of step and mesh potential in safe grounding system design. There are two main design goals to be achieved by any substation ground system under normal as well as fault conditions. These are 1) to provide means to dissipate electric currents into the earth without exceeding any operating and equipment limits and 2) to assure that a person in the vicinity of grounded facilities is not exposed to the danger of critical electric shock. The design procedures described in the following sections are aimed at achieving safety from dangerous step and touch voltages within a substation. It is possible for transferred potential to exceed the GPR of the substation during fault conditions.

DOI: 10.4018/979-8-3693-2969-6.ch006

6.1 INTRODUCTION

Since mankind has used high and medium voltage networks for the transmission and distribution of electrical energy, the danger is increasing as a result of direct exposure to excessive overvoltage's of these networks on one hand, and to lightning strokes that may strike workers in these networks on the other hand, which may cause death or injuries in the best case. Effects of current on human beings and livestock are investigated by IEC/TR 60479-1-4 (IEC/TR 60479-4, (2004)), (IEC/TS 60479-1, (2005)). Safety current limits for human beings are presented in IEEE Std. 80, Guide for Safety in AC substations (IEEE Std 80, (2000))

Lightning is one of the most powerful natural phenomena that happen on earth (Koont D. R., (2015)), and it can be harmful to any devices on the rooftops and/or outdoor areas, such as deterioration in photovoltaic systems (Hetita I., et al, (2022)) and wind turbines (Zalhaf A. S., et al, (2022)) where it is necessary for providing effective lightning protection structures. Also, the lightning occurrence can cause human death and harm electrical devices and the environment (Andrews CJ, (2007)). Thousands of injuries produced by lightning have been reported from the start of humankind, resulting in fatalities and damage to humans with both acute and long-term consequences (Misbah N.R., Ab Kadir M.Z.A., Gomes C., (April, 2011)). Further, to evaluate the risk caused by lightning-human interaction in the vicinity of humans, structures, and the surrounding earth, a sufficient Lightning Protection System (LPS) is designed and analyzed. The effectiveness of the grounding system and complex earthing models during lightning are utilized to evaluate the role performed by aspects such as earth electrode dimensions, soil resistivity, and the equivalent circuit nature elements. Moreover, the role of lightning strikes on humans caused by touch and step potential is determined by devising a lumped electrical model of human to evaluate its role and influence on wet and dry skin (Sreedhar S. and Indragandhi V., (2022)).

It is well known that influences of the amount of electrical current going through the human body are mostly determined by the duration, frequency, and amplitude of the event (Andrews CJ, (2007)). Over the last few decades, various efforts have been made to define reference limits and do studies relating to the impacts of electric current on humans and animals. However, the majorities of these researches only look at power frequencies and are mostly concerned with the study of sinusoidal current with frequency in the range of 15–100 Hz, which ended in the preparation of international standards actually in the use (Misbah N.R., Ab Kadir M.Z.A., Gomes C., (April, 2011)), Sreedhar S. and Indragandhi V., (2022), (IEC/TR 60479-4, (2004)), (IEC/TS 60479-1, (2005)). Since there are no specified criteria for determining the maximum electric current, a human may suffer from lightning strokes. The IEEE Std 80(IEEE Std 80, (2000)) is currently the most widely used standard for determining human-tolerated step and touch potentials and currents, although it is crucial to note that this standard is based on steady-state analysis at power frequencies, which is not ideal for transient investigations.

Unlike studies on the influence of alternating current at power frequencies and low magnitudes, research on lightning is extremely difficult to do because of the numerous experiments that must be completed. There is also minimal thought given to grounding systems, the effects of soil resistivity, and the variance in lightning characteristics. Furthermore, when lightning current flows through the network, the performance of the grounding electrodes in uniform soil is studied, where the frequency impact on the surrounding soil is considered (Cavka D., Mora N., and Rachidi F., (2014)). Recently, Gouda *et al.* investigated the impact of lightning strokes in uniform and two-layer soils on the step and touch potentials (Gouda, Osama E., et al., (2022)). Based on the literature review and the authors' knowledge, most existing studies lack concerns with non-homogenous soil resistivity parameters that effectively change the touch and step voltages.

Based on this remark and to cover this gap in the literature, this paper investigates the effect of lightning current strokes on human safety. Further, this article focused on the current passing in the heart caused by step and touch (contact) voltages generated by first and subsequent lightning strokes when hit the transmission tower in case of non-homogenous soil resistivity. Also, the study of the parameters that affect the step and touch voltages such as different reflection factors, and different peaks and front times of lightning stroke currents are investigated.

6.2 MATHEMATICAL MODELS FOR THE PROBLEM FORMULATION

In this part of the article, the simulation of the grounding system of the high-voltage transmission line towers is presented. The simulations of lightning current waveforms and the human body are also presented. The soil ionization phenomenon caused by lightning is also presented in this part of the article.

6.2.1 Modeling of HVTL Towers Grounding

The grounding system of the HVTLs tower is shown in Fig. 6.1. Hence, Fig. 6.1.a illustrates the dimensions of the transmission line tower of 220 kV. The data of this line are illustrated in Table 6.1. The grounding system of the tower contains four vertical grounding electrodes, each electrode has a length of L=3 m and a radius of r=0.02 m. To model the transient behaviour of those vertical grounding electrodes, a transmission line method (TLM) with the aid of the ATP program (ATP rule book,

(1995)) is used to compute the grounding impedance when lightning current is injected into those vertical grounding electrodes of the tower guiding system. In the TLM technique, any earthing electrode can be subdivided into M segments each one containing earthing resistance, inductance and capacitance elements as shown in Fig. 6.1.b. The grounding resistance, $R_{g,}$ capacitance C_g, as well as specified electrode inductance L_g of the vertical rod are computed by using the following equations (6.1) to (6.3), respectively Leonid, (G., (2009)), (Nahman, J., Paunovic, L., (2006)).

$$\mathbf{R}_i = \frac{\rho_a}{2\pi}\ln\left(\frac{4l}{a} - 1\right) \tag{6.1}$$

$$\mathbf{C}_i = \frac{\rho_a \varepsilon_g}{\mathbf{R}_g} \quad (6.\ 2)$$

$$\mathbf{L}_i = \frac{\mu}{2\pi}\left[\ln\frac{2l}{a} - 1\right] \quad (6.\ 3)$$

where μ represents the soil permeability (H/m), a depicts the radius of the grounding electrode, which is replaced by r_i when the ionization process is began, followed by ρ means the soil resistivity, ε_g represents the soil dielectric constant and l means the grounding electrode length.

6.2.2 Simulation of Lightning Current Wave Forms

In this article, two lightning current waveforms equivalent to the first and subsequent lightning strokes are employed. The following Heidler's formula of lightning current function (HF) is selected to represent the current waveforms (Rachidi F., et al., (Aug.2001)).

$$i(t) = \frac{I_0}{\eta}\frac{\left(\frac{t}{\tau_1}\right)^n}{1 + \frac{t}{\tau_1}^n} e^{(-t/\tau_2)} \tag{6.4}$$

$$\eta = e^{-(\tau_1/\tau_2)(n(\tau_2/\tau_1))^{1/n}} \tag{6.5}$$

Where the time in second represents t, I_0 means the current pulse amplitude, followed by τ_1 that depicts the front time constant. Further, τ_2 means the decay time constant, as well as n refers to the exponent having values within range of 2 to

10, finally η means the amplitude correction factor (Heidler F., Cvetic J., (March/ April 2002)).

6.2.3 Human Body Simulation

The impedances of human body elements depend on the degree of moisture of the skin, skin temperature, current path duration, touch (contact) or step voltages, and the body's physical properties. According to IEC/TS 60479-1 and other standards and studies, the human body has been simulated by two impedances (internal and external) (Andrews CJ, (2007)). The human body can be simulated by a combination of resistive and capacitive elements. Simplified simulations are carried out by many researchers (Gazzana, Daniel S., et al., (2014)); one of these is shown in Figure 3. (Gazzana, Daniel S., et al., (2014)). Some adjustments have to be made for the evaluation of the electric current passing through the heart. The contact resistance $R_{contant}$ in between the feet of a person and the soil surface can be estimated by using equation (6. 6) (Becerra M., Cooray V., (Aug.2009)).

$$R_{contact} \cong 3.\rho \quad (6.\ 6)$$

The impedances of human body elements based on reference (Gazzana, Daniel S., et al., (2014)) are illustrated in Figure 2. The simulated equivalent circuit of the human body is used to examine the behavior of the current passing through the heart considering the step and touch (contact) potentials. This current is created from the change of potential in the soil between two specified points on its surface in case of step potential and between the touch point and point on the soil surface in case of touch (contact) voltage. The simulation of the human body model in the case of step potential is illustrated in the equivalent circuit given in Figure 3. Hence, Figure 4 illustrates the simulation of the touch (contact) potential mechanism. The models given in Figure 3 can be used to estimate the current in the heart.

Figure 1. Geometric configuration of transmission line tower dimensions of 220 kV

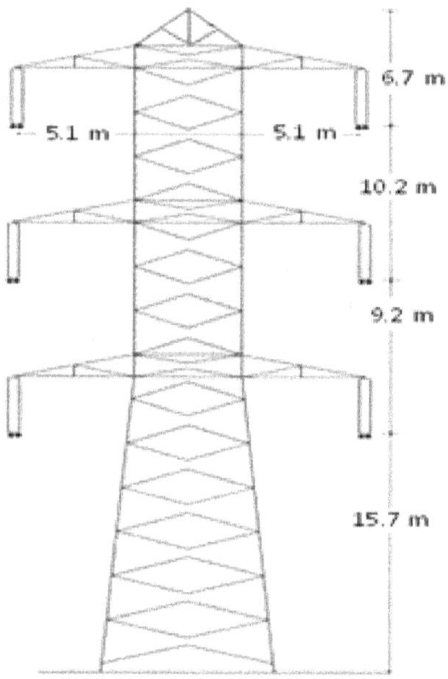

Figure 2. Geometric configuration of grounding electrode and its equivalent circuit.

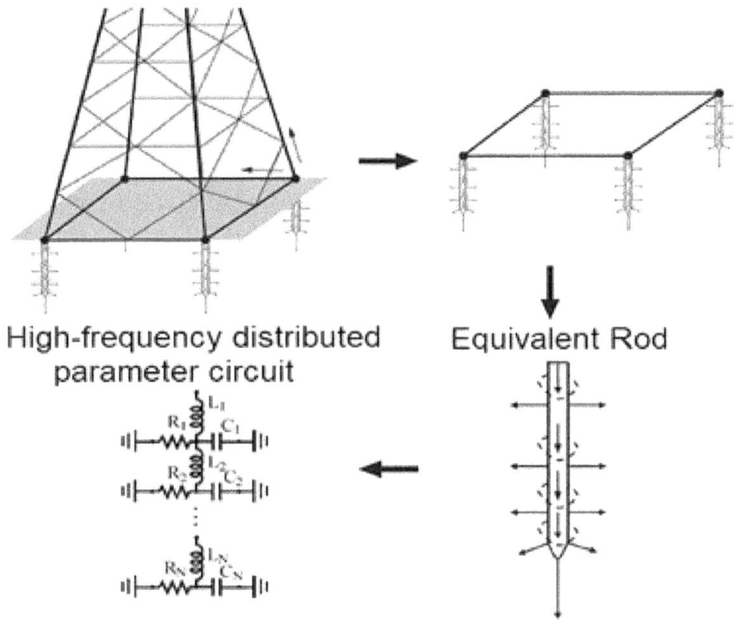

Table 1. Data of 220 kV power line

tem	Units	value
Transmission line capacity	MVA/circuit	158
Line voltage of the transmission line	kV(r.m.s)	220
Transmission line length	km	90
Positive and negative sequence impedance	per phase in ohm	3.6+ j 15.3
Zero sequence impedance	per phase in ohm	12.2+j50.95
Number of circuits per tower	-	2
Number of conductors per phase	-	2
Number of ground wires	-	1
Single conductor diameter	mm	27
Spacing between a conductor in the bundle	cm	30
Span between each two towers	meter	360

6.3 SOIL IONIZATION

Due to lightning current, soil ionization happens when the dissipated current in the soil surpasses its critical value. The soil ionization occurrence by lightning current indicates an increase in the grounding electrode radius and subsequently; it decreases the ground resistance, step and touch (contact) potentials, and transient impedances. Manna and Chowdhuri in reference (Manna, T.K., Chowdhury, P., (2007)), which is one of many that dealt with this topic, proposed a relation between the critical breakdown field strength (E_c) as well as the soil dielectric constant ε_g and its conductivity σ_g as presented in equation (6. 7).

$$E_c = 8.6083\,\varepsilon_g^{-0.0103}\,\sigma_g^{-0.15264} \quad (6.\ 7)$$

The new effective grounding electrode radius r_i in meter is suggested by equation (6. 8) (Mokhtari, M., Abdul-Malek, Z., and Gharehpetian, G. B., (2016)).

$$r_i = \frac{\rho I_m}{2\pi E_c l} \quad (6.\ 8)$$

6.4 CALCULATION OF GROUND POTENTIAL RISE

This section explains gauss–Sedial method (Gauss-Seidel Method, (2022)) that is used to estimate earth surface potential (ESP) in non-homogenous soil that is injected into human feet when walking near the grounding system of the 220 kV tower. Also, in this paper Gauss–Sedial method is used also in the calculations of step and touch (contact) potentials that are affecting the person near the towers of HVTLs. After calculating all the aforementioned parameters, the transient behavior of current passing through the human heart can be obtained by the use of ATP-EMTP (ATP rule book, (1995)).

Figure 3. Equivalent circuit of the human body in case of (a) step potential (Gazzana, Daniel S., et al., (2014))

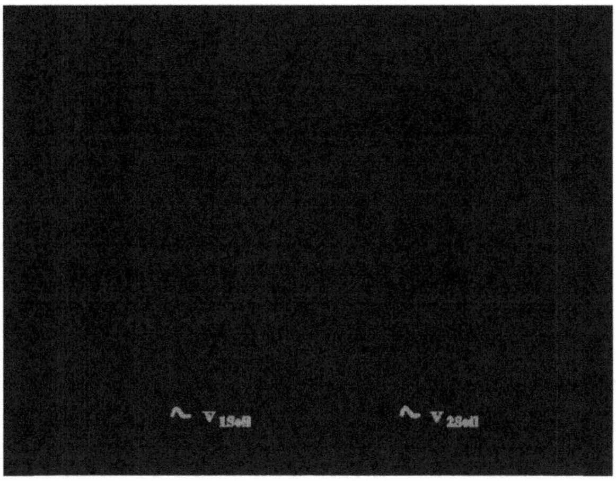

Figure 4. Equivalent circuit of the human body touch (contact) potential (Gazzana, Daniel S., et al., (2014))

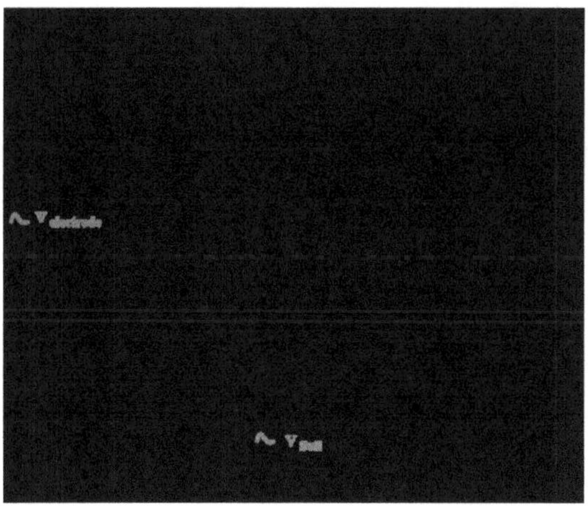

The grounding fault current is i_g, which is discharged from the grounding system into the earth according to equation (6. 9):

$$i_g = \sum_{m=1}^{nn} i_m \tag{6.9}$$

i_m is the current of element m= 1, ……, nn

Two lightning current waveforms as given in equations (6.4) and (6.5) are used. The resistance of the grounding system will be:

$$R = \frac{u}{i_g} = \left(\sum_{k=1}^{n} \frac{i_m}{u}\right)^{-1} \quad (6.10)$$

The potential of the grounding system is:

$$u = Ri_g \text{ (6. 11)}$$

The electric potential u_T at any point (T), caused by the grounding system element buried in non-homogenous soil can be calculated by super-position of potential from an infinite sequence of images as follows:

$$u_T = \sum_{k=1}^{n} r_{mT} i_m \quad (6.12)$$

In the above equation r_{kT} is the mutual resistance of grounding system element m and point (T)is on the surface of the upper soil layer. The reflection factor g is defined as:

$$g = (\rho_2 - \rho_1) \cdot (\rho_1 + \rho_2)^{-1} \quad (6.13)$$

Whereρ_1: is the resistivity of the upper soil, and ρ_2 is the lower soil resistivity. The mutual resistance of element m and point T is given by:

$$r_{mT} = \frac{\sum_{s=0}^{\infty} Q_{mT}(s)}{i_m} \quad (6.14)$$

The integer s represents the general index of the image sequences. $Q_{mT}(s)$ is representing the voltage potential caused by terms s of the image sequences.

$$Q_{mT} = \begin{cases} \frac{\rho\, i}{4\pi l} ln \frac{G+1}{G-1}, G > l + \frac{d^2}{2l} \\ \frac{\rho\, i}{4\pi l} ln \left(\frac{2l}{d}\right)^2, G \leq l + \frac{d^2}{2l} \end{cases} \quad (6.15)$$

Where $G = \sqrt{(P,T)} + \sqrt{(Q,T)}$

$$r_{mT} = \frac{\rho_1}{4\pi l_m} \{A_{mT}(0) + B_{mT}(0) + \sum_{s=1}^{\infty} k^s \left[A_{mT}(s) + B_{mT}(s) + C_{mT}(s) + D_{mT}(s)\right]\} \quad (6.16)$$

P ,Q are being the endpoints of the conductor. Symbols i , l , d denote the conductor's current, length, and diameter of grounding electrode respectively and ñ is the soil resistivity.

6.5 THE IMPACT OF SOIL IONIZATION ON THE STEP AND TOUCH POTENTIALS AND CURRENT PASSING THROUGH THE HUMAN HEART

The stochastic characteristics of a lightning surge and current peak values in the range of 4 to 250 kA are considered. Based on the mathematical simulation and Gauss–Sedial method considering the soil ionization impact it is possible to estimate the passing current through the human heart. The steps of the calculation of touch (contact) and step potentials as well as the current passing through the human body are as follows:

1) Insert equivalent circuit of the human body model in case of step potential in the ATP program, ρ_1, ρ_2, start point and end point of each element, and effective radius of electrode considering ionization.
2) Calculate $V_{1\ soil}$, $V_{2\ soil}$, V_{soil}, and $V_{electrode}$ illustrated in Figures 6.2(a) and (b) by the use of equation (6. 12).
3) Insert in ATP $V_{1\ soil}$ and V_{2soil} in the human body model as given in Figure 6. 2(a) in case of step potential and V_{soil} and $V_{electrode}$, as given in Fig. 6.2(b) in case of contact potential.

Figure 5 illustrates the behaviour of step potential that a person suffers from grounding potential rise when located above the grounding system of the tower at two-layer soils due to the first lightning stroke when the lightning surge current peak was 100 kA. The dielectric constants of the soils are considered between 6 and 20 depending on the soil composition, and the thickness of the upper layer was 10 meters (Mokhtari, M., Abdul-Malek, Z., and Gharehpetian, G. B., (2016)). The thickness of the upper layer is considered 10 meters. The figure shows the step voltages with and without considering the soil ionization.

Figure 5. Step potential above the grounding system affecting the human body at a first lightning model with and without soil ionization

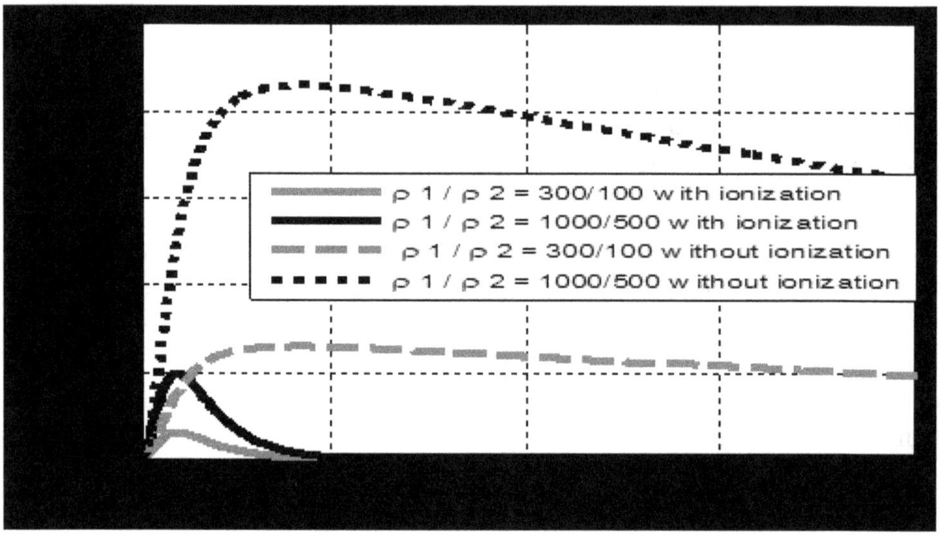

It is seen that step potential affects a person decreases when the soil ionization phenomenon is considered by values of 71% and 80% at $\rho_1/\rho_2 = 300/100$ *and* $\rho_1/\rho_2 = 1000/500$, respectively.

Figure 6 illustrates the influence of soil ionization phenomena on the current that passes through the heart when step potential affects the human body at a first lightning stroke in two-layer soils. It is observed that soil ionization has helped in reducing the current value that affects the human heart if the person is stationed above the tower grounding system in the event of a lightning-stroke earth fault, and this is expected due to the reduction of the step voltage in this case.

It is observed that the soil ionization phenomenon helps in the decrease of the current by about 74% - 73.5% depending on ρ_1/ρ_2 .It is also noted that with the increase in the resistivity of the upper layer of soil, the value of the current pass-through heart of humans increases.

The touch or contact voltage is calculated by the use of the human body model given in Figure 3. The calculations of touch or contact voltages are carried out considering soil ionization. The contact potential, and the ground potential in cases of first lightning and subsequent return strokes for different reflection factors considering the ionization procedure with the first peak wave current was 100 kA are given in Figures 7 and 8.

The corresponding currents passing through the heart are illustrated in Figures 6 and 7. It is noticed that the positive reflection factors gave higher peak values of touch or contact voltage and current pass through the human heart compared with the negative reflection factor.

Figure 6. Current passing through the heart when step potential affects the human body at first lightning stroke with and without soil ionization

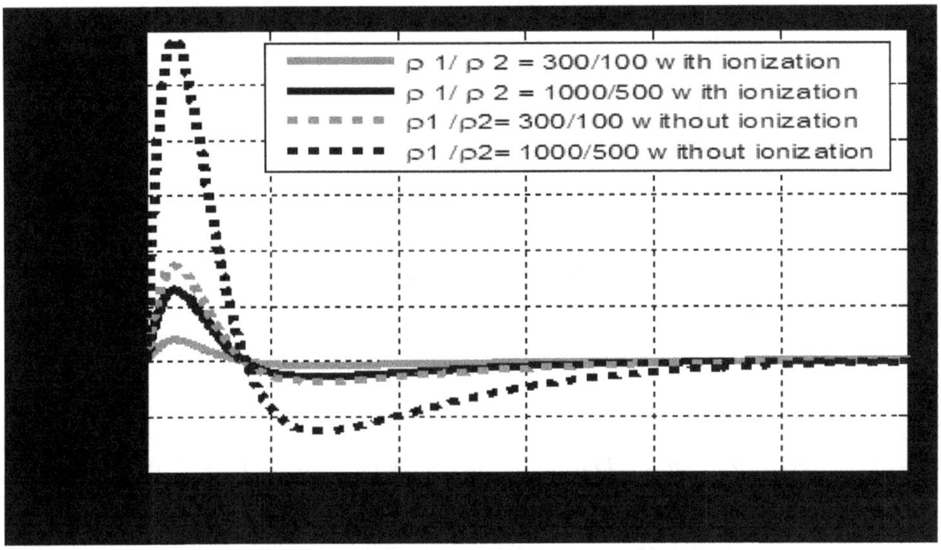

Figure 7. Contact potential submitted in the body model at different reflection factors first stroke

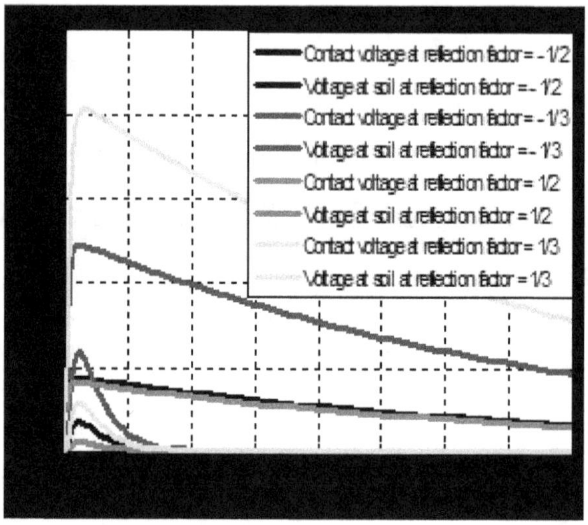

Figure 8. Contact potential submitted in the body model at different reflection factors subsequent return strokes

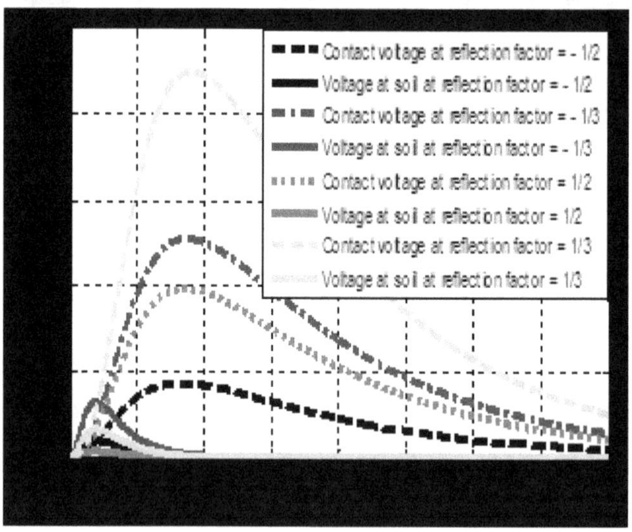

Figure 9. Current passing through the heart when contact potential is submitted to the human body at first lightning stroke

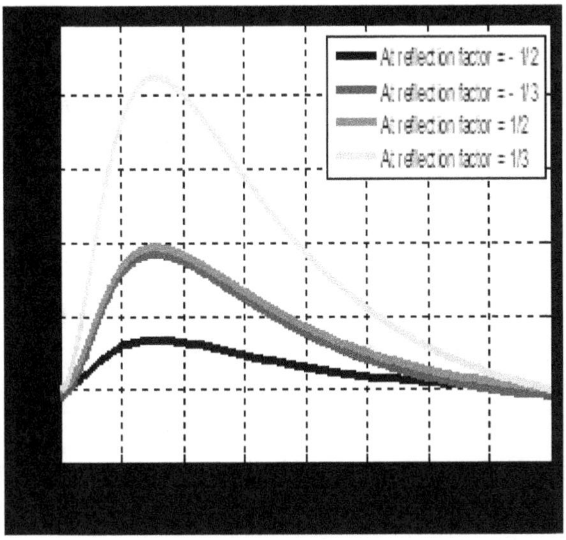

Figure 10. Current passing through the heart when contact potential is submitted to the human body at subsequent return strokes

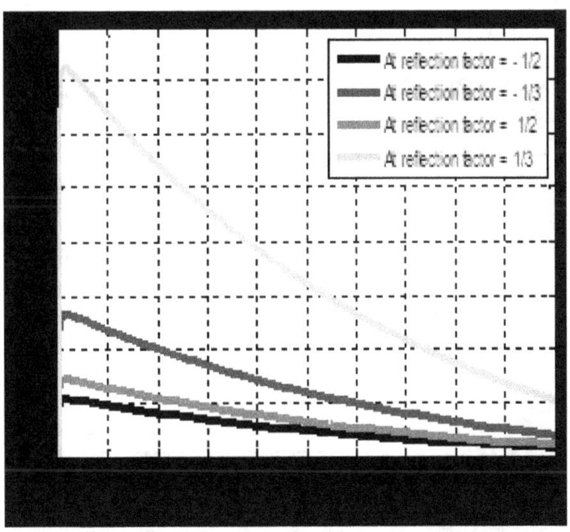

The obtained results are in agreement with those that were reached in reference (Gazzana, Daniel S., et al., (2014)), and the differences are due to the fact that in this article the study was done in the case of two-layer soils and that the ionization state in the soil was considered.

By comparing the results of the step voltage and those of the touch voltage, we find that in the first case, the step voltage increases with a negative reflection coefficient, while the opposite is limited in the case of the touch voltage. Perhaps this is due to the fact that the current in the latter case tends to pass through the second layer due to the necessity of passing through the ground.

6.6 THE INFLUENCE OF THE STROKES FRONT TIME ON THE CURRENT PASSING THROUGH THE HUMAN HEART

Figures 11 and 12 show the influence of the front time of the first lightning strokes on the contact potential submitted in the body model at negative and positive reflection factors and the current peak of lightning stroke = 100 kA. The corresponding currents passing through the heart are illustrated in Figures 8 and 9 when soil ionization is taken into account. The created current in the body is progressing from the difference of potential between the soil surface and contact point, considering the feet is together, besides 1 m apart from the touch tower.

Figure 11. Contact potential submitted in the body model at and current peak of lightning stroke = 100 kA negative reflection factor

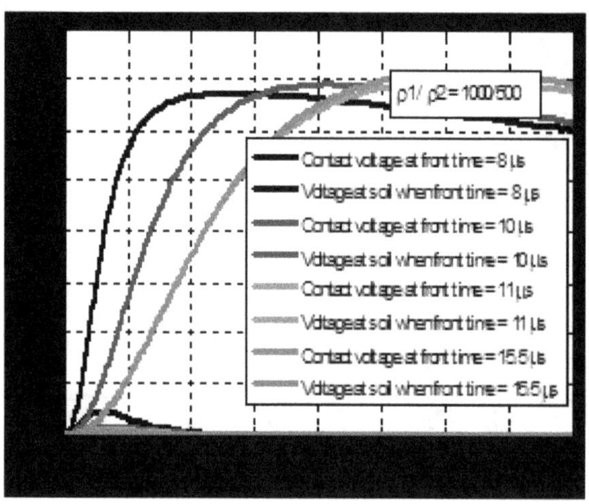

Figure 12. Contact potential submitted in the body model at and current peak of lightning stroke = 100 kA positive reflection factor

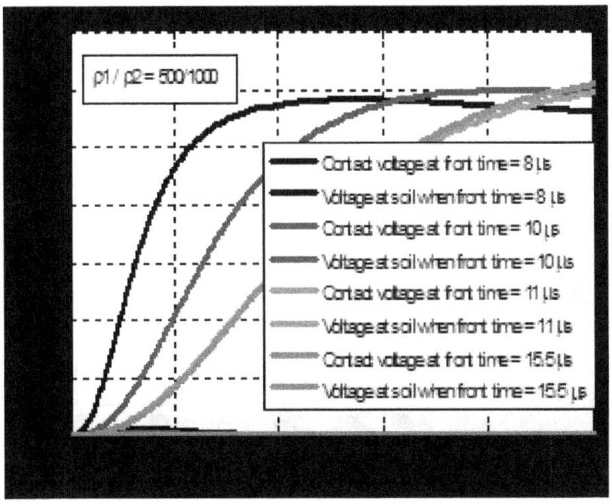

From Figures 13 and 14, it is observed that when contact potential is submitted to the human body current pass through the heart increases when the resistivity of the second layer of soil is higher than the upper layer's resistivity. Also, it is noted that when the front time of lightning current increases, the current pass-through the heart takes a long time to reach its peak value.

Figure 13. Current passing through the heart when contact potential submitted to the human body at negative reflection factor

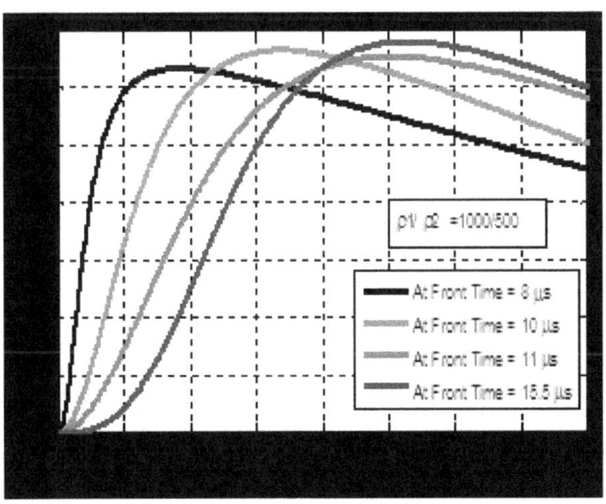

Figure 14. Current passing through the heart when contact potential submitted to the human body at positive reflection factor, the current peak of lightning =100 kA

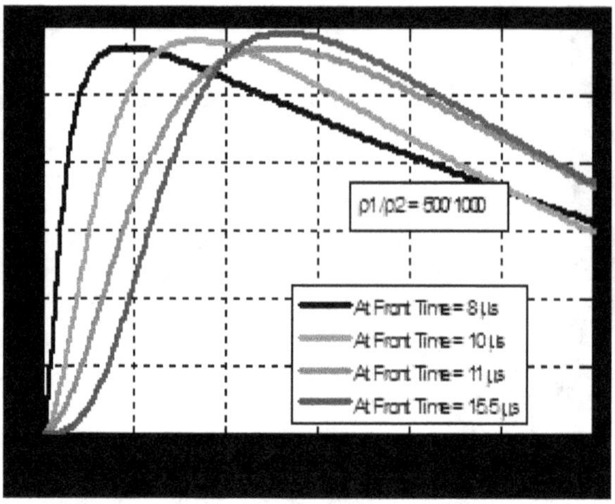

It is also observed that the current magnitudes are higher than that produced by step potential. Moreover, it is observed that in the positive reflection factor the current passing in the human heart is much higher than the similar current in the case of negative reflection factors. Again, it is monitored that the soil resistivity of the second layer plays an effective role in increasing the contact voltages and currents. Figures 15 and 16 show the behaviors of the step potential at a person located above the grounding system of the 220 kV tower in non-homogenous soil due to first and subsequent lightning strokes, respectively. It is noticed that the step potential of the first stroke is higher than that of the subsequent lightning strike by about 35%.

Figure 15. Step potential submitted in the body model at a first lightning stroke

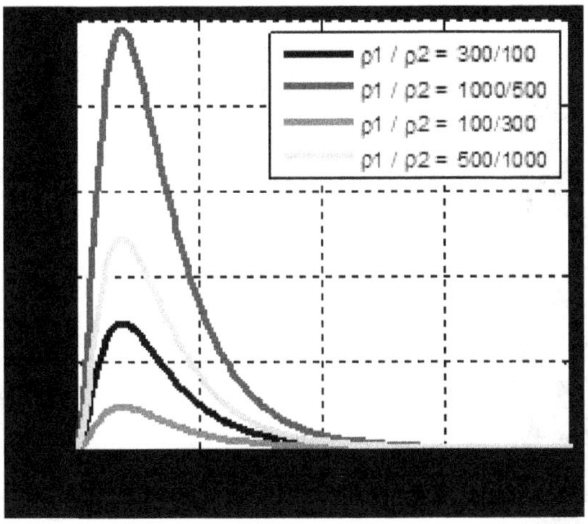

Figure 16. Step potential submitted in the body model at subsequent return stroke

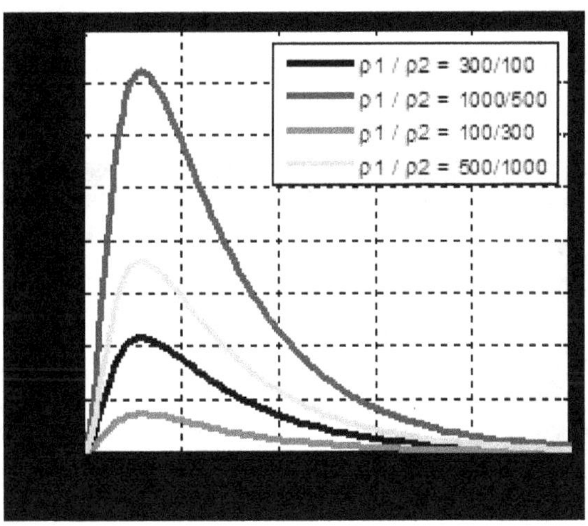

Figure 17. Current passing through the heart when step potential at first lightning stroke

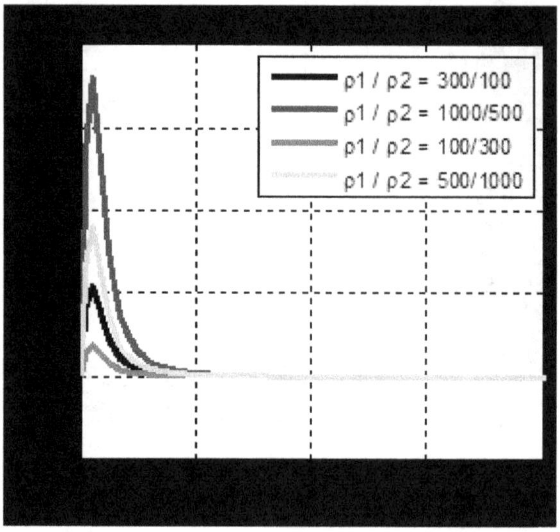

Figure 18. Current passing through the heart when step potential at subsequent return stroke

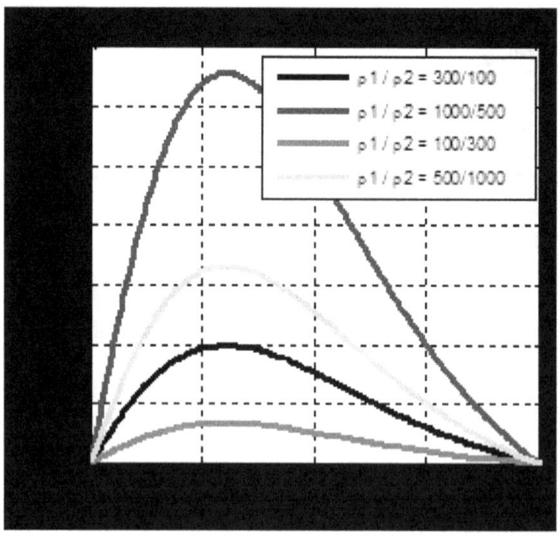

Figures 19. Step potential submitted into the body mode

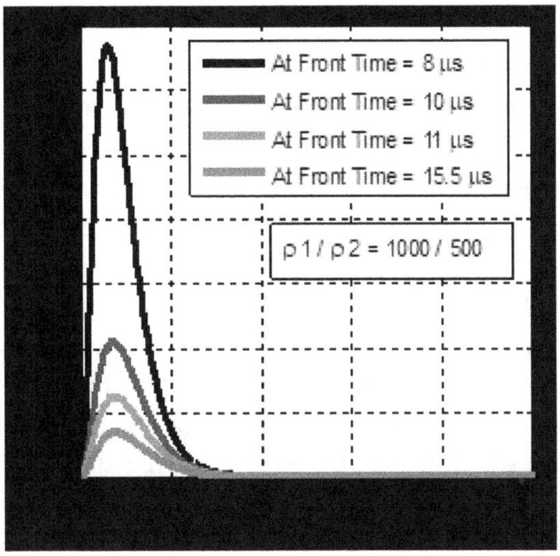

Figure 20. Current passing through the heart when the current peak of lightning =100 kA and has a negative reflection factor

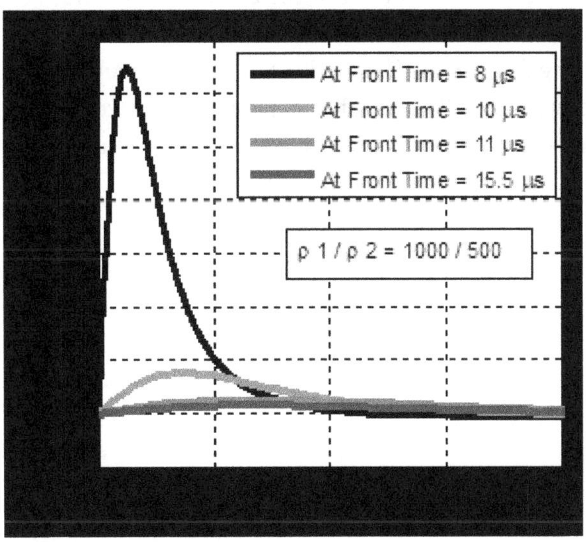

Figures 19 and 20 illustrate the influence of the first and subsequent lightning strokes on the current passing through the heart of a human when step potential is submitted into the body model in non-homogenous soil when the soil ionization phenomenon is taken into account. It is seen that the current passing through the heart of humans decreases at subsequent return strokes much higher than its value at the first stroke due to the high values of the step potential. Also, it is seen that when the reflection factor is negative the current pass-through of the heart increases. Finally, it is noticed that the time of current pass-through the heart reaches zero quickly in the first lightning stroke.

Figures 19 and 20 illustrate the behavior of step potential of a person stationed above the tower grounding system at non-homogenous soil resistivity when a soil ionization phenomenon is taken into account. The front time of the lightning waves is changed as shown in the figure. It is seen that the step potential that is submitted to a person increases when the front time of lightning stroke is decreased. Again, it is seen that with the increase of the resistivity of the upper layer of soil, the step potential that is submitted to a person increases.

6.7 THE INFLUENCE OF THE WAVE PEAKS ON THE CURRENT PASSING THROUGH THE HUMAN HEART

Figures 21 and 22 show the effect of the current peak values of lightning pulse on step potential at a person stationed above the tower grounding system on non-homogenous soil resistivity with a positive reflection factor when soil ionization phenomena are considered. It is seen that step potential as well as the current increase when the current peak of the lightning pulse increases. Similar results are obtained in Figure 7. 13 with a negative reflection factor when a soil ionization phenomenon is considered a soil negative reflection factor.

Figure 21. Step potential submitted into the body model

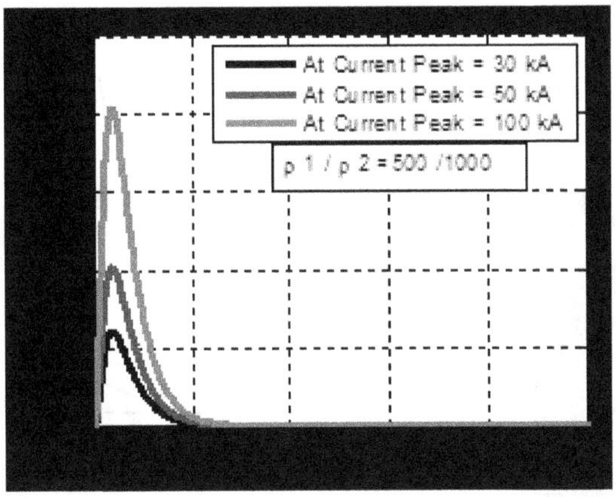

Figure 22. Current passing through the heart with different values of the current peak of lightning and the soil layers has a positive reflection factor

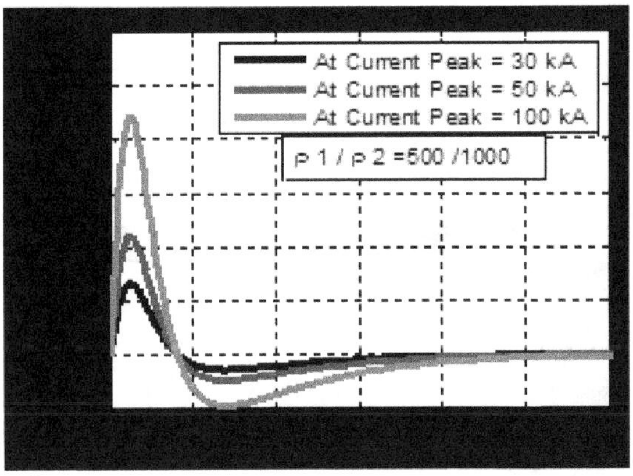

Comparing Figures 23 and 24 with Figures 25 and 26, it can be seen that the current pass in the heart of the human as well as the step potential values are higher in case of negative soil reflection factor (ρ_1/ρ_2 = 1000/500) in comparison with the positive flection factor $(\rho_1/\rho_2$ = 500/1000).This reflects the effect of the soil resistivity of the upper layer. Also, it is seen that when the upper layer of soil resistivity is high, the current passing in heart of a human is increased. In the end, it is seen at the different current peaks of lightning pulse that the current pass through the heart reaches zero at a time equal to 10 μsec. this means the current peak of lightning non-effectible for the duration of a current pass in the heart.

Figure 23. Step potential submitted into the body model with soil negative reflection factor

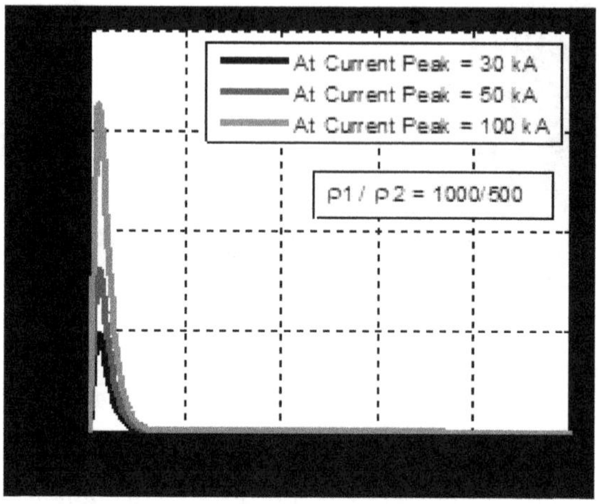

Figure 24. Current passing through the heart with different values of the current peak of lightning, and the soil layers have a negative reflection factor

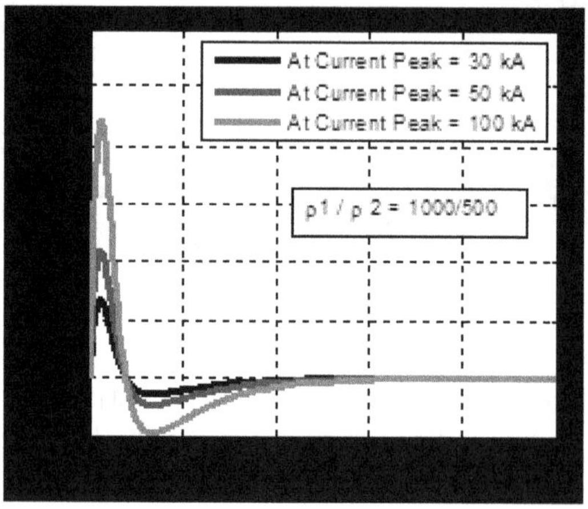

The confirmation of the obtained results cannot be achieved practically, as it is not possible to expose people to the influences of lightning strikes for the electric shock measurements, because these people will be exposed to certain death. Therefore, the modeling process is the only method for conducting such studies. In any case, it is possible to compare the results obtained with that reported by others (Gazzana, Daniel S., et al., (2014)), (Andrews C., (2003)); especially those who used the same model of human representation (Gazzana, Daniel S., et al., (2014)).

The contact potentials with negative and positive reflection factors for different wave peaks considering soil ionization are given in Figures 25 and 26. The corresponding currents passing through the heart are illustrated in Figures 27 and 28.

By comparing the results obtained in this study with those obtained in a reference, it is found an agreement in the case of touch voltage. Also, the quantitative comparison of the step current, the touch voltage, and the current proved to match the results between what was obtained in this article and the previous study in the reference (Gazzana, Daniel S., et al., (2014)), taking into account that in this article the study was done in the case of non-homogenous soils considering the ionization process in the soil.

As a validation process for the article results, calculations are done considering uniform soil and other data that are reported by Ref. (Gazzana, Daniel S., et al., (2014)). In the simulations produced here, a standard horizontal electrode with 7.5 mm of radius, with 9 m length, as well as buried at 0.5 m depth in homogeneous soil is implemented. A slow wave shape, double exponential with 5 kA (8 × 20) s is used in the calculations. The soil resistivity was 500 Ω. meter. An agreement is noticed between the paper results as well as Ref. (Gazzana, Daniel S., et al., 2014)) results as illustrated in Figure 16. The differences are related to considering the soil ionization procedure in this article.

Figure 25. Contact potential submitted in the body model when the lightning stroke has a different peak current at negative reflection factor

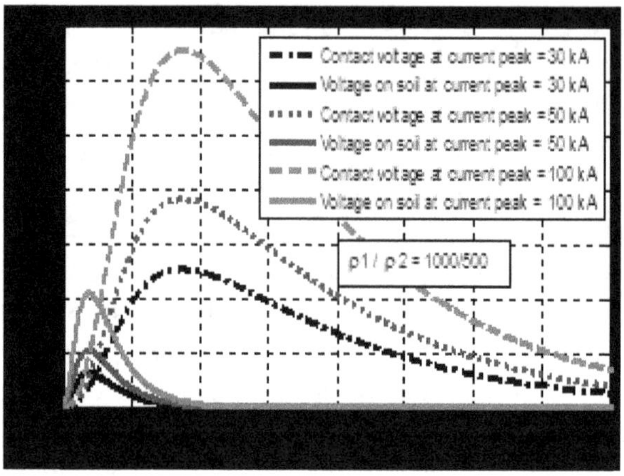

Figure 26. Contact potential submitted in the body model when the lightning stroke has a different peak current positive reflection factor

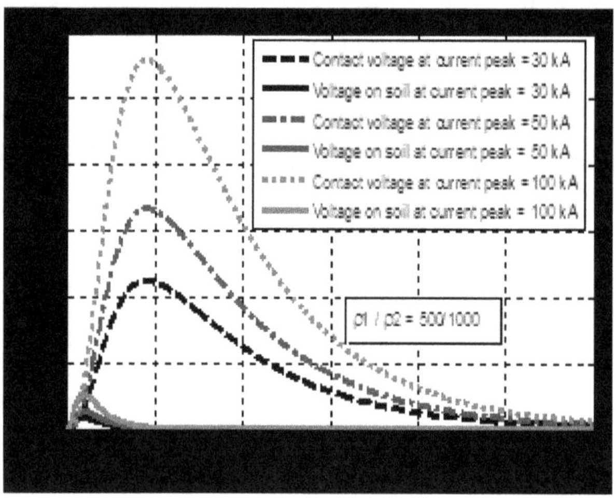

Figure 27. Current passing through the heart when contact potential submitted into the human body at negative reflection factor

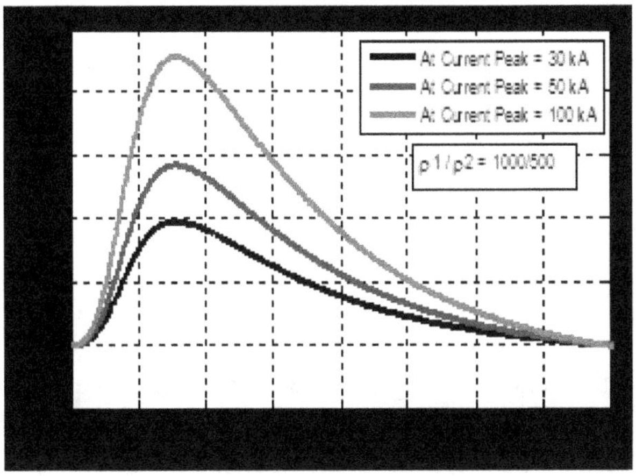

Figure 28. Current passing through the heart when contact potential submitted into the human body at positive reflection factor

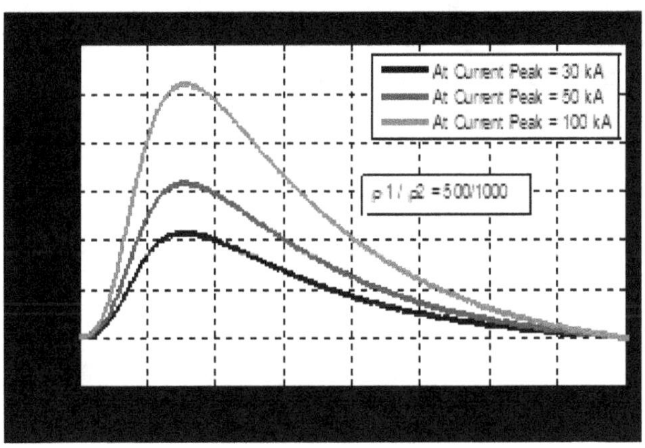

6.8 ELECTRICAL ENERGY GENERATED IN THE HUMAN BODY

To evaluate the impact of soil construction and lightning characteristics on the risky electrical energy that affects the person during step and contact voltages due to lightning strokes, the estimated current passing through the heart can be used. Equation (6.17) can be utilized in the estimation of such this energy per unit ohm of the current path:

$$E_e = \int_0^t i^2\, dt \text{ (j/}\Omega\text{)} \quad (6.\ 17)$$

Where E_e (j/Ω) is the energy per unit ohm and i is the current passing through the heart in Ampere during time t in seconds.

Table 2 shows the quantities of generated energy due to the step potential procedure at first and subsequent return strokes when soil ionization takes place. Table 3 shows the quantities of the electrical energy generated due to the contact potential procedure at first and subsequent return strokes when soil ionization is considered.

It is noted that the ratio between the soil resistivities for each of the first and second layers plays a sensitive role in the values of the current passing through the heart and, accordingly, the amount of generated electrical energy for each unit of its resistance. Considering that the heart resistance is in the range between 25 and 150 ohms, it can be concluded that the generated energy has significant value, especially in the case of touch (contact) potential. It can be concluded that the touch (contact) potential will be fatal (Bronzino J.D., (2000)).

Figure 29. Comparing results between Ref. (Gazzana, Daniel S., et al., (2014)) and ATP as the validation process

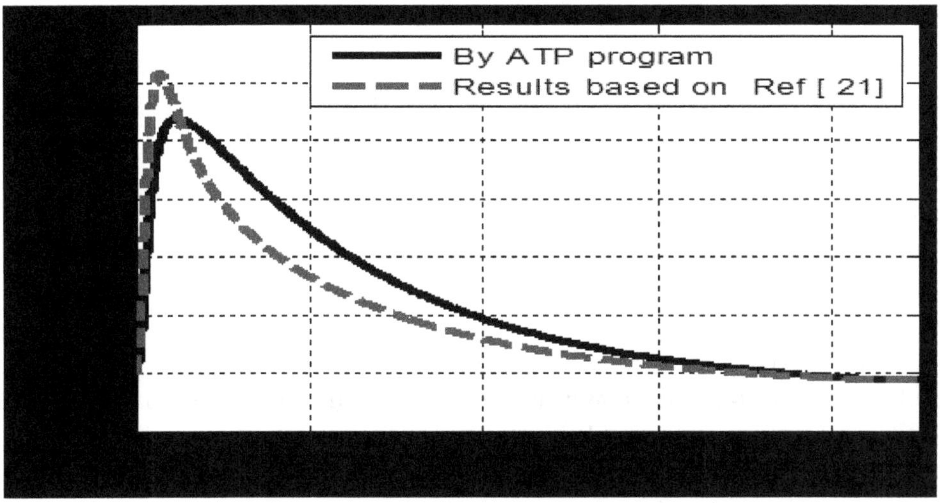

Table 2. Electrical energy generated due to step potential mechanism at first lightning and subsequent return strokes when soil ionization takes place

ρ_1/ρ_2	I_{Max} (A) pass in heart due to first lightning	E_e (j/Ω) due to first lightning	$I_{Max(A)}$ pass in heart due to subsequent return lightning	E_e (j/Ω) due to subsequent return lightning
300/100	0.0395	6.1028e-09	1.0641	8.5049e-08
100/300	0.1312	6.7133e-08	0.3566	9.5504e-09
1000/500	0.0132	6.8371e-10	3.5909	9.6843e-07
500/1000	0.0662	1.7105e-08	1.7780	2.3743e-07

Table 3. Electrical energy generated due to contact potential mechanism at first lightning and subsequent return stroke when soil ionization takes place

ρ_1/ρ_2	I_{Max} (A) pass in the heart due to the first lightning	E_e (j/Ω) due to the first lightning	$I_{Max(A)}$ pass in the heart due to subsequent return lightning	E_e (j/Ω) due to subsequent return lightning
300/100	340.7706	1.4000	107.9977	0.1785
100/300	975.1190	11.4631	144.9161	0.3215
1000/500	972.3712	11.3986	267.3263	1.0940
500/1000	2.1290e+03	54.6424	725.6813	8.0616

This chapter investigates the impact of direct lightning strikes on human safety near towers of HVTLs considering soil ionization influence. The article studies the current pass–through the heart of the human in cases of step and touch voltages in non-homogenous soil. In the present study, the first stroke and subsequent return stroke were considered, and the evaluation is carried out for lightning with a short duration. It is concluded that the step and touch potentials as well as the current pass-through heart in non-homogenous soil depend mainly on the resistivity of the upper layer, i.e., the negative reflection factors gave higher values in step potential and current more than positive reflection factors, while in touch potential the opposite has happened. This may be due to the difference in the path of the current in both cases.

It is observed that soil ionization plays an influential role in reducing the step and touch (contact) potentials as well as the values of the current pass-through heart. It is seen that the front time of lightning stroke affects the time to reach by the current pass-through heart to its beak value. The current pass-through of the heart of humans increases when the current peak of the lightning pulse increases, also when the reflection factor is positive the current pass–through the heart is decreased in case of step potential and increased in touch potential.

REFERENCES

Andrews, C. (2003), Electrical aspects of lightning strike to humans. In The Lightning Flash. IEEE Press.

Andrews, C. J. (2007). Lightning caused injuries in humans. *Journal of Electrostatics*, 65(5-6), 386–394. 10.1016/j.elstat.2006.09.016

ATP rule book, (1995), Can/Am EMTP User Group.

Becerra, M., & Cooray, V. (2009, August). on the interaction of lightning upward connecting positive leaders with humans. *IEEE Transactions on Electromagnetic Compatibility*, 51(4), 1001–1008. 10.1109/TEMC.2009.2033265

Bronzino, J. D. (2000). *The Biomedical Engineering Handbook* (2nd ed.). CRC Press.

Cavka, D., Mora, N., & Rachidi, F. (2014). A comparison of frequency dependent soil models: Application to the analysis of grounding systems. *IEEE Transactions on Electromagnetic Compatibility*, 56(1), 177–187. 10.1109/TEMC.2013.2271913

Effects of lightning strikes on the human body, (2019). Cardiac Arrest, Brain Injury, Seizures, Spinal Cord Damage, Amnesia, https://www.prevention.com/health/a32851873/struck-by-lightning-effects

Gauss-Seidel Method. (2022), https://math.libretexts.org/Bookshelves/Linear_Algebra/Introduction_to_Matrix_Algebra_ (Kaw)/01%3A_Chapters/1.08%3A_Gauss-Seidel_Method

Gazzana, D. S., Bretas, A. S., Dias, G. A. D., Telló, M., Thomas, D. W. P., & Christopoulos, C. (2014). A study of human safety against lightning considering the grounding system and the evaluation of the associated parameters. *Electric Power Systems Research*, 113, 88–94. 10.1016/j.epsr.2014.03.015

Gouda, O. E., Mohamed, A. Z. E. D., Al-Harthi, M. M., Omar, S. Y., & Ghoneim, S. S. M. (2022). Performance of grounding electrodes under lightning strokes in uniform and two-layer soils considering soil ionization. *IEEE Access : Practical Innovations, Open Solutions*, 10, 76855–76869. 10.1109/ACCESS.2022.3192394

Heidler F., & Cvetic J. (2002), A Class of Analytical Functions to Study the Lightning Effects Associated with the Current Front. ETEP, 12(2).

Hetita, I. D., Mansour, A., Han, Y., Yang, P., & Zalhaf, A. S. (2022). Experimental and Numerical Analysis of Transient Over voltages of PV Systems When Struck by Lightning, IEEE Transactions on Instrumentation and Measurement.

IEC/TR 60479-4, (2004), Effects of Current on Human Beings and Livestock – Part 4: Effects of Lightning Strokes on Human Beings and Livestock", Edition 1.0.

IEC/TS 60479-1, (2005), Effects of Current on Human Beings and Livestock – Part 1, General Aspects, Edition 4.0.

IEEE Std 80, (2000), Guide for Safety in Ac Substation Grounding, IEEE, New York.

Koont, D. R. (2015). *Lightning*. Headline Publishing Group.

Leonid, G. (2009). Modeling of grounding electrodes under lightning currents'. *IEEE Transactions on Electromagnetic Compatibility*, 51(3), 559–571. 10.1109/TEMC.2009.2025771

Manna, T. K., & Chowdhury, P. (2007). Generalized equation of soil critical electric field based on impulse tests and measured soil electrical parameters'. *IET Generation, Transmission & Distribution*, 1(5), 811–817. 10.1049/iet-gtd:20060559

Misbah, N. R., Ab Kadir, M. Z. A., & Gomes, C. (April, 2011), Modeling and analysis of different aspect of mechanisms in lightning injury, *Proceedings of the 4th International Conference on Modeling, Simulation and Applied Optimization*

Mokhtari, M., Abdul-Malek, Z., & Gharehpetian, G. B. (2016). A critical review on soil ionization modeling for grounding electrodes. *Archives of Electrical Engineering*, 65(3), 449–461. 10.1515/aee-2016-0033

Nahman, J., & Paunovic, L. (2006). Resistance to earth of earthing grids buried in multi-layer soil. *Electrical Engineering*, 88(4), 281–287. 10.1007/s00202-004-0282-y

Rachidi, F., Janischewskyj, W., Hussein, A. M., Nucci, C. A., Guerrieri, S., Kordi, B., & Jen-Shih Chang, . (2001, August). Current and electromagnetic field associated with lightning return strokes to tall towers. *IEEE Transactions on Electromagnetic Compatibility*, 43(3), 356–367. 10.1109/15.942607

Sreedhar, S., & Indragandhi, V. (2022). A Comprehensive Framework for Direct Lightning-Structure-Human Interaction Modeling in Heritage Monuments and Safety Assessment. *Energies*, 15(19), 7053. 10.3390/en15197053

Zalhaf, A. S., Mansour, D. E. A., Han, Y., Yang, P., & Darwish, M. M. F. (2022). Numerical and Experimental Analysis of the Transient Behavior of Wind Turbines When Two Blades are Simultaneously Struck by Lightning, IEEE Transactions on Instrumentation and Measurement.

Chapter 7
Step and Touch Voltages and Grounding Potential Rise (IEEE and IEC)

Osama El-Sayed Gouda
Faculty of Engineering, Cairo University, Egypt

ABSTRACT

This chapter explores set and touch voltages and grounding potential rice (IEEE and IEC). This includes safety criteria according to IEEE standards, step and touch voltage criteria, reduction factor, simplified equations for mesh and step voltage, application of step and mesh potential in safe earthing system design, Safety Criteria-IEC 479-1 std, permissible touch and step voltage IEC 479-1, optimum compression ratio (OCR), test and verify, approximated analysis to calculate voltage profile using apparent soil resistivity, scaling factor, test and verify the approximated method, and ground potential rise (GPR) of faulty substations having equal and unequal spacing grounding grids conductors.

7.1 INTRODUCTION

Since the early days of the electric power industry, safety of personnel in and around electric power installations has been a prime concern.

Over the years and after many investigations on the effects of electric current on humans, safe limits have been established and standards have been developed which provide permissible values of body currents to avoid electrocution. Two such standards are IEEE 80 (1961) and the IEC60 479-1(1984).

The purpose of both standards is to establish safe (permissible) body currents.

DOI: 10.4018/979-8-3693-2969-6.ch007

Since IEEE Standards 80 publication in 1961, Guide for Safety in AC Substation Grounding, has been favorably received by the industry and gained a broad acceptance worldwide, this chapter presents the improvements in a broader context of the Guide's development to this date.

This chapter is organized as follows: first the safety criteria as stated in the two documents, IEEE Standards (2000) and IEC 60479-1 (1984), are described in detail and evolution of IEEE standard. A common basis for comparison is developed in terms of permissible touch and step voltages.

A detailed analysis of potential profile at a surface above energized unequally spaced grounding grid has been carried out, for soils with multiple layers. Exact and approximated algorithms are used to obtain the voltage profile and then to calculate the step and touch voltages.

7.2 SAFETY CRITERIA ACCORDING TO IEEE 80 STANDARDS

7.2.1 Step and Touch Voltage Criteria

7.2.1.1 Allowable Body Current Limit

An electric shock accident is determined by the magnitude, duration, and frequency of the current passing through the heart. When a current passes through the heart it causes ventricular fibrillation, and when the current flows through the nerve center it suppresses the respiratory as reported by IEEE 80 (2000).

Therefore, safety design of grounding system takes into account the critical value of an electric shock current causing ventricular fibrillation. Many Researchers have studied the permissible body current limit as reported by Schon. J.H (1998) and ESBTSD (1994). For an AC current with a frequency of 50 Hz, it can be classified into three categories according to human responses:

- Sensory current: this is the minimum current a human can feel. For a male adult it is 1.1 mA, and for a female adult, it is 0.7 mA.
- Extrication current: this is the maximum current that can be extricated by human after electroshock. It is also related to gender, for a male adult it is 16 mA and for a female adult, it is 10.5 mA.
- Fatal current: this is the minimum current that threatens life in the shortest time, by causing ventricular fibrillation or asphyxiation as reported by Schon. J.H (1998).

It is concluded that the value of current play an important role in electric shock on the human body, and also the duration of the current passing through the body is considered. If the duration is very short, the current that a human can tolerate and which does not cause ventricular fibrillation as reported by Dalziel, C. F. and Massogilia, F. P, (1956) is:

$$I_K^2 t = K \tag{7.1}$$

Where the unit of I_K is A, t is the time passing through human in second, and K is an energy coefficient related to the weight of the human body. The energy coefficients of 50 and 70 kg humans are $K_{50} = 0.0135$ and $K_{70} = 0.0247$, respectively. Therefore, we can get the allowable body current I_{K50} for a 50 kg human

$$I_{K50} = 0.116/\sqrt{t} \tag{7.2}$$

According to the result from Equation 2, the tolerable current corresponding to 1 s duration is 116 mA, while that corresponding to 0.1 s is 367 mA and that corresponding to 0.05 s is 519 mA.

The allowable body current I_{K70} for a 70 kg human is

$$I_{K70} = 0.157/\sqrt{t} \tag{7.3}$$

7.2.1.2 Allowable Body Voltage

Once allowable body current I_K is known the allowable body voltage calculated by the multiplying the allowable body current I_K and the body resistance R_B is obtained:

$$U_{SV} = I_K R_B \tag{7.4}$$

By setting Equation (7.3) into Equation (7.4), the short-time allowable body voltage for a 50 kg human is:

$$U_{SV} = 0.116 R_B/\sqrt{t} \text{ (7.5 a)}$$

Also, the allowable body voltage for a 70 kg human can be obtained:

$$U_{SV} = 0.157 R_B/\sqrt{t} \text{ (7.5 b)}$$

7.2.1. 3 Touch and Step Voltage Criteria

The safety of a people in substation depends on preventing the cirticl amount of shock energy from being absorbed before the fault is cleared by protection system.

The maximum voltage of any accidental circuit should not exceed the limits defined as follows.

The limit of step voltage:

$$E_{Step} = (R_B + 2R_f) . I_B \tag{7.6}$$

For body weight of 50 kg

$$E_{Step} = (1000 + 6C_s . \rho_s) . \frac{0.116}{\sqrt{t_s}} \tag{7.7}$$

For body weight of 70 kg

$$E_{Step} = (1000 + 6C_s . \rho_s) . \frac{0.157}{\sqrt{t_s}} \tag{7.8}$$

Similarly, the limit of touch voltage:

$$E_{touch} = \left(R_B + \frac{R_f}{2}\right) . I_B \tag{7.9}$$

For body weight of 50 kg

$$E_{touch} = (1000 + 1.5C_s . \rho_s) . \frac{0.116}{\sqrt{t_s}} \tag{7.10}$$

For body weight of 70 kg

$$E_{touch} = (1000 + 1.5C_s . \rho_s) . \frac{0.157}{\sqrt{t_s}} \tag{7.11}$$

Where;

E_{step} is the step voltage in V, E_{touch} is the touch voltage in V

ρ_s is the resistivity of the surface material in Ω.m.

t_s is the duration of shock current in seconds, C_s is the correction coefficient of the foot grounding resistance.

The goal of a safe grounding design is to prevent the possibility of situations when a person can be vulnerable to absorbing a dangerous level of electrical shock energy, before the fault is cleared and the system de-energized. Hence,

$V_{sc} >= Vc$, is always required for safety,

V_{sc} is safe voltage limit; Vc is voltage of accidental circuit.

The expression (7.12) is obtained when the depth of soil of resistivity is very large. Since the depth of the crushed rock or gravel layer is finite and its resistivity is more than that of soil underneath, ground resistance of the foot is calculated from (7.13)

$$R_f = \frac{\rho_s}{4b} \tag{7.12}$$

$$R_f = \frac{\rho_s}{4b} C_s \tag{7.13}$$

b=0.08m

Application of Cs Factor in **1986** IEEE Edition: Without the use of computer, Cs is rather tedious to evaluate, especially when a protective surface layer of small thickness h_s, is covering the switchyard and its resistivity is greater than that of the underlying soil. **A** set of curves, was therefore provided, (Figure 7.1). The curves were based on the following formula:

$$C_s = \frac{1}{0.96}[1 + 2\sum_{n=1}^{\infty} \frac{k^n}{\sqrt{1 + (2n h_s/0.08)^2}}] \tag{7.14}$$

$$K = \frac{\rho - \rho_s}{\rho + \rho_s}$$

Figure 1. Reduction factor Cs as a function of reflection factor K and crushed rock layer thickness hs

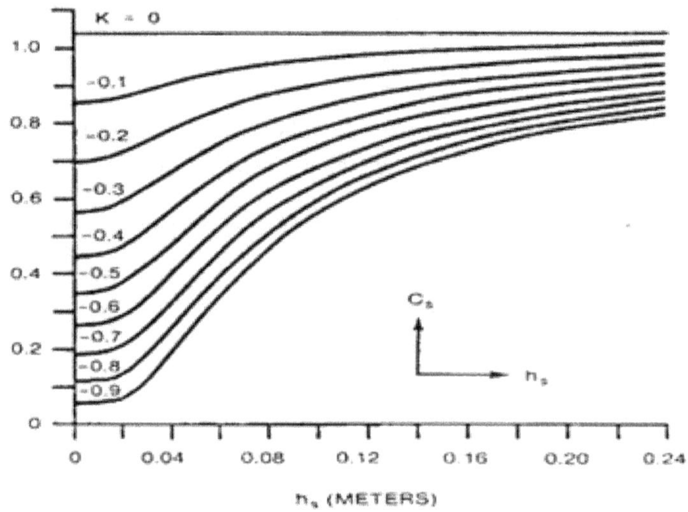

As a footnote option left to designer discretion, the 1986 Guide also suggested a very simple semi-empirical equation for alternative estimate of Cs shown here as C_{sest} in equation (7.15)

$$C_{sest} = 1 - a\left[\frac{1 - \frac{\rho}{\rho_s}}{2h_s + a}\right] \quad (7.15)$$

a=0.106 m. which avoids the infinite summation series, are also possible;

However, analytical studies done have revealed that equation 7.14 contains a minor error. The shape of de-rating curves is symptomatic of this oversight: There is an initial convex-to-concave curvature for depths between 0 and 0.04 m, which causes that values of *Cs* are too conservative for hs < 0.15 m.

On the other hand, as also documented by Dawalibi et al. (1975), the seemingly unsophisticated equation 15 which was adapted from Sverak (1984) just for the mentioned "footnote approximation" - has proven itself to yield not only reasonably correct results, but quite unexpectedly, for very thin layers to do so within **a** smaller error range than other formulas.

Application of Cs Factor in 1996 Edition: In view of the above described findings, it is likely that the following equations will be included in the 1996 edition, giving the value of *Cs* with very good accuracy as given in equation (7.16):

$$C_s = C_1 - C_2 \tan^{-1}\left(\frac{2h_s}{b}\right) - C_3\left[e^{-7h_s} - e^{-30h_s}\right] \tag{7.16}$$

$$C_1 = (1+k)/(1-k), C_2 = 4k/[\pi(1-k)],$$

$$C_3 = 0.21\,k^2, \text{ hs and b in meters}$$

Equation 16 has been derived as a polynomial fit by Thapar et al. **(1991), to** give values of *Cs,* very close to the results obtained with their accurate but rather complex analytical method. The error is less than 5%. This equation is applicable for hs varying from 0 to 0.3 m and K varying from 0 to -0.98. **A** set of Cs curves for b = 0.08 m will also be provided, Figure 2, replacing Figure 1.

The older equation 15 is more accurate for very thin surface layers between 0.005 m and 0.02 m, and for the rest of parameter range is about 2 -5% more conservative than equation (7.16).

Ultimately, using the factor 1/0.96 from equation 14 makes its results practically identical to those of equation (7.16). Thus,

$$C_s = C_{sest}/0.96 \tag{7.17}$$

Figure 2. Reduction factor Cs as a function of reflection factor K and crushed rock layer thickness hs

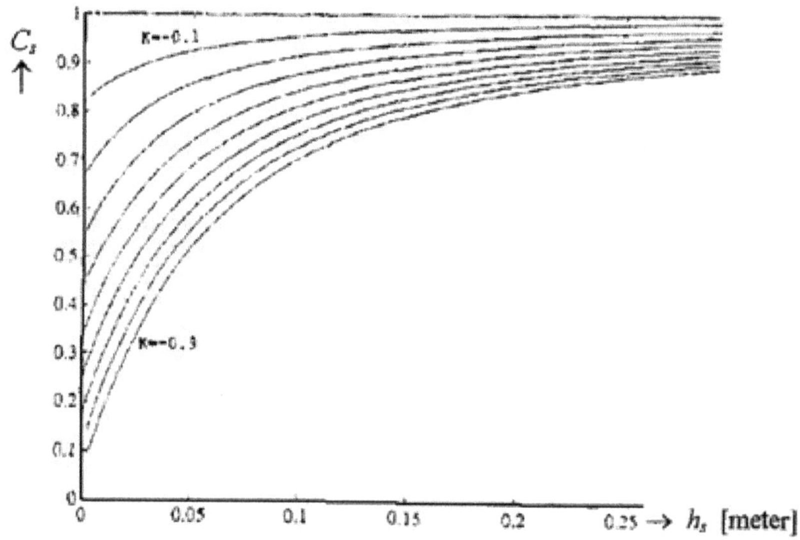

7.2.1.4 Simplified Equations for Mesh and Step Voltage

In 1959, a small working group led by Stevens developed a practical mathematical model, allows approximating the maximum step and touching voltages in the critical area of a corner mesh. For its time, this was a remarkable achievement. The model, which so far has been used in all editions of the Guide, is based on a simple abstraction of n parallel conductors in depth h. The mesh voltage, in volts, is obtained as.

$$E_m = \frac{\rho . K_m . K_i . I_G}{L_m} \tag{7.18}$$

An analogous formula for calculation of the maximum step voltage, Es in volts, is

$$E_s = \frac{\rho . K_s . K_i . I_G}{L_s} \tag{7.19}$$

Note: A formal distinction between L=Ls and L=Lm indicates that the effective buried conductor length may be defined differently for the step and the touch calculations.

In either formula, the calculated voltage is obtained as a product of the soil resistivityρ, of an average current density per unit of the effective buried length, IG/Lm or IG/Ls, and of two factors:

Km or Ks characterizing the grid geometry, and

Ki, an irregularity factor, which accounts for some of the errors introduced by the assumptions made in deriving Km and Ks

The effective length of buried conductors includes ground rods connected to the grid. The entire grounding system is energized and conducts current IG into the earth. A homogeneous soil of uniform resistivity is assumed.

AIEE No. 80 (1961) and IEEE 80 (1976) Editions reported equations to calculate the constants *Km, Ks* and *Ki* for square grid consisting of *n* parallel conductors evenly spaced *D* as follows:

$$K_m = \frac{1}{2\pi}\cdot\left[\ln\left(\frac{D^2}{16\cdot h\cdot d}\right)\right] + \frac{1}{\pi}\ln(\lambda) \qquad (7.20)$$

$$K_s = \frac{1}{\pi}\left[\frac{1}{2h} + \frac{1}{D+h} + \frac{1}{2D} + \frac{1}{3D} + \ldots.. + \frac{1}{(n-1)D}\right] \qquad (7.21)$$

$$K_i = 0.65 + 0.172n \qquad (7,22)$$

n is number of parallel grid conductors in one direction
D spacing between parallel conductors in m
d is diameter of grid conductor, m
h is depth of grid burial, m
λ is a finite series (3/4)(5/6)(7/8)…..((2n-3)/(2n-2))

The effective buried length L, used both in IEEE 80 (2000), was simply defined as

$$L = L_c + L_r \qquad (7.23)$$

L_c=total length of grid conductors, in m,

Lr= total length of all ground rods, in m

Factors *Km, Ks* and *Ki* perform the necessary corrections due to the difference in geometries of a configuration of *n* parallel conductors representing the grid and of the actual grounding system. *Km* simulates the influence of currents flowing in *n* -2 conductors of the *n* conductor model upon the current density in two most outward parallel conductors representing the corner mesh, whereas *Ki* compensates for the difference in performances of the whole model and of a complete grid.

To assist the designer in choosing a convenient square grid representation of rectangular or L-shaped grids and in interpreting the results, a figure was provided for six different grid spacing patterns, depicting the distribution of mesh voltages within each grid. These auxiliary data reflected the outcome of Koch's (1950) measurements on miniature grid models in an electrolytic tank. A complementary example, describing the development of a grounding design for an L-shaped grid from a rough idea to a viable concept, was also provided. Otherwise, implicitly assuming the presence of one's good engineering sense and prudent judgment, no explicit applicability limits were mentioned.

In practical use, however, some deficiencies emerged. It has been found that the application of *Km* as defined by equation 20, causes optimistic errors in the calculated values of the mesh voltage and, for very dense grids with high *n* and *D* approaching the order of *h* parameter, the result becomes negative. In contrast, *Ks* defined by equation (7.21) often yields overly conservative values for grids buried deeper than *0.25* m below the earth surface.

IEEE 80 Edition (1986) improved the accuracy of calculations for a broad class of rectangular grids with and without ground rods, Sverak. J.G. (1984) reassessed the effect of simplifying assumptions used in AIEE No. 80 (1961) and established a deferent modeling approach, to develop new simplified equations for Km and Ks, which corrected most of the deficiencies of the previous equations.

Given below as equation (7.24), the resulting equation for Km is more accurate and versatile than the previously used

$$K_m = \frac{1}{2\pi} \cdot \left[\ln\left(\frac{D^2}{16 \cdot h \cdot d} + \frac{(D+2h)^2}{8 \cdot D \cdot d} - \frac{h}{4 \cdot d} \right) + \frac{K_{ii}}{K_h} \ln \frac{8}{\pi(2n-1)} \right] \tag{7.24}$$

$K_{ii}=1$ for grids with ground rods along perimeter or with ground rods both in the grid corners and inside the grid area, or, $K_h=\sqrt{1+\frac{h}{h_0}}$, $h_0=1$

$K_i = \frac{1}{(2n)^{\frac{2}{n}}}$ for grids with a few inner ground rods, or no rods.

The approximation of a current density distribution pattern includes two weighing terms, K_{ii} and K_h The first term adjusts the influence of n - 2 conductors upon the current density in the first two conductors forming the corner mesh, according to the presence (K_{ii} = 1), or the absence (K_{ii} = 2n (exp)-2h) of peripheral grounding rods. The second term, K_h, corrects for non-zero depth of the remaining n – 2 conductors. The finite series λ = (3/4) (5/6) ... etc., which had been previously used in equation 20 to represent these conductors, is replaced by an asymptotic sum.

Ks Model: For the calculation of step voltages, two versions of factor *Ks* were used:

- The older formula given in equation 7.21, applicability of which was limited to shallow depths between 0 and 0.25 m; and
- Sverak's formula below, applicable to depths greater than 0.25 m.

$$K_s = \frac{1}{\pi}\left[\frac{1}{2h} + \frac{1}{D+h} + \frac{1}{D}(1 - 0.5^{n-2})\right] \tag{7.25}$$

Other Changes: Compared to the value given in equation 7.22, the value of a starting point in the definition of irregularity factor *Ki* was slightly adjusted, yielding exactly *Ki* = 1 for a singular mesh (n = 2);

$$K_i = 0.656 + 0.172n \tag{7.26}$$

An important novel detail in the application of *Km* and *Ks* was a more precise definition of parameter *n* for rectangular grids, characterized by an actual number of the parallel conductors in each direction, n_A, n_B. For the calculation of mesh voltages,

$$n = \sqrt{n_A n_B}$$

Rounded to the nearest integer (7.27)

For the calculation of step voltages,

$$n = Max.(n_A, n_B) \tag{7.28}$$

Only one value of the effective buried conductor length *L*, was used both in the general formula for mesh and step voltage, that is, $L_M = Ls = L$. However, it was defined differently for grids with peripheral ground rods and for other grids. For grids with ground rods predominantly around perimeter

$$L = L_c + 1.15L_r \tag{7.29}$$

The use of older definition presented in equation 23, that is, of *L* = *Lc* + *Lr*, was restricted to grids with no ground rods, or with only a few rods located in inner parts of the grid, away from the perimeter.

Finally, the following, conservative applicability limits were recommended for square and rectangular grids:

$n < 25 \quad D > 2.5$ m

d < 0.25 *h*   0.25 m < *h* < 2.5 m

In IEEE Edition (**1996**) to extend the applicability of simplified calculations to T-shaped, L-shaped and triangular grids, as well as to further improve the overall accuracy of results, it is proposed to introduce a new definition of parameter *n,* which will broaden the applicability of existing mesh voltage and step voltage equations Additional minor refinements concern various empirical co-factors in definitions of the effective conductor length and an adjustment of *Ki* formula., reflecting the result of bench-mark tests performed by the Working Group. The specific refinements are:

1. Equation (7.24) for calculations of the mesh voltage, retain without change the formulation of *Km* itself, but broaden its applicability by a new definition of parameter *n.*
2. Equation (7, 25) for calculations of the step voltage, retain without change this more recent formulation of *Ks* itself (as given for grids buried in depths >= 0.25 m), but broaden its applicability by the new definition of parameter *n,*
3. Equation (27.1) abandons this older version of *Ks,* which has been already limited by the 1986 edition to grids in a shallow depth < 0.25 m.
4. Using four grid shape components developed by Thapar et al (1991) the newly extended formulation of parameter *n,* is:

$$n = n_I . n_{II} . n_{III} . n_{IV} \tag{7.30}$$

Where
$n_{II} = 1$ for square grids
$n_{III} = 1$ for square and rectangular grids
$n_{IV} = 1$ for square, rectangular and L-shaped grids
And

$$n_I = \frac{2L_c}{L_p}; n_{II} = \sqrt{\frac{L_p}{4\sqrt{A}}}; n_{III} = \left[\frac{L_x L_y}{A}\right]^{\frac{0.7A}{L_x L_y}}; n_{IV} = \frac{D_m}{\sqrt{L_x^2 + L_y^2}}$$

A=area of grid, m^2
Lc=total length of conductors forming the grid, m
Lp=peripheral length of the grid, m
Lx=maximum length of the grid in the x direction, m
Ly=maximum length of the grid in the y direction, m
Dm=maximum diagonal of the grid, m

5.Equations 7.18 and 7.19 modify the definitions of effective buried conductor lengths, $L = Lm$ and $L = Ls$, using different co-factors for the mesh and step voltage formula, as follows:

i) For the calculation of mesh voltages, the effective buried length, L_M, applicable to grids with ground rods in the corners, as well as along the perimeter and inside the grid, is:

$$L_m = L_c + \left[1.55 + 1.22\left(\frac{L_r}{\sqrt{L_x^2 + L_y^2}}\right)\right] \cdot L_R \qquad (7.31)$$

Where
Lc = total length of grid conductors, in m
L_R = total length of all ground rods, in m
L_r = length of each ground rod, in m

i) For the calculation of step voltages, the effective buried length Ls, applicable to grids with or without ground rods, is:

$$L_s = 0.75L_c + 0.85L_r \qquad (7.32)$$

6. Implement a modification of the irregularity factor Ki, which has been derived by Thapar et al (1991) in conjunction with the extended formulation of parameter n. Compared to previous definitions, there is a slight adjustment both in the slope and in the starting point of the Ki curve:

$$K_i = 0.644 + 0.148n \qquad (7.33)$$

7.3 APPLICATION OF STEP AND MESH POTENTIALS IN SAFE GROUNDING SYSTEM DESIGN

In IEEE 80 standard (2013) it is concluded for grounding system design that

$$K_m . K_i . \rho . \frac{i_G}{L_m} \langle ((1000 + 1.5C(h_s, k)\rho_s) \cdot \frac{0.116}{\sqrt{t}}$$

This relation could be used to calculate the safe conductor length to ensure safe step and touch voltages as follow. The influence of soil resistivity on step and touch potential Figure 3 gives the relation between touch potential of the grounding system and total length of rods and grid conductors, Figure 4 shows relation between step potential of the grounding system and total length of rods and grid conductors. Grid dimension 50 m.50 m with 11 parallel grid conductors at each direction and rods located at perimeter, buried at 0.5 meter depth, crushed rock of 3500 Ω.m resistivity and thickness 10 cm used, grid current assumed to be 1 kA for 1 second fault clearance duration. The calculations are done taking three values of soil resistivity 100,500 and 1000 Ω.m. The safety values (step and touch) are shown also in the figures. The safety design length decreases with decreasing the earth resistivity.

Figure 3. Touch potential of the grounding system as function of grounding system length

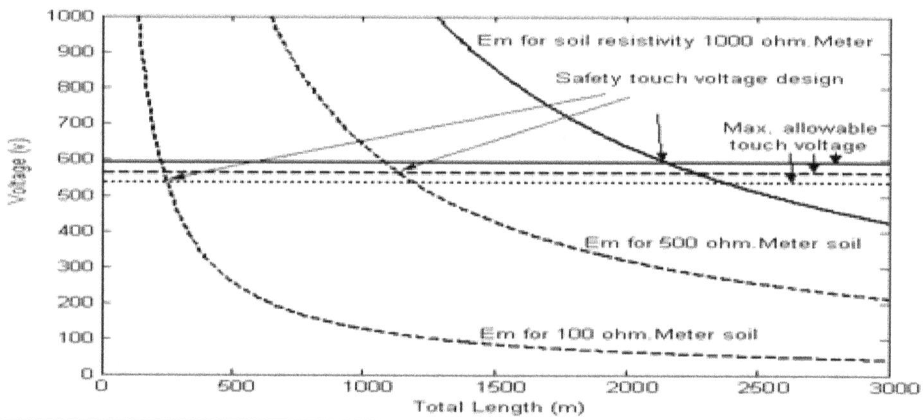

Figure 4. Step potential of the grounding system as function of grounding system length

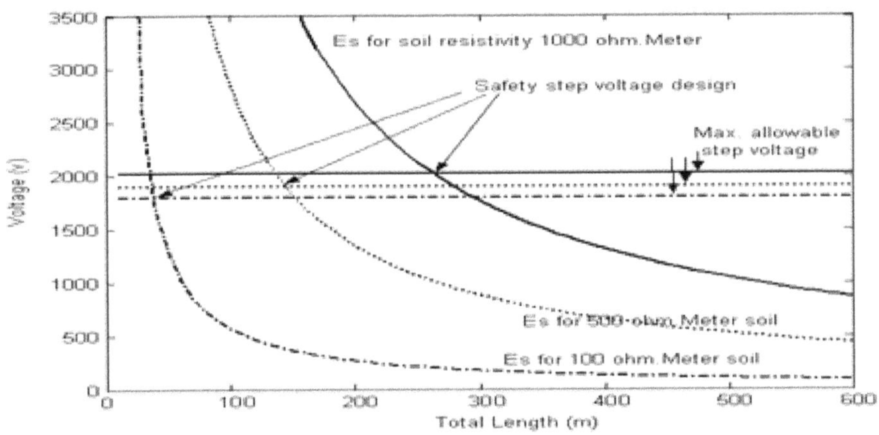

In case of step and mesh potential are very high than the tolerable step and touch potential and the ground system design must be modified, it is possible to increase crushed rock thickness or resistivity as it will affect the maximum allowable step and touch potential. Figure 5 shows the relation between step voltage and maximum allowable step voltage for different crushed rock resistivity. The influence of crashed rock thickness on touch potential is shown in Figure 6. This figure shows the relation between mesh potential and maximum allowable touch voltage for different crushed rock thickness (0.1, 0.15 and 0.2 meter).

Figure 5. Step potential of the grounding system for different crushed rock resistivity (h_s=0.1m)

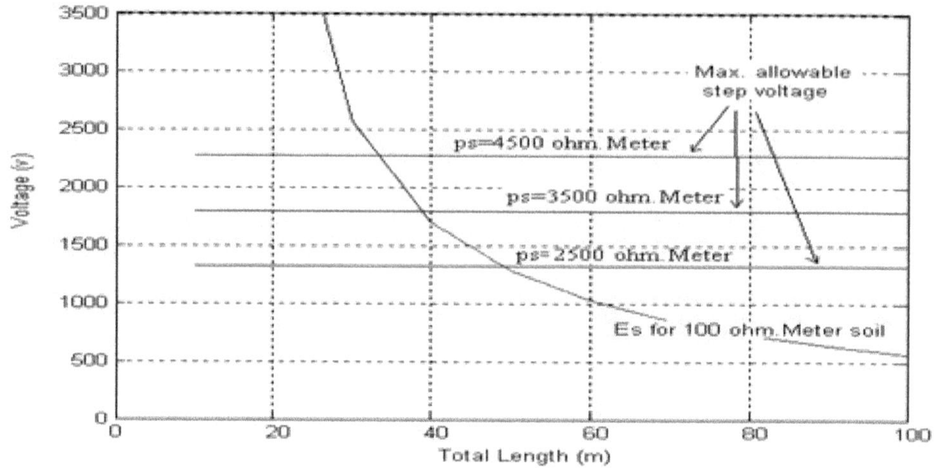

Figure 6. Touch potential of the grounding system for different crushed rock thickness (ρ_s =3500 Ω.m)

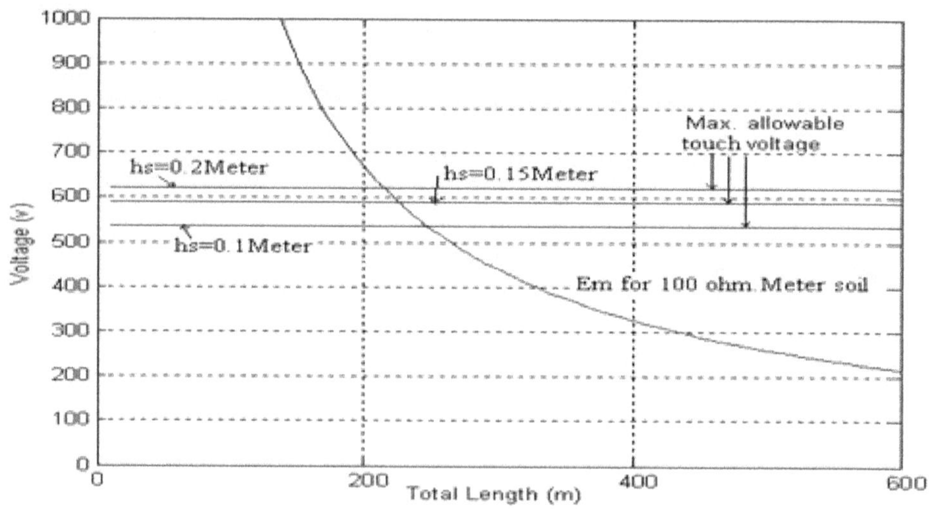

7.4 SAFETY CRITERIA-IEC 60479-1 STD

The parameters of the electric shock model are listed in Figure 7 showing differences in IEEE and IEC. This model is usually translated into permissible touch (or step) voltages.

IEC 60479-1*(1984)* is less specific than IEEE Std 80 (2000) for design purposes. The standard provides data for body resistances as a function of body voltage, which are illustrated in Figure 8, and data of body resistance as a function of path, which are illustrated in Figure 8. In Figure 8, the 5% curve indicates body resistance values which were not exceeded by 5% of the population; the 50% curve indicates body resistance values which were not exceeded by 50% of the population, etc. All the values in (Figure 8) are for hand-to-hand resistance. In Figure 9, the numbers not in brackets indicate the impedance of several paths in the body as a percentage of the hand-to-hand impedance. The numbers in brackets are for current paths between both hands and the corresponding part of the body. IEC 60479-1*(1984)* also provides values of permissible body current versus electric shock duration as shown in Figure 10. Points on curve C1 represent 0.14% probability of ventricular fibrillation, points on curve C2 represent 5% probability of ventricular fibrillation and points on curve C3 represent 50% probability of ventricular fibrillation. These curves separate the space of body current and shock duration into zones. As an example, Zone 4 represents all the combinations of body current and shock duration which will lead to ventricular fibrillation with probability more than 50%.

Figure 7. Electrical shock model differences between IEEE 80 Std. and IEC 60479-1

	IEEE Std 80	IEC 479-1
Body Resistance	1000 ohms	Voltage Dependent and Path Dependent
Thevenin Equivalent Resistance	$1.5c_s\rho_s$ for touch voltage $6.0c_s\rho_s$ for step voltage	no guidance
Thevenin Equivalent Voltage	Simplified Equations k_t k_r L I or use of computer models is suggested	no guidance
Permissible Body Current	$0.116A/\sqrt{t}$ for 50 kg person $0.157A/\sqrt{t}$ for 70 kg person	S-curves independent of human size

Figure 8. Human body resistance as a function of body voltage

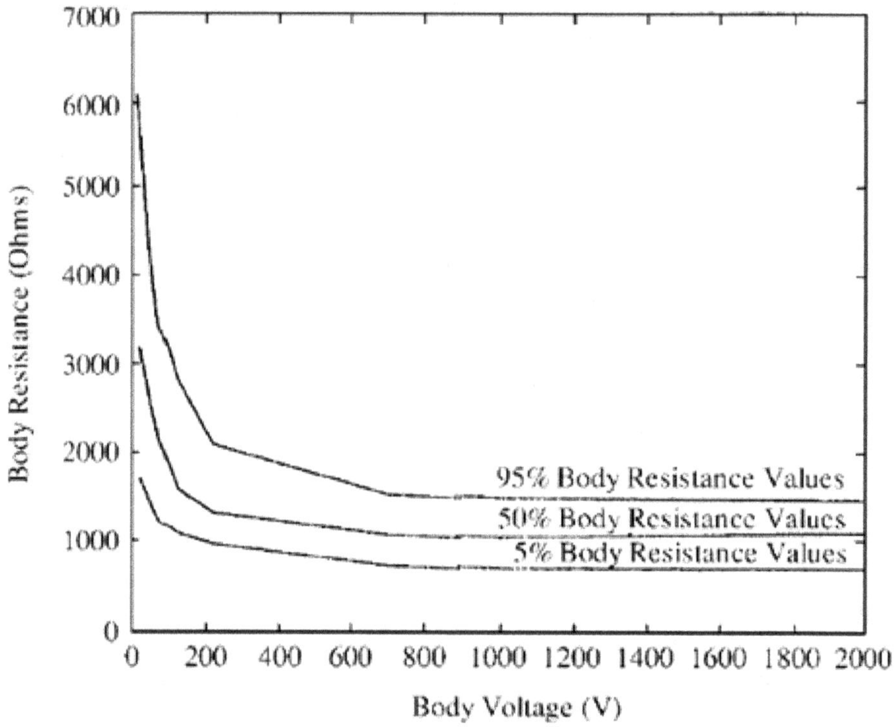

Figure 9. Internal impedance of the human body as a function of the current path

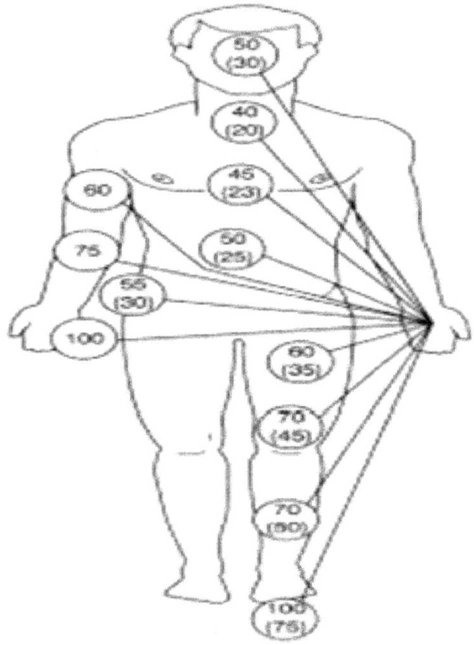

The data of IEC 60 479-1 ***(1984)*** can be utilized in two ways:

(1) Actual body currents can be computed for an individual subjected to touch or step voltage in a specific system and under specific conditions, and
(2) Permissible touch and step voltages can be computed for a specific system.

7.5 PERMISSIBLE TOUCH AND STEP VOLTAGES IEC 60479-1(1984)

The permissible (or allowable) touch, V_T^a, and step, V_S^a, voltages are computed from the following equations:

$$V_T^a = i_{b,perm}(t)\left[R_b^T\left(i_{b,perm}(t)\right) + r_{eq,T}\right] \qquad (7.34)$$

$$V_S^a = i_{b,perm}(t)\left[R_b^S\left(i_{b,perm}(t)\right) + r_{eq,S}\right] \qquad (7.35)$$

Figure 10. Permissible body current per IEC 60479-1

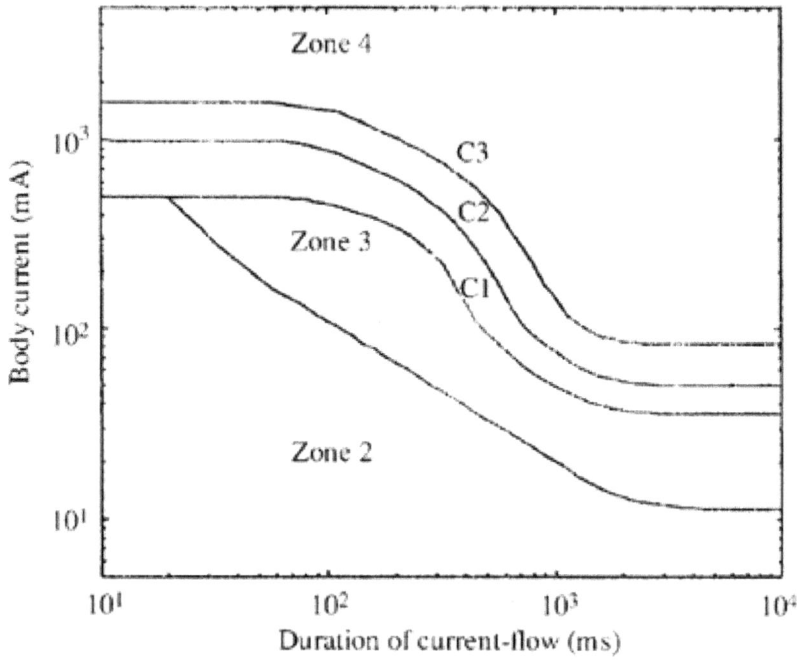

Where

$i_{b,\mathrm{perm}}(t)$	**is the permissible body current per IEC 60479-1 for an electric shock duration *t*. This current is obtained from the data shown in Figure 8.**
R_b^T (*ib*,perm(*t*))	is the body resistance for the path specified by the touch voltage (typically, hand to two feet) and for a body current equal to $i_{b,\mathrm{perm}(t)}$. This value can be obtained from the data shown in Figures 8 and Figure 9.
R_b^S (*ib*,perm(*t*))	is the body resistance for the path specified by the step voltage (foot to foot) and for a body current equal to $i_{b,\mathrm{perm}(t)}$. This value can be obtained from the data shown in Figure 8 and Figure 9.
$r_{\mathrm{eq},T}$	is the feet to soil resistance for touch voltage; i.e., the two feet to soil resistances are in parallel.
$r_{\mathrm{eq},S}$	is the feet to soil resistance for step

IEC 60479-1 does not provide any data for $r_{\mathrm{eq},T}$ or $r_{\mathrm{eq},S}$. For this reason, we shall use the data of IEEE 80 Std., $req_s = 6C_s\rho_s$, $req_T = 1.5C_s\rho_s$

Note that application of above equations to obtain the permissible touch and step voltages is straightforward and involves the following steps:

Step 1: For a given electric shock duration t and a given assumed probability of ventricular fibrillation, determine the value of permissible body current, $i_{b,\text{perm}(t)}$, from Figure 10.

Step 2: For the current $i_{b,\text{perm}(t)}$, determine the body resistances R_{bT} and R_{bS} from the data shown in Figure 8 and Figure 9. For touch voltage, it is expedient to assume that the path will be one hand to two feet (75% of the body resistance given in Figure 9, and for step voltage that the path is foot to foot (100% of the body resistance given in Figure 9.

Step 3: Compute $r_{\text{eq},T}$ and $r_{\text{eq},S}$ per IEEE 80 Std.

Step 4: Compute permissible touch and step voltages using Equations 7.34 and 7.35.

For purposes of comparison with IEEE Std 80 (2000), the permissible body current $i_{b,\text{perm}(t)}$ is computed for probability 0.5% of ventricular fibrillation using proper interpolation between curves C1 and C2 in Figure 10.

7.6 COMPUTATION OF POTENTIAL PROFILE AT A SURFACE ABOVE ENERGIZED UNEQUALLY SPACED GROUNDING GRID

Studying the effect of voltage distribution is an important parameter to design the grounding systems of substations to ensure the safety of people above the substation ground. This can be achieved by equalizing the potential distribution of the ground surface and reducing step and touch potentials. This is reported by Zhongdong, Du., Yao, Zhenyu, Wen, Xishan, Xu, Hua, (2005)

The earth surface potential distribution above the equally spaced grid is extremely non-uniform. In this section the voltage distribution is calculated for unequally spaced by using the suggested image method and also by using approximated method in case of multi-layer soil. The obtained results are compared with the IEEE method as reported by ANSI/IEEE Std. 80-1986. IEEE guide has proved to be a very popular guide for substation grounding applications. IEEE gradient analysis method allows a recursive point by point integration of surface gradients through all parallel grid conductors .The analytically most significant equations of the IEEE gradient analysis method are given by Dawalibi, F., Mukhedkar, D. (1975).

In the suggested algorithm the expressions of potential at any point due to a point current source in multi-layer soils can be obtained by image technique or by solution of Laplace's equations. These are in the form of infinite summation series. The rate of convergence of these series is mainly dependent on the reflection factor between layers of the soil. For a reasonable accuracy, up to a few thousand terms of series may have to be computed. Determination of potential from such expressions, therefore, forms significant part of computational effort in analysis and design of a grounding system in three-layer soil.

A new simplified method for calculating the grounding voltage distribution of grounding grid in three-layer earth structure (multi-layers) has been presented.

7.6.1 Analytical Analysis

(Figure 7.11) shows the grounding system model under study, which consists of horizontal three layers. The grid is placed in the first layer of grounding system which its resistivity (ρ_1), over a second layer of resistivity (ρ_2), which is over the third layer of resistivity (ρ_3). Thickness of the first layer and second layer are (d_1), and (d_2) respectively. The depth of the buried grid in the first layer is (b), while the distance between the grid and surface of the second layer is (h). The ground voltage profile of the grid can be determined with the use of image technique.

The three boundary planes between air, first layer, second layer, and third layer produce an infinite sequence of images. All these images discharge current into the first layer of its resistivity (ρ_1). Whenever an image is taken on the boundary between air and first layer, the current of the image remains the same. However, when an image is taken on the boundary between the first and second, the current of the image will be reduce by a ratio.

$$k_{12} = \frac{\rho_2 - \rho_1}{\rho_2 + \rho_1} \tag{7.36}$$

Whenever, an image is taken on the boundary between the second and third, the current of the image reduces by a ratio

$$k_{23} = \frac{\rho_3 - \rho_2}{\rho_3 + \rho_2} \tag{7.37}$$

Where K_{12}: Reflection factor between first and second layer, K_{23}: Reflection factor between second and third layer

Figure 11. Grounding system model under study

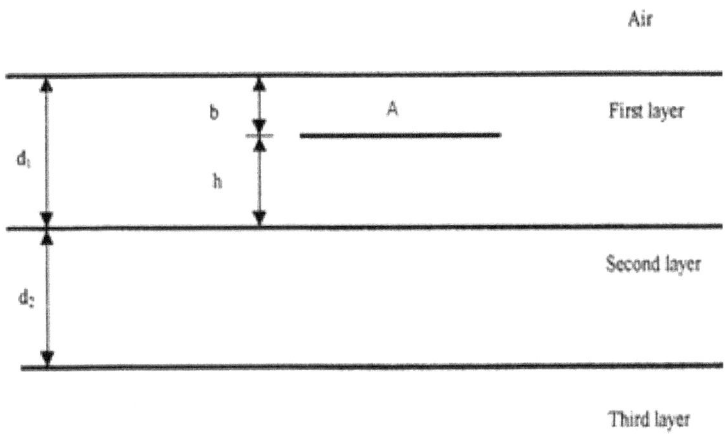

7.6.2 Infinite Series Potential (I.S.M.) Expressions in Three Dimensions

(Figure 11) shows the three-layer model under study in case of calculating the voltage distribution on the grid surface where: A: Area of grid, m^2, b: Burial depth of grid, m, For any grounding system in three-layer earth structure, the current in each conductor element produces a potential at a point (x, y, z) in the surrounding medium in three dimension with taking the reflection factors between first, second and third layer in account will be as follows as reported by Heppe, R., J. (1979).

$$v(x,y,z)=\frac{I\rho_1}{4\pi l}[\ln f(x,y,z)+\ln f(x,y,2b)$$

$$+\sum_{n=1}^{\infty}(k_{12}^{n}.\ln f(x,y,z+n2h+(n-1)2b)+$$

$$2k_{12}^{n}\ln f(x,y,z+n2h+n2b)+$$

$$k_{12}^{n}\ln f(x,y,z+n2h+(n+1)2b)+$$

$$\sum_{n=1}^{\infty}(k_{23}^{n}.\ln f(x,y,z+n2(d_2-h)+(n-1)2(d_1+h))+$$

$$2k_{23}^{n}\ln f(x,y,z+n2(d_2-h)+n2(d_1+h))$$

$$+k_{12}^{n}\ln f(x,y,z+n2(d_2-h)+$$

$$(n+1)2(d_1+h)))$$

$$] \qquad (7.38)$$

Where,

$$f(x,y,z)=\frac{\sqrt{(x+l/2)^2+y^2+z^2}+x+l/2}{\sqrt{(x-l/2)^2+y^2+z^2}+x-l/2} \qquad (7.39)$$

l: Length of a grid side, (m), r: Radius of grid conductor, (m),, n: Number of images
Computer program has been developed based on the I.S.M. The flowchart is given in Figure 12.

7.7 OPTIMUM COMPRESSION RATIO (OCR)

In unequally spaced grid when the conductors are arranged according to an exponent regularity, the span from the center could be calculated by the equations

$$d_n = d_{max} \cdot C^n \tag{7.40}$$

$$d_{max} = \frac{L(1-c)}{1+c-2c^{(N/2+1)}}$$

N even (7.41)

$$d_{max} = \frac{L(1-c)}{2(1-2c^{(N-1)/2})}$$

N odd (7.42)

Where C is the compression ratio, L is the side length of grounding system, N is the conductors number arranged on it and d_{max} is the central span .Empirical expression is obtained by Hung, L., Chen, X., Yan, H. (1995) to calculate (OCR).

7.7.1 Testing and Verification

The analytical analysis in three-dimension is applied for a grid grounding system consists of 96 meshes with dimensions given in Table 1 and soil structure given in Table 2 to obtain the voltage profile on the earth surface in the different position along the grid length.

The short circuit current of substation 220 kV grounding system is 10 kA. The (I.S.M.) method suggested in this section is applied on example of IEEE 80 (2000) and proves that it is very accurate. The maximum percentage difference is approximately 5%. Figure 13 to Figure 16 show the voltage profile for the ground grid with different compression factors.

The number of images that is used has been chosen to be 30 which will give acceptable result accuracy for different reflection factors (k=0.1 to 0.9).

Figure 12. I.S.M program flowchart

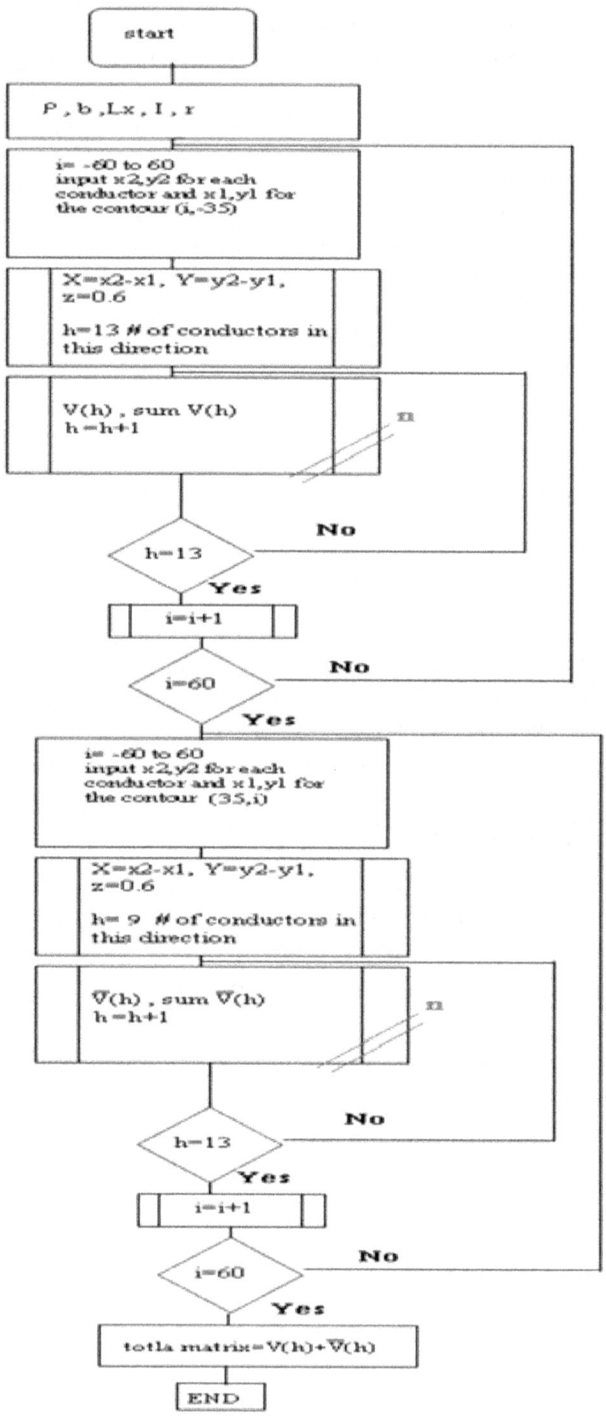

Table 1. Grid dimension parameters

Grid dimension parameters	
Grid area, m²	**120X80**
burial depth, m	0.6
Radius of grid conductors, m	0.0175
number of meshes	96

Table 2. Soil structure parameters

Soil structure parameters	
thickness of the first layer, m	**2**
Thickness of the second layer. m	2
resistivity of the first layer Ωm p1=30	30
resistivity of the second layer Ωm p2=40	40
resistivity of the third layer Ωm p3=250	250
Number of images that vary to infinity.	30

Figure 13. Voltage profile along the unequally spaced grid in case of C=0.4

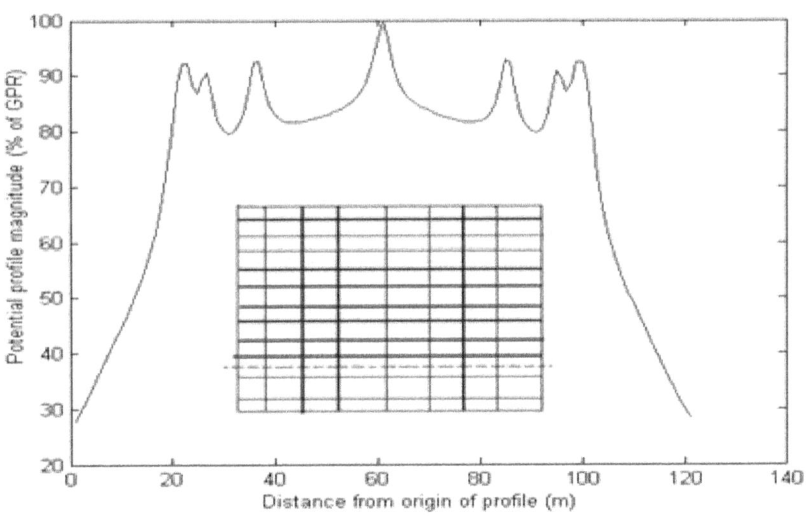

Figure 14. Voltage profile along the unequally spaced grid in case of C=0.6

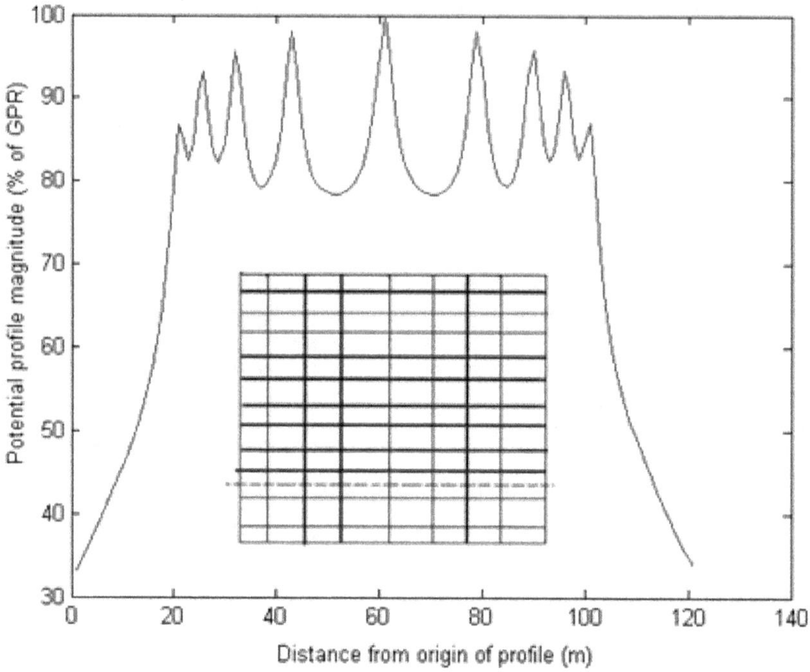

Figure 15. Voltage profile along the unequally spaced grid in case of C=0.8

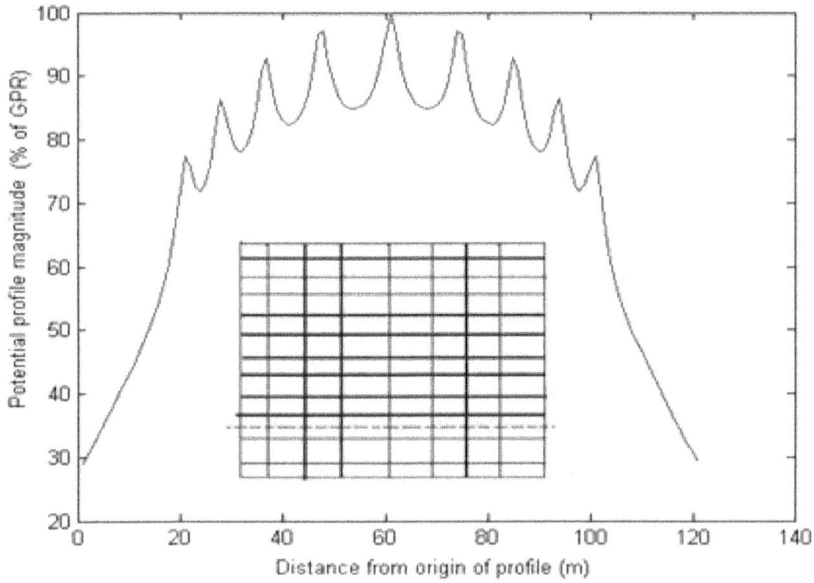

Figure 16. Voltage profile along the unequally spaced grid in case of C=1.0

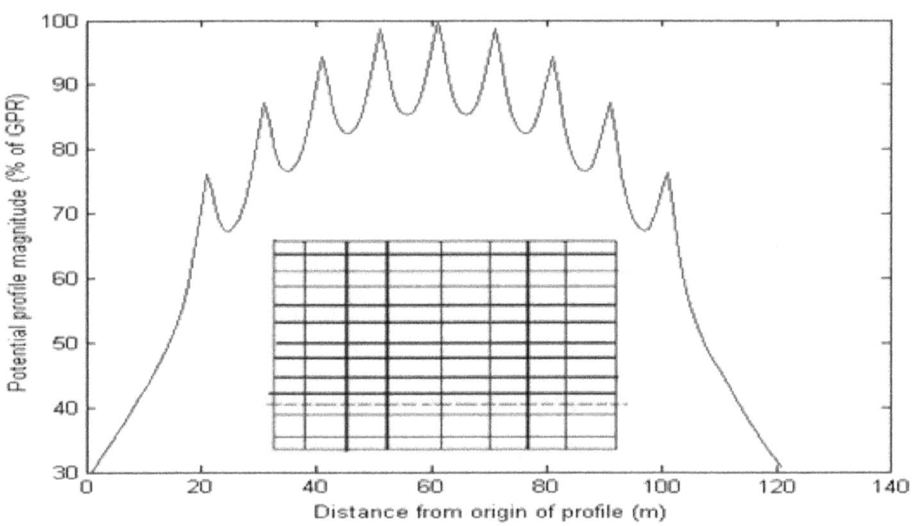

From the obtained results given in Figures 13 to 16 it is noticed that when C≈ 0.6 the potential different between any two conductors is minimum and this may lead to minimum difference between the maximum and minimum touch potential, this is in agreement with the obtained results given by Hung, L., Chen, X., Yan, H. (1995) in calculating OCR. Calculating the voltage profile by the used method consumes long calculation time to give accurate result for this reason the following method may be used.

7.7.2 Approximated Analysis to Calculate Voltage Profile Using Apparent Soil Resistivity

The apparent soil resistivity of the three layers is calculated by using IEEE 80 (2000) formula.

In this case the potential at a point (x,y,z) in surrounding medium in three-dimension for any grounding system in three-layer earth structure will be as follows:

$$v(x,y,z) = \delta \frac{I\rho_a}{4\pi l}[\ln f(x,y,z) + \ln f(x,y,z+2b)] \tag{7.43}$$

Where,

$$f(x,y) = \frac{\sqrt{(x+l/2)^2+y^2+z^2}+x+l/2}{\sqrt{(x-l/2)^2+y^2+z^2}+x-l/2} \tag{7.44}$$

δ Scaling factor

7.7.2.1 Scaling Factor

Scaling factor δdepends on geometry of the grid, distance from the point of origin and compression ratio C.

Table 3 shows the values of scaling factor that gives voltage profile closer to that calculated by (I.S.M.) method using different Compression ratio.

Table 3. The influence of scaling factor with compression ratio

Compression ratio C	Scaling factor δ
1	4.1
0.8	4.72
0.6	4
0.4	5.571

In order to ascertain the scaling factor when we know the compression ratio, we must find a general law, using the method of least squares to fitting curves.

The fitting equation is as follows:

$$\delta = -76.088\,C^3 + 165.67\,C^2 - 115.71C + 30.215 \tag{7.45}$$

7.7.2.2 Testing and Verification of the Approximated Method

A) Effect of compression ratio on voltage profile

The designed computer program requires various parameters of the grounding system, which are divided as follows:

The same grid and soil parameters given in Tables 1 and 2 are used to calculate the voltage profile by using the approximated method using ρ_a=220 Ωm. Computer program has been developed based on the approximated method Figure 17 shows flowchart for the program. The obtained results are given in Figures 18 to 21.

From the obtained results given in Figure 21 the maximum difference between the peaks values of the two methods is about 12% in case of compression ratio=0.4 and minimum difference is 0%.

When C=0.6,0.8 and 1 Figure 20, Figure 19, and Figure 18 the maximum difference between the peak values of the two methods is about 4%, 8% and 13% respectively and the minimum difference in all cases is 0%.

The minimum value of the maximum percentage difference between the I.S.M. and the approximated method occurs at C=0.6

Figure 17. Approximated method program flowchart

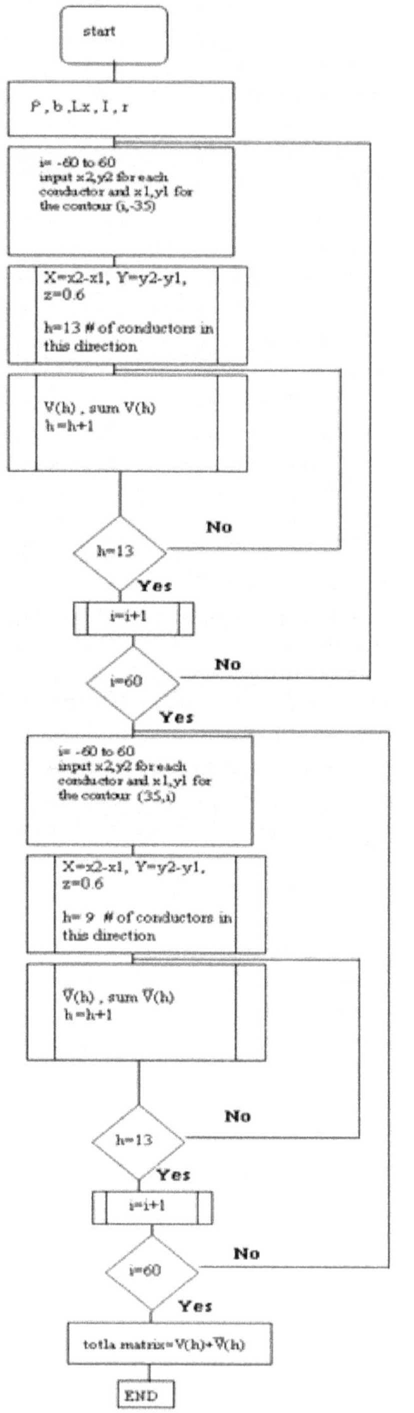

Figure 18. Voltage profile along the unequally spaced grid in case of C=0.4

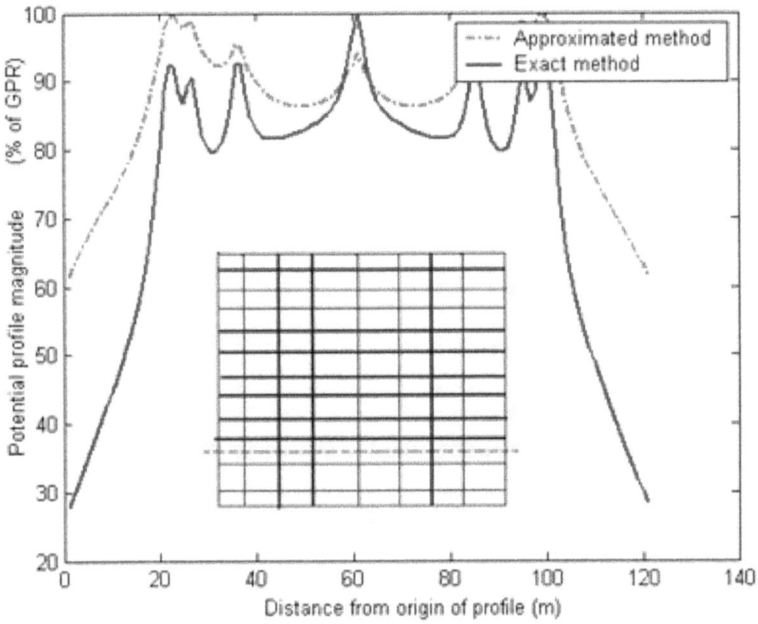

Figure 19. Voltage profile along the unequally spaced grid in case of C=0.6

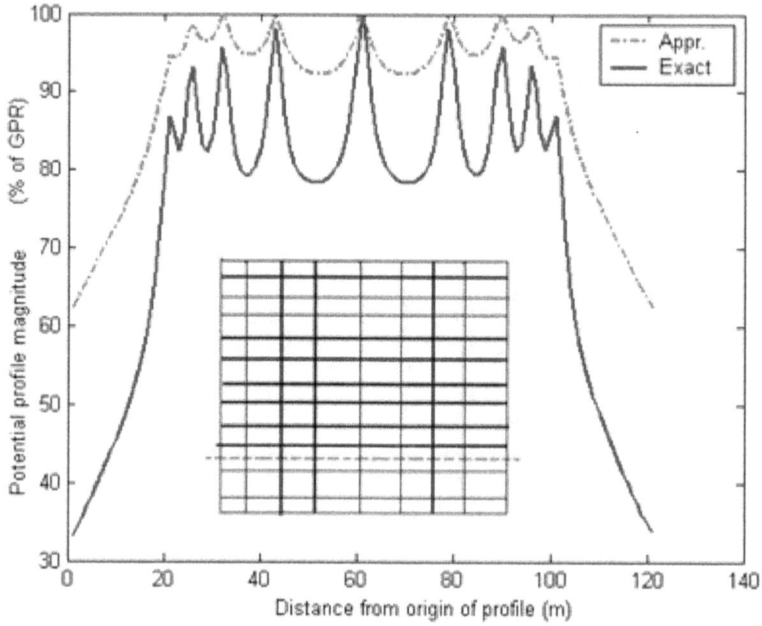

Figure 20. Voltage profile along the unequally spaced grid in case of C=0.8

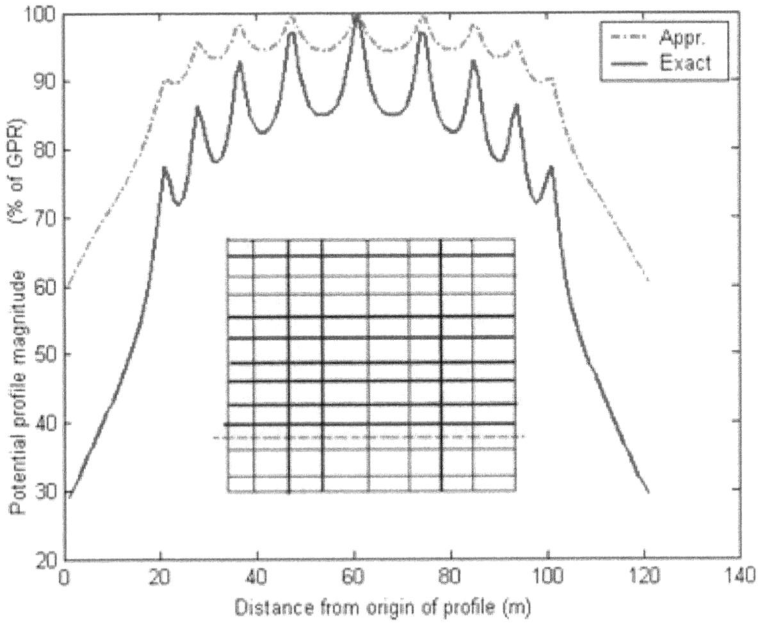

Figure 21. Voltage profile along the unequally spaced grid in case of C=1

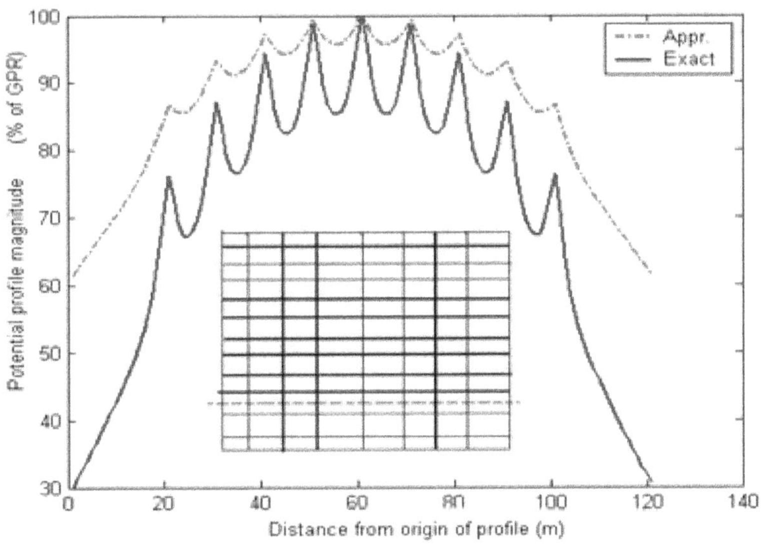

B) Effect of number of meshes on voltage profile

a) For square grid

Four different grids are used for testing and verification CS01, 04, 16 and 64 given in Figure 22. Table 4 and Table 5 give the parameters of the four square grids and soil structure parameters. Figures 23, 24, 25 and 26 show the voltage profiles for the grids given in Figure 22 for the purpose of comparison between the exact and approximated calculations.

Figure 22. Different numbers of meshes for square grids

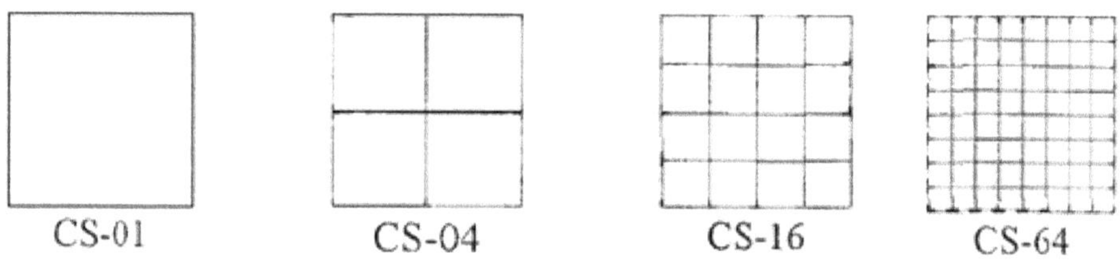

Table 4. Parameters of the four square grids

	Area m*m	**n (number of parallel conductors)**	**N (number of meshes)**	**D (space between conductors)**
CS01	10×10	2	1	10
CS04	10×10	3	4	5
CS16	50×50	5	16	12.5
CS64	100×100	9	64	12.5

Table 5. Soil structure parameters

soil structure parameters	
thickness of the first layer, m	**2**
thickness of the second layer. m	2
resistivity of the first layer Ωm	100
resistivity of the second layer Ωm	300
resistivity of the third layer Ωm	500
Number of images that vary to infinity.	30

Figure 23. Voltage profile along the CS01 grid

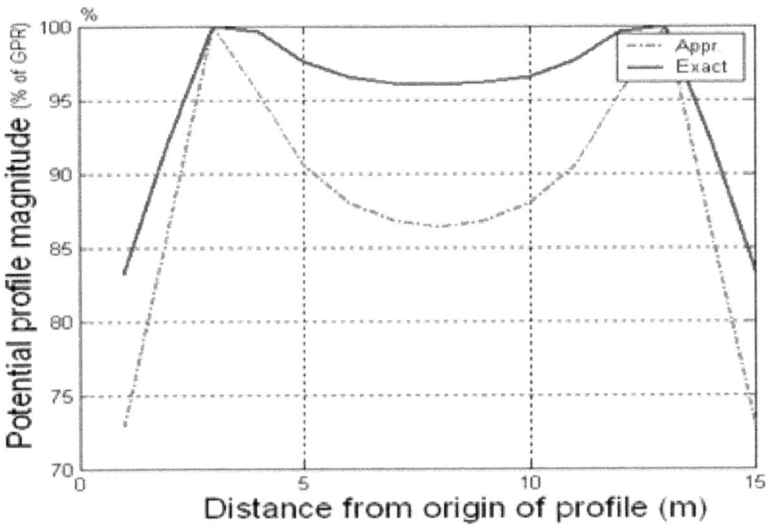

Figure 24. Voltage profile along the CS04 grid

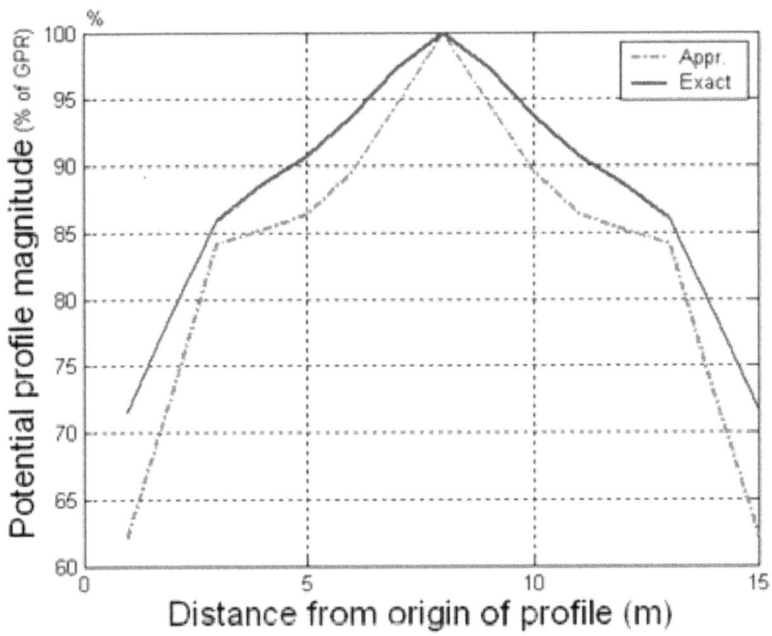

Figure 25. Voltage profile along the CS16 grid

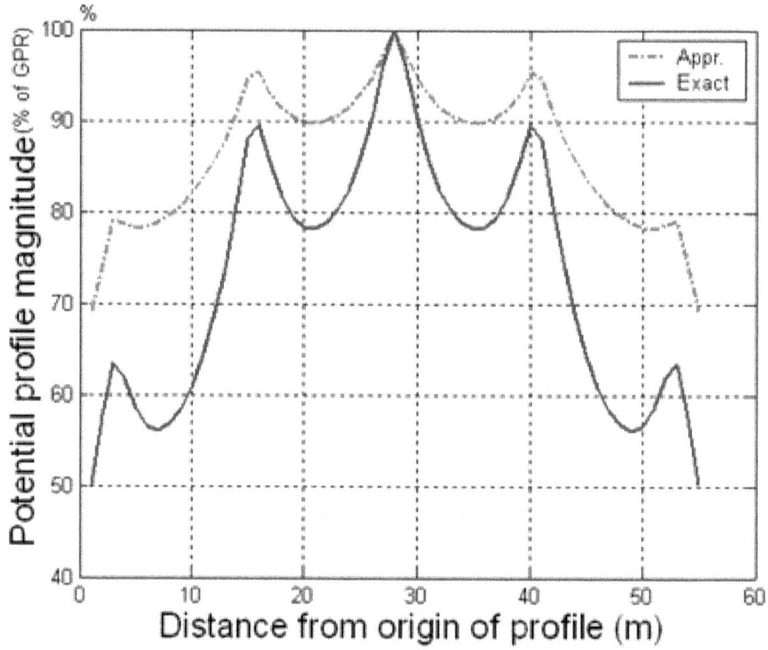

Figure 26. Voltage profile along the CS64 grid

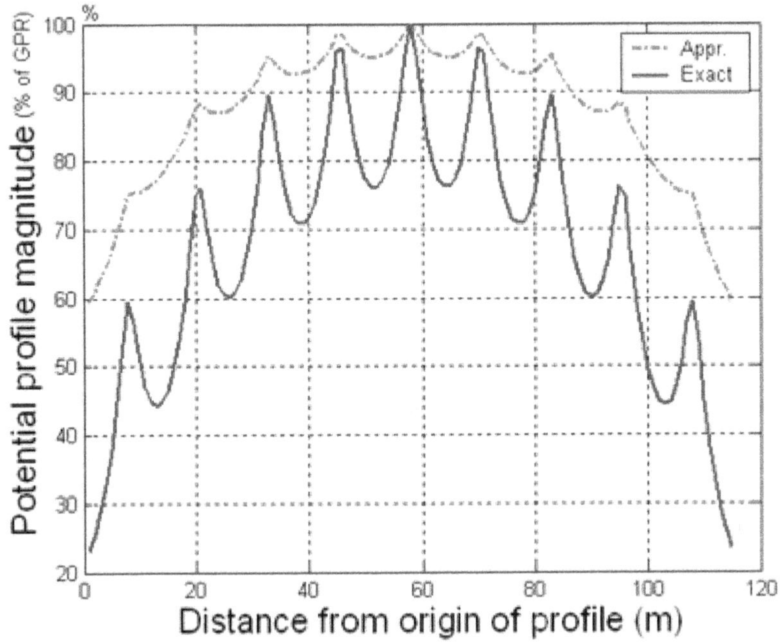

From the obtained results given in Figure 23, it is notice that when number of meshes=1 the potential difference of the both methods (I.S.M. & approximated) are closed to each other at the perimeter and differ at the center of the grid

When numbers of meshes equal 4, 16 and 64 (Figures 24, 25 and 26) both methods are much closed at the center of the grid and with potential difference does not exceed 15% at perimeter of the grid.

b) For rectangular grid

Four different rectangular grids are used for testing and verification CR01, 04, 16 and 64 given in Figure 27. Table 7 gives the parameters of the four rectangular grids.

Figure 27. Different number of meshes for rectangular grids

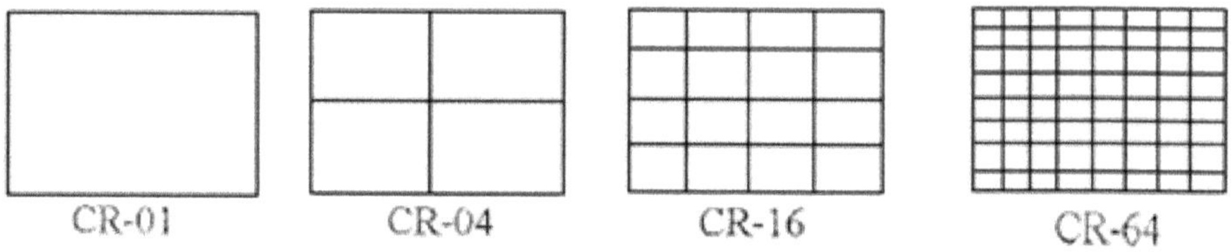

Table 6. Parameters of the four rectangular grids

	Area m×m (m^2)	n (number of parallel conductors)	N (number of meshes)	D
CR01	10×20	2	1	14.14
CR04	10×20	3	4	7.07
CR16	50×100	5	16	17.67
CR64	100×200	9	64	17.7

Figures 28- 31 give the voltage profile along the grids under study

Figure 28. Voltage profile along the CR01 grid

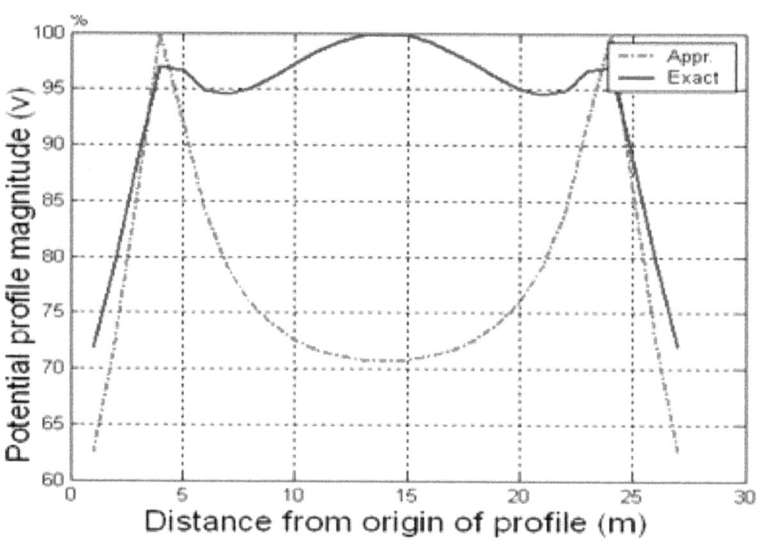

For the grid with number of meshes equal one, potential difference at the center of the grid is difference for the two methods.

Figure 29. Voltage profile along the CR04 grid

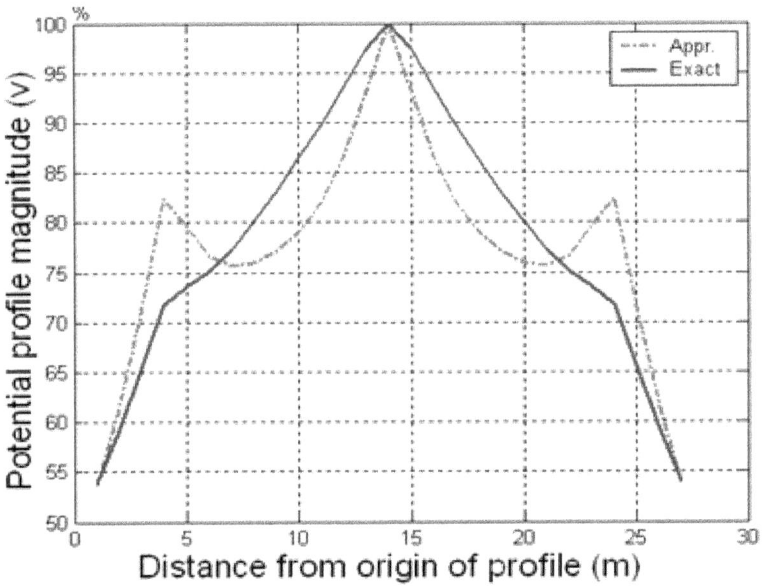

Figure 30. Voltage profile along the CR16 grid

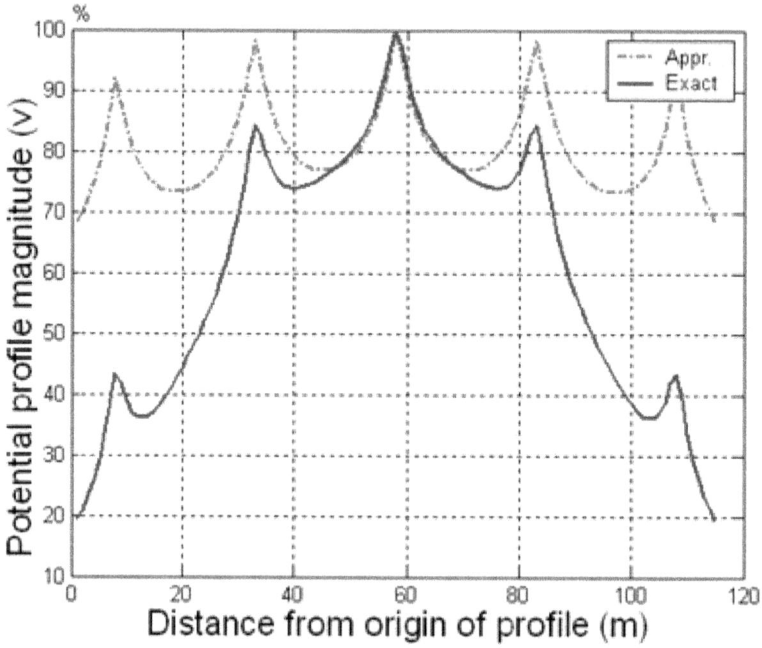

Figure 31. Voltage profile along the CR64 grid

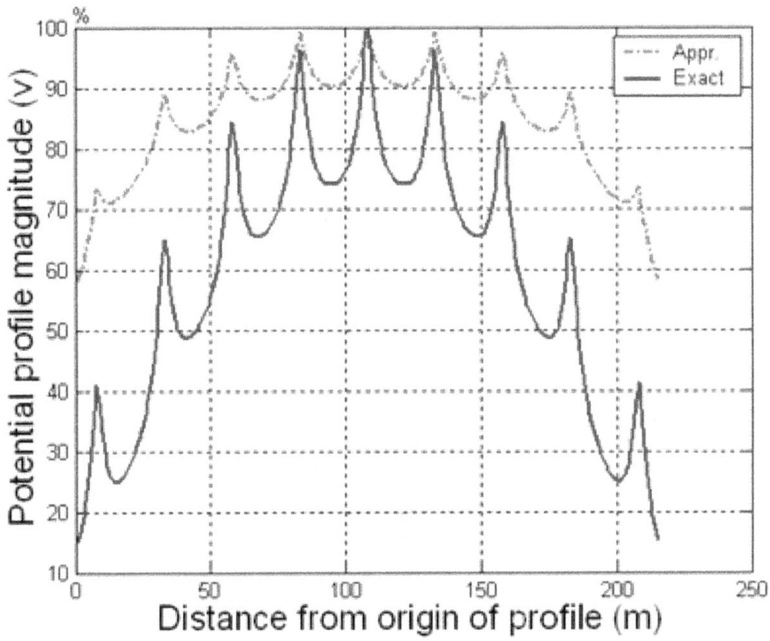

The results of the two methods are much closed for number of meshes 4, 16 and 64 and it is noticed that the potential difference between the two methods vary from center to perimeter.

7.8 CALCULATION OF THREE DIMENSIONS POTENTIAL USING THE APPROXIMATED

Four different square grids are used for testing and verification of CS01, CS04, CS16 and CS64 described in Table 4, the potential calculated for every meter in both x and y direction and saved in 3D matrix using Matlab program. Results shown in Figures 32, 33, 34 and 35 are so close to that obtained by Bendito, E., Carmona, A., Encinas, A., M., Jimenez, M., J., (2000).

For grid consists of one mesh the difference between the center potential and the perimeter potential is 12% and for grid consists of four mesh the difference between the center potential and the perimeter potential is 15% and for grid consists of sixteen mesh the difference between the center potential and the perimeter potential is 35% and for grid consists of 64 mesh the difference between the center potential and the perimeter potential is 40%.

Figure 32. 3D Earth surface potential over CS01 grid

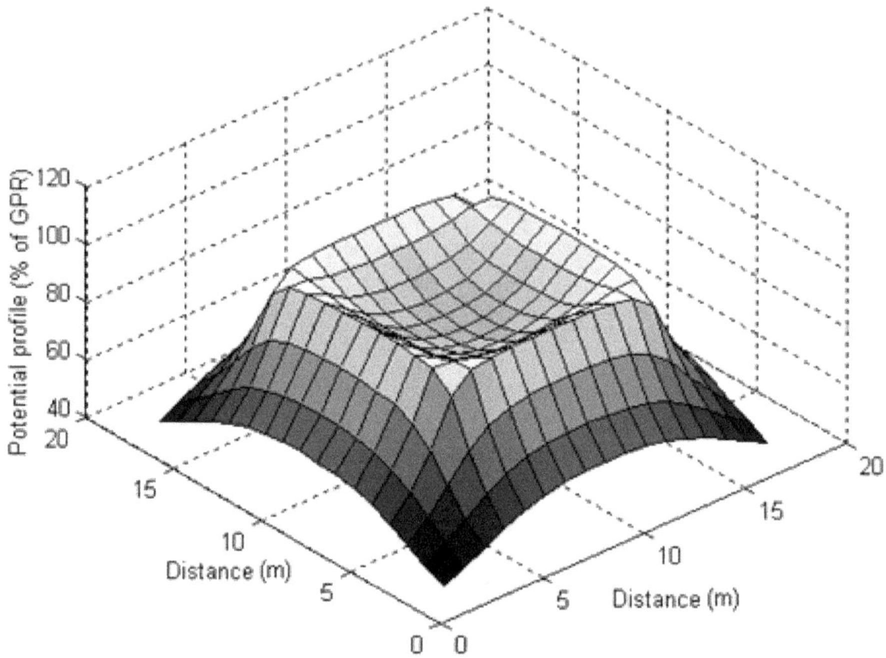

Figure 33. 3D Earth surface potential over CS04 grid

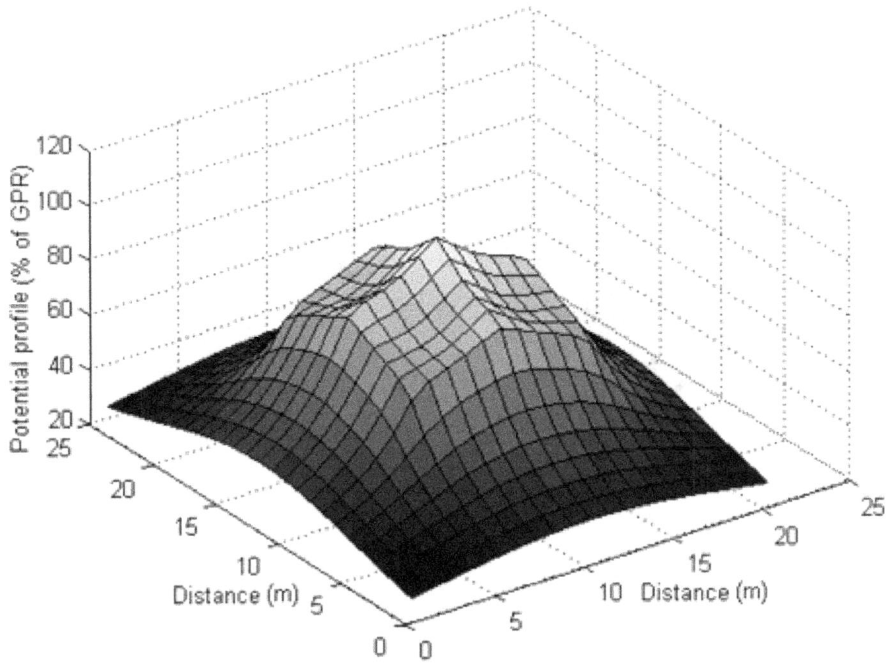

Figure 34. 3D Earth surface potential over CS16 grid

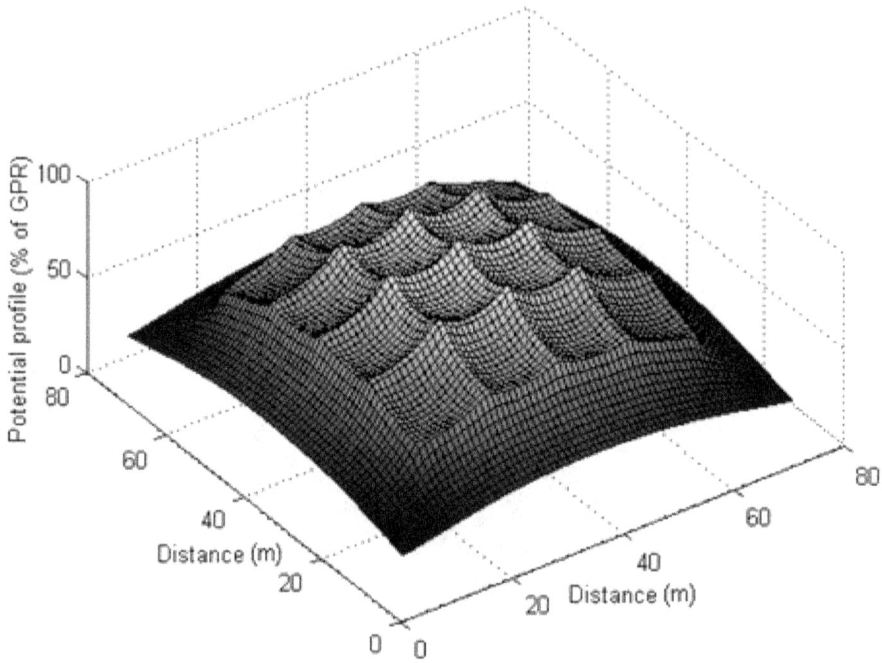

Figure 35. 3D Earth surface potential over CS64 grid

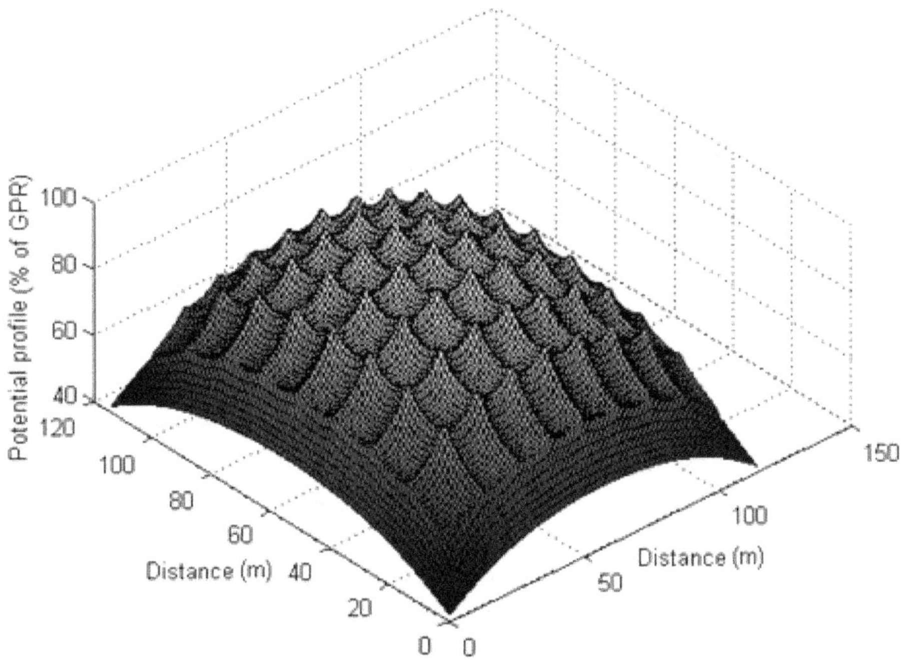

7.9 GROUND POTENTIAL RISE (GPR) OF FAULTY SUBSTATIONS HAVING EQUAL AND UNEQUAL SPACING GROUNDING GRIDS CONDUCTORS

Grounding of power substations is to provide safety for personnel during normal and fault to ground conditions by limiting step and touch potentials, to dissipate lightning strokes and to stabilize voltage during transient conditions. One important goal of this paper is to present simplified approaches for the earth surface potentials and ground potential rise (GPR) of substation grounding in case of uniform and non-uniform soil. The equalization of the current density and then decreasing of GPR is achieved. The Infinite Series Potential Method (I.S.M) is used to calculate the GPR and the earth surface potentials with accurate results for the grids buried in uniform and non-uniform soil structure. The voltage profiles have been designed in 3D to make the selection of suitable grounding system easier.

There are two main design goals to be achieved by any substation grounding system under normal as well as faulty cases. One of them is to provide means to dissipate electric currents into the earth without exceeding any operating and equipment limits, and the other is to assure that a person in the vicinity of grounded facilities is not exposed to the danger of critical electric shock. Step voltages are inherently less dangerous than mesh voltages. However, safety within the grounded area is achieved with the assistance of a high resistivity surface layer (surface material), which does not extend outside the fence. In any event, the computed step voltages should be compared with the permissible step voltage after a grid has been designed that satisfies the touch voltage criterion.

Studying the effect of voltage distribution on the grid surface is an important parameter to design the grounding system of substations to ensure the safety of people above the substation ground. This can be achieved by equalizing the potential distribution of the ground surface and reducing step and touch potentials. As it is well known the person's safety is affected by the ground potential rise (GPR) of grounded structures during electric power faults. In order to limit these voltages to safe values, it is necessary to design a right grounding system that effectively connects all metallic structures of the substations to ensure the safety criteria for both human and substations taking into account the economical purposes.

However, the unequally spaced grid can make current density which is emanating from grid conductors more uniform, and therefore remarkably decrease of potential gradients on the earth's surface. Then, the maximum touch voltage for this grid will reach its lowest values and thus the safety conditions for human above the substation earth surface will be ensured. The unequally spaced grid saves about 34% of grounding grid material. It also costs less to install the grid.

7.9.1 Unequally Spaces Grids Design Procedures

The spacing between the grounding grid conductors should be so arranged that the touch and step voltages will be smaller than the safe value for the personnel. The optimal design of grounding systems for substations is to suitably arrange the conductors of grounding grid systems to keep the maximum touch voltages within the safe limit; this would ensure making all grounding grid conductors sufficiently utilized so it can be certified as a safe and economic design method.

The conductor compression ratio "C" is related to the grid dimensions and the maximum conductor separation d_{max} which occurs toward the center of the grid and is given by Zhongdong, Du., Yao, Zhenyu, Wen, Xishan, Xu, Hua, (2005):

$$d_{max} = \frac{L(1 - C)}{1 + C - 2C^{(N/2+1)}}$$

, with N even number (7.46)

Or,

$$d_{max} = \frac{L(1 - C)}{2(1 - C^{(N-1)/2})}$$

, with N odd number (7.47)

$$d_i = d_{max} C^i$$

. (7.48)

Where N is the desired number of conductors perpendicular to grid axis of length L. In unequally spaced grid when the conductors are arranged according to an exponent, the span from the center could be calculated according to equations 7.46, 7.47 and 7.48, where i, is the conductor number.

For the optimal design of grounding grid system, the optimal compression ratio (OCR) could be determined, in which the touch voltage reaches its minimum if the grounding grid is designed using this compression ratio. An empirical expression is suggested by Zhongdong, Du., Yao, Zhenyu, Wen, Xishan, Xu, Hua, (2005) to calculate OCR as follows:

$OCR=a_0+a_1exp\ (0.0001h)\ +a_2\ exp\ (bh)$ (7.49)

Where

$b=-0.3503-9.6311exp\ (-0.03666L)$ (7.50)

$a_0=a_{01}+a_{02}k+a_{03}k^2$
(7.51)

$a_1=a_{11}+a_{12}k+a_{13}k^2$
(7.52)

$$a_2 = a_{21} + a_{22}k \quad (7.53)$$

In the above equations, k is the number of segments of the conductor in any direction; h is the grid laying depth. The relationship between a_0, a_1... and L is given in Table 7.

Table 7. Relation between a_0, a_1... and grid length

Coefficients	L≤100 m	100 m ≤ L ≤ 175 m	175 m ≤ L ≤ 250 m	L > 250 m
a_{01}	0.44	0.38	-0.51	0.32
a_{02}	-77.43	-50.65	-33.18	-15.44
a_{03}	15.63	13.88	18.49	13.42
a_{11}	0.033	0.19	1.15	0.38
a_{12}	76.9	50.21	32.82	15.16
a_{13}	-15.56	-13.83	-18.44	-13.38
a_{21}	-0.067	-0.037	-0.029	-0.022
a_{22}	0.5	0.41	0.34	0.26

7.9.2 Effect of Conductors' Distributions on Voltage Profile of Grids Buried in Uniform Soil

A. Infinite Series Potential Method (I.S.M)

Software is designed by the authors to illustrate the effect of equal and unequal grid conductor's distribution on the earth surface potentials above the grid. A square grid of 105×105 m^2 with number of meshes 64 equally and unequally spaces parallel conductors in each direction is taken as an example as shown in (Figure 36). The grid material is copper having 5 mm radius, calculated according to IEEE standard 80.

By using Infinite Series Potential Method (I.S.M), the potential difference in volts can be calculated in two dimensions using the following equation.

$$V(x,y) = \frac{I\rho_1}{4\pi l}\ln f(x,y)$$

(7.54)

Where,

$$f(x,y) = \left(\sqrt{(x+l/2)^2+y^2}+x+l/2\right)/\left(\sqrt{(x-l/2)^2+y^2}+x-l/2\right)$$

Where I is the short circuit current, L is grid side length in meter and $\rho_1$is the soil resistivity in Ω.m. The following similar relation is used to obtain the grounding grid voltage profile in three dimensions.

Figure 36. Examples for grid module of N=64 with equally, unequally spaced having denser conductors at the edges and at the middle, (A), (B) and (C) respectively.

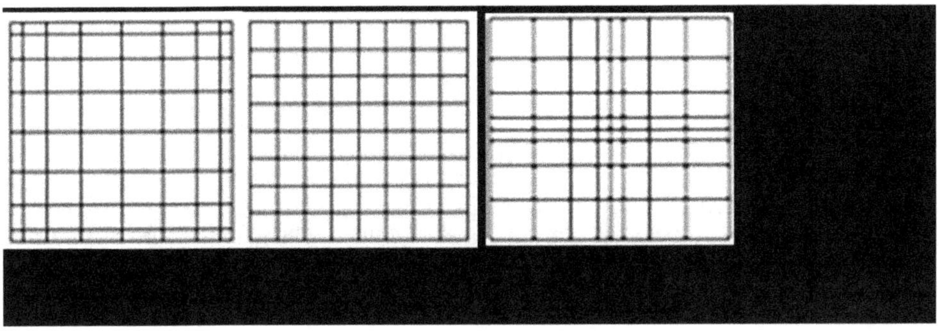

$$V(x, y, z) = \frac{I\rho_1}{4\pi l}\ln f(x, y, z) \tag{55}$$

Where,

$$f(x, y, z) = \left(\sqrt{(x + l/2)^2 + y^2 + z^2} + x + l/2\right)/\left(\sqrt{(x - l/2)^2 + y^2 + z^2} + x - l/2\right)$$

To achieve the effect of grid conductors distribution on the ground potential rise, assuming the current emanating from the faulty substation into the grounding grid at short circuit condition of 220 kV substations is I=10 kA, grid burial depth is 0.6 m. The earth surface in volts and as percentage of GPR for grid modules A, B and C are calculated. The obtained results are given in Table 8 when the soil resistivity was 100 Ω.m.

From the obtained results, it is concluded that the GPR decreases in unequally spaces grids having denser conductors at the grid edges compared with equally spaces grid conductors and unequally spaces having denser conductors at the grid middle, in general using grounding grids decreases the potential difference between each two conductors and more uniform voltage profile can be obtained. (Figure 37) represents the three dimensions earth surface potentials, for every one meter in both x and y directions, above the mentioned grids of different conductors' distribution configurations.

Table 8. Earth surface potentials in volts and as percentages of GPR along center conductor for equally and unequally spaced grid of (A=105×105 m²)

Equally spaces grid conductors									
x, (m)	**-52.5**	**-39.38**	**-26.25**	**-13.13**	**0**	**13.1**	**26.3**	**39.4**	**52.5**
V, (volt)	4003	4732	5076	5249	5297	5249	5076	4732	4003
V, (% GPR)	0.76	0.89	0.96	0.99	1	0.99	0.96	0.89	0.76
Unequally spaces having denser conductors at the grid edges									
x, (m)	-52.4	-46.8	-35.1	-19.1	0.0	19.1	35.1	46.8	52.4
V, (volt)	4080	4431	4629	4762	4799	4762	4629	4431	4080
V, (% GPR)	0.85	0.92	0.96	0.99	1	0.99	0.96	0.92	0.85
Unequally spaces having denser conductors at the grid middle									
x, (m)	-52.4	-33.3	-17.3	-5.6	0.0	5.6	17.3	33.3	52.4
V, (volt)	4156	5265	5785	6138	6197	6138	5785	5265	4156
V, (% GPR)	0.66	0.85	0.93	0.99	1	0.99	0.93	0.85	0.66

Figure 37. Voltage profile along grid conductors in case of equally spaced grid (grid A), the number of meshes =64, short circuit current 10 kA, soil resistivity 100 Ω.m

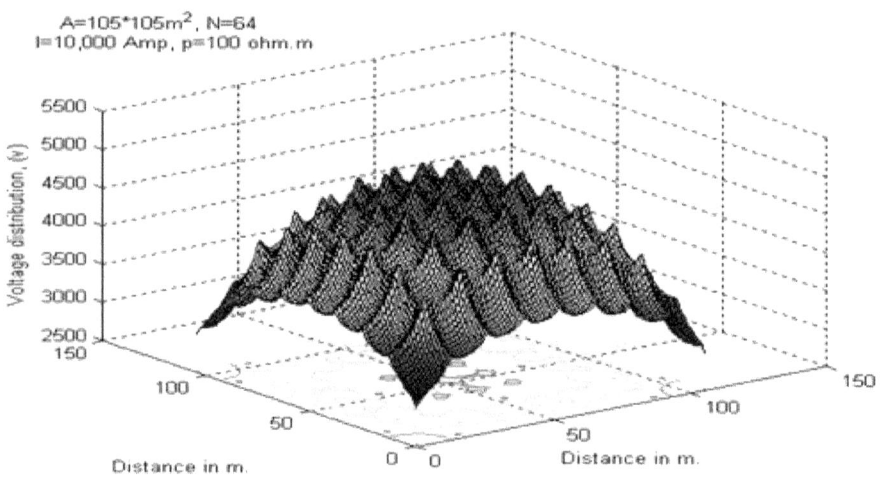

Figure 38. Voltage profile along grid conductors in case of unequally spaced grid having denser conductors at the middle, where the number of meshes =64, short circuit current 10 kA, and soil resistivity 100 Ω.m

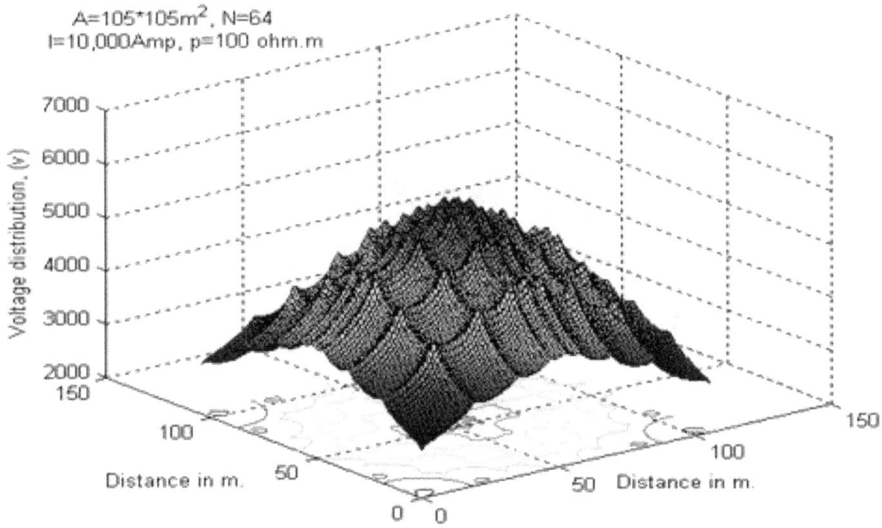

Figure 39. Voltage profile along grid conductors in case of unequally spaced grid having denser conductors at the edges, the number of meshes = 64, short circuit current 10 kA, soil resistivity 100 Ω.m

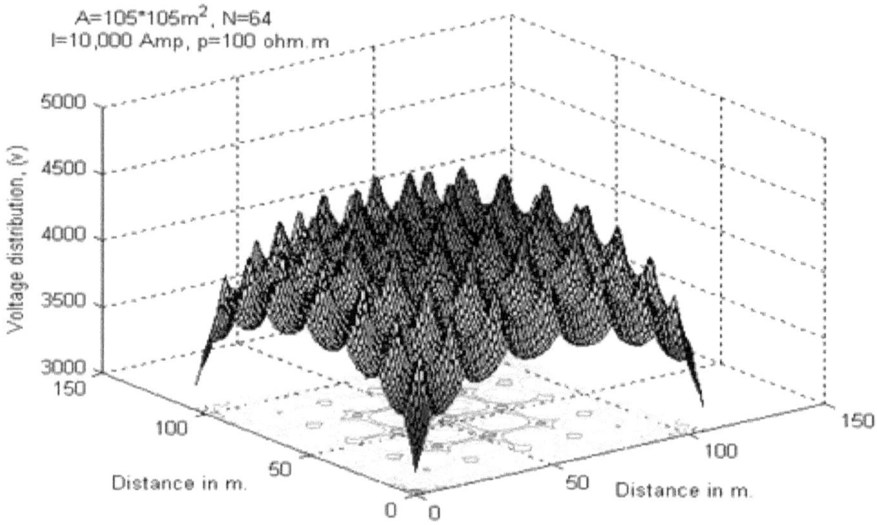

Regarding to grid conductors' distribution with equally and unequally spaces, the percentages of the potential rise at the corner conductors with respect to the grid center potential rise in case of number of meshes 9×9, 13×13 and 17×17, are given in Table 9.

Table 9. GPR of meshes corners for different number of meshes, when the grid is loaded by 10 kA short circuit current and 100 Ω.m soil resistivity, grid area equals 105×105 m^2

Grid number of meshes	Equally space grid corner (A) GPR	Unequally grid (B) corner GPR	Unequally grid (C) corner GPR
9×9	0.76	0.85	0.66
13×13	0.743	0.850	0.64
17×17	0.733	0.860	0.624

From this table it is noticed that increasing the number of grid meshes at the same area reduces the grid GPR regardless of grid configuration.

Noteworthy Tables 8 and 9 and (Figure 39) confirm the previous conclusion, that the increasing of unequally grid conductors having denser conductors at the grid middle increases the GPR for this grid. While that having denser conductors at the edges has more flat earth voltage distribution.

As it is noticed from Table 9, the GPR whatever number of grid conductors and meshes for the same grid area, for unequally grid denser conductors at the grid edges has lowest GPR and the lowest potential differences between center and corner conductors compared with the other two grid configurations. It provides the most efficient grounding grid design. This is in agreement with according to IEEE 80, 2000

(Figure 40) gives the earth surface potential along the center of the square grounding grids of various meshes A, B and C. (Figures 41 and 42) give the touch and step voltages for different shapes of the square grounding grids.

B. Optimal Compression Ratio (OCR)

In unequally spaced grid, when the conductors are arranged according to an exponent regularity, the span from the center can be calculated by equations 46 to 48. Empirical expression is used to calculate OCR as given in equations 49 to 55 and Table 7. To illustrate the effect of unequally space conductors' distribution on the earth surface potentials and touch voltage profiles for the grid buried in uniform soil and to satisfy the optimal grounding grid design, the grid module area, 105×105 m^2, $N=8\times8$, as an example is used. The grid is buried in a depth of 0.6 meter of uniform soil has 100 Ω.m resistivity. The earth surface potentials and touch voltages are calculated along the center conductor for this grid under different values of compression ratios (OCR) as presented in (Figure 43).

Figure 40. Earth surface potential along the center of the square grounding grids of various shapes of meshes, short circuit current 10 kA, soil resistivity 100 Ω.m

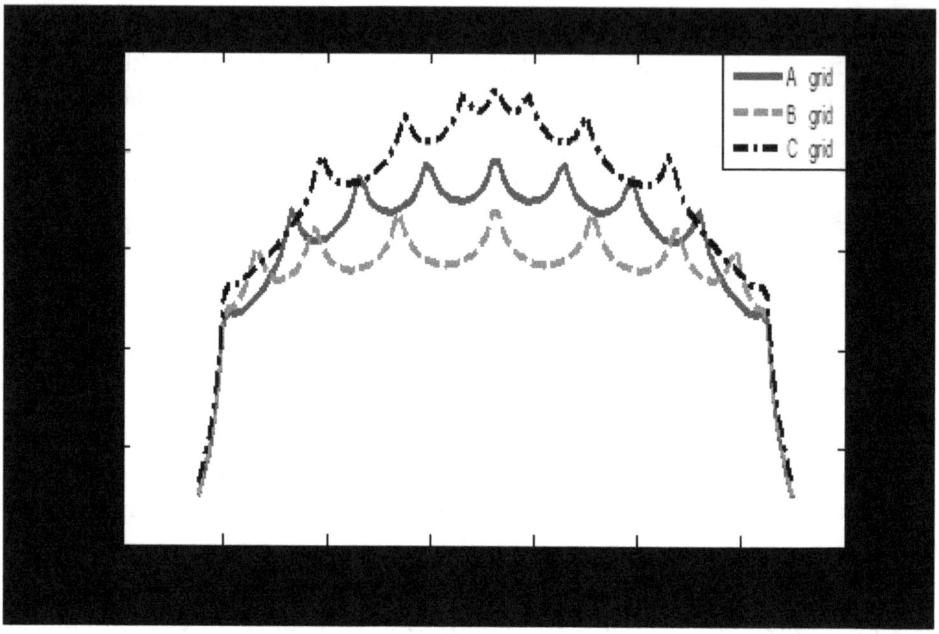

Figure 41. Effect of the different shapes of meshes on the touch voltage of the square grounding grids, short circuit current 10 kA, and soil resistivity100 Ω.m

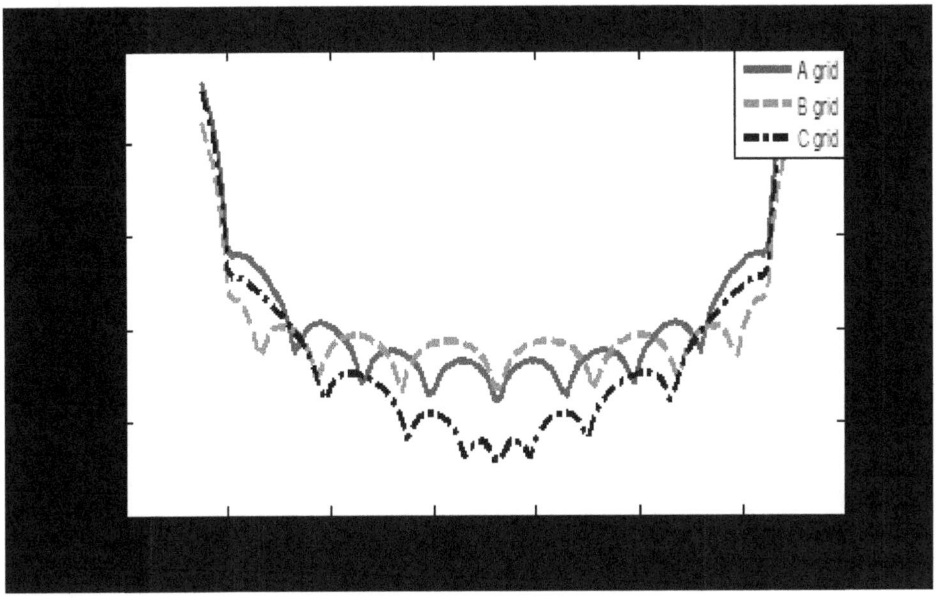

Figure 42. Effect of different shapes of meshes on the step voltage of the square grounding grids, short circuit current 10 kA, soil resistivity100 Ω.m

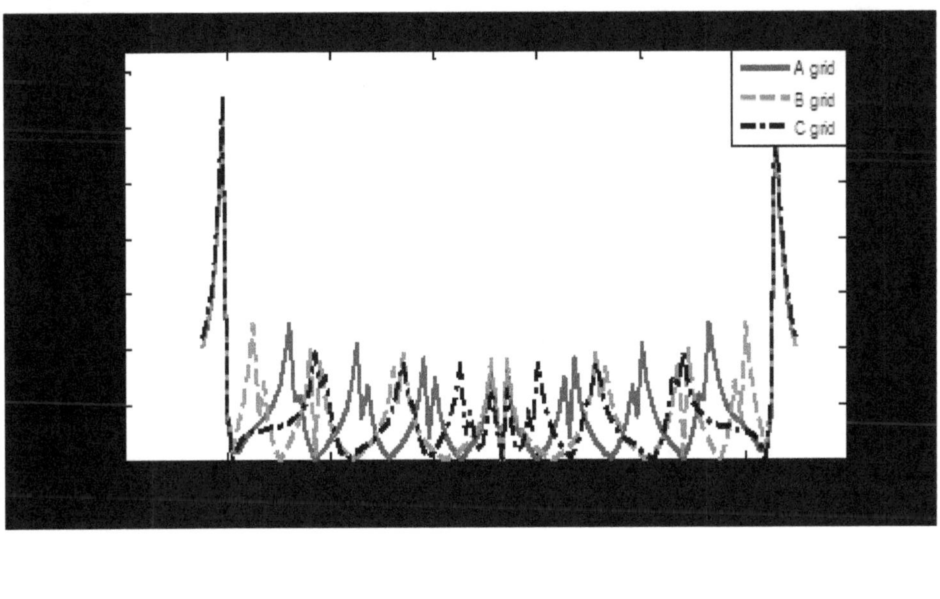

By the analysis of the calculated data represented in this figure, it is concluded that the touch voltage along the center conductor reaches its lowest value when the compression ratio for this grid reaches 0.538, it means that, the optimal compression ratio obtained gives the most save and economical grid design, and the best arrangement of grounding conductor span is achieved, so the minimum touch voltage and thus, the optimal design for grounding grid is satisfied. These findings are in agreement with results obtained by Hung, L., Chen, X., Yan, H. (1995). Figure 42 illustrates the relationship between the touch voltages over center conductor for the unequally spaced grid as a function of conductor compression ratio. Remarkable, this relationship usually has an obvious "U" shape. As it is noticed from this figure the minimum touch voltage is obtained when the compressed ratio (OCR) is 0.538 in case of the grid is laid in uniform soil.

Samples of voltage distribution in 3D for different values of compressed ratios are given in (Figure 45). From this figure it is clear that at C=0.54 the earth surface potential distribution is more uniform.

Figure 43. Earth surface potentials and touch voltages over the center conductor for A=105×105 m², N=8×8, grid having different compression ratios

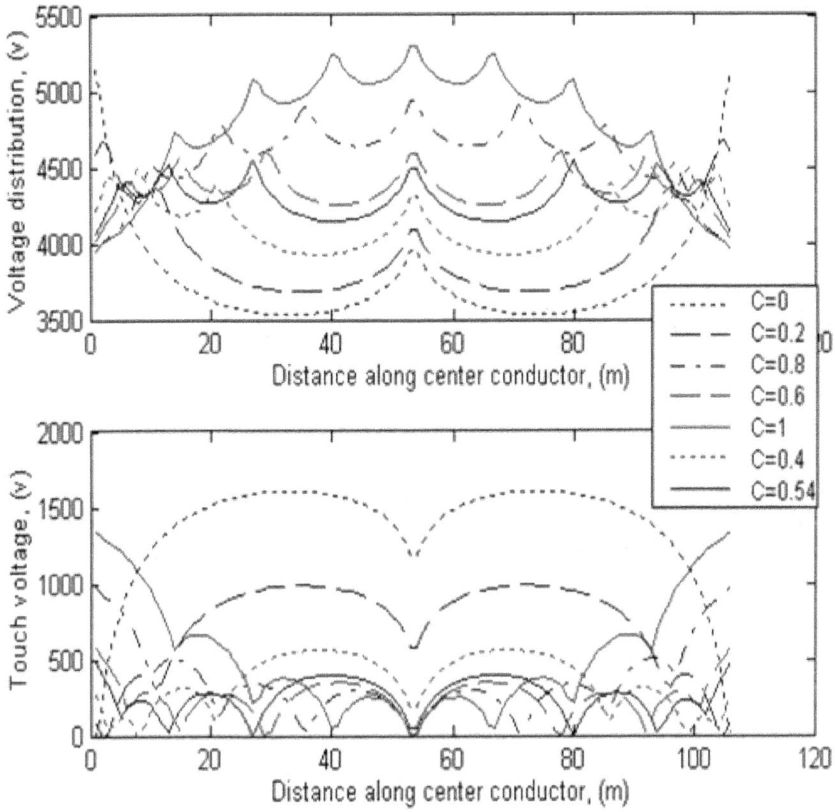

Figure 44. Touch voltages over center conductor for grounding grid module (A) as a function of conductor compression ratio in uniform soil of 100 Ω.m

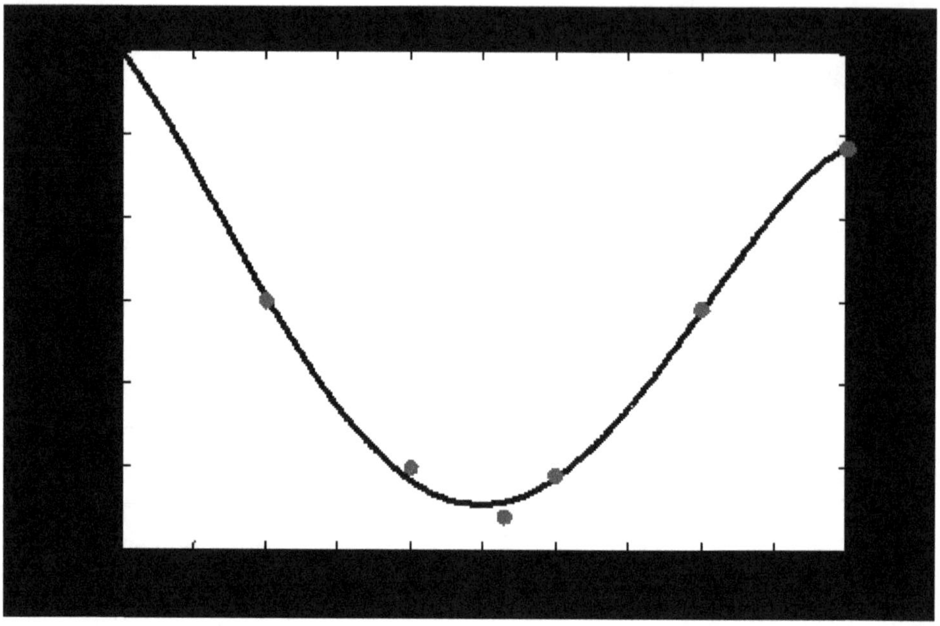

Figure 45. Earth surface potentials over all grid conductors for grounding grid of module (A) (A=105×105 m², N=8×8) in case of C=0.2, short circuit current 10 kA, soil resistivity 100 Ω.m

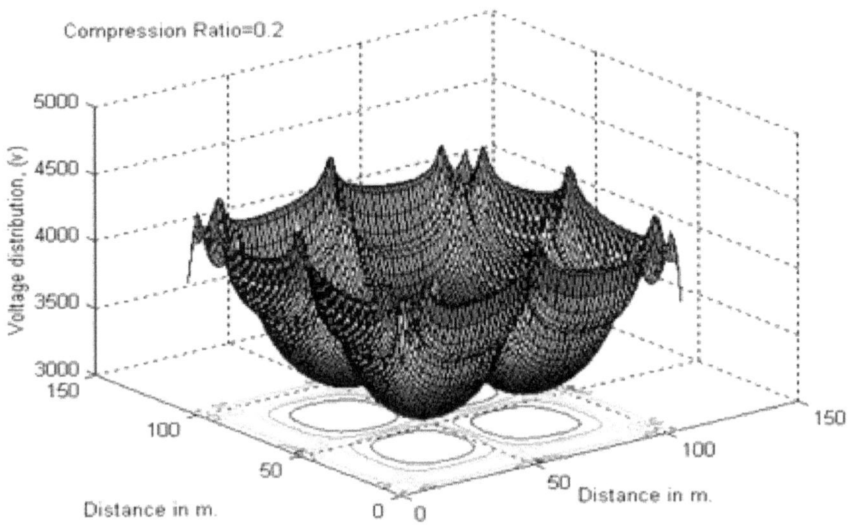

Figure 46. Earth surface potentials over all grid conductors for grid of (A=105×105 m², N=8×8) in case of C=0.54, short circuit current 10 kA, soil resistivity 100 Ω.m

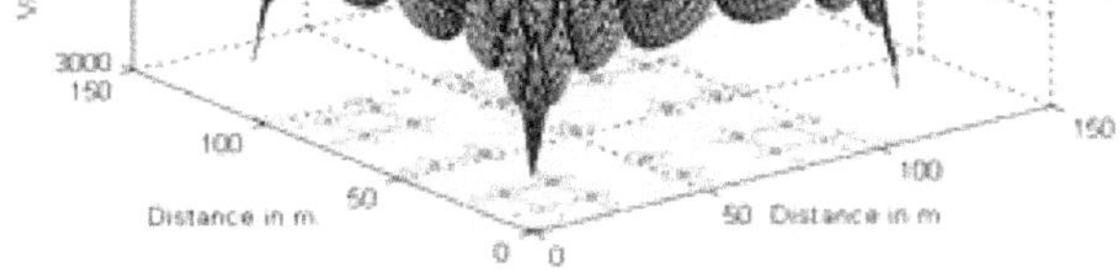

Figure 47. Earth surface potentials over all grid conductors for grid of (A=105×105 m², N=8×8) in case of C=0.6, short circuit current 10 kA, soil resistivity 100 Ω.m

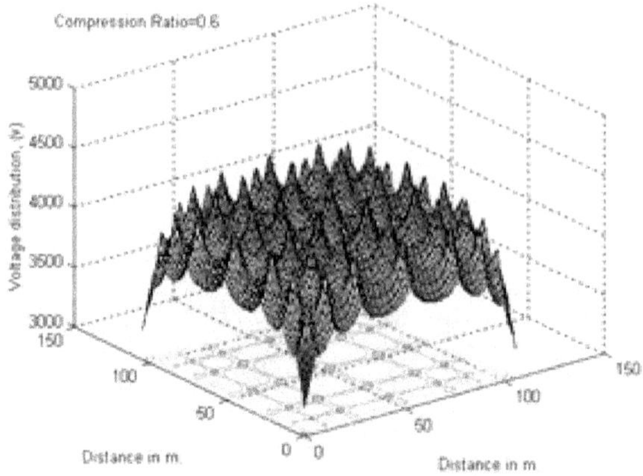

Figure 48. Earth surface potentials over all grid conductors for grid of (A=105×105 m², N=8×8) in case of C=1, short circuit current 10 kA, soil resistivity 100 Ω.m

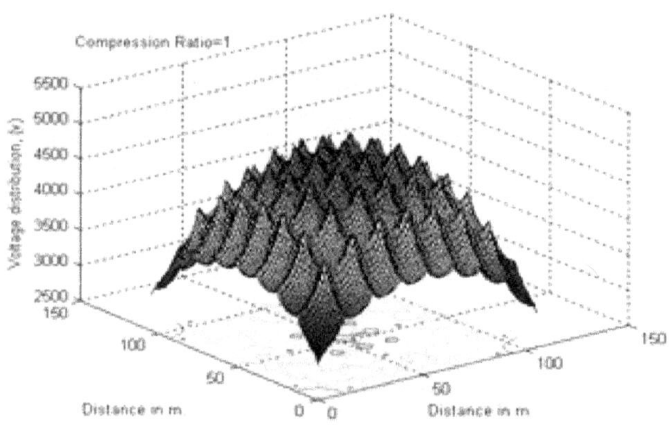

7.9.3 Voltage Profile of Grids Buried in Double Layers Soil

Often the installation of grounding grids of electric power stations and electrical networks is carrying in heterogeneous soil contains two or more layers. To illustrate the effect of changing the upper layer thickness in double layers soil structure with the variation of the upper and lower soil resistivity on the voltage profile and step and touch voltages, I.S.M method is used.

The graphical representation for earth surface potentials is plotted along the center conductor for 105×105 m^2 grid area with eight equally spaces conductors buried in non-uniform soil having specifications that given in Table 10. The current emanating the grounding grid is 10 kA, and the grid laying depth is 0.6 meter. (Figure 50) shows the earth surface potential along center conductors for equally spaced grid with 105×105 m^2, N=8x8 (module A) with various soil structures. From this figure, it is noticed that the GPR voltage of the grid buried in upper layer of double layers soil structure in which the second layer has low resistivity, is lower than that having high soil resistivity of second layer. It is noticed also that at the same upper layer soil resistivity in which the grid is buried, the upper layer thickness has little effect on GPR.

Table 10. Specifications of used non-uniform soil

ρ_1 (upper layer resistivity), Ω.m	**100**	**100**	**100**	**100**	**100**
ρ_2 (lower layer resistivity), Ω.m	100	40	40	250	250
H (upper layer thickness) in meters	infinity	10	20	10	20

Figure 49. Earth surface potentials along center conductor for equally spaced grid (A= 105×105 m2, N=8×8) with various soil structures

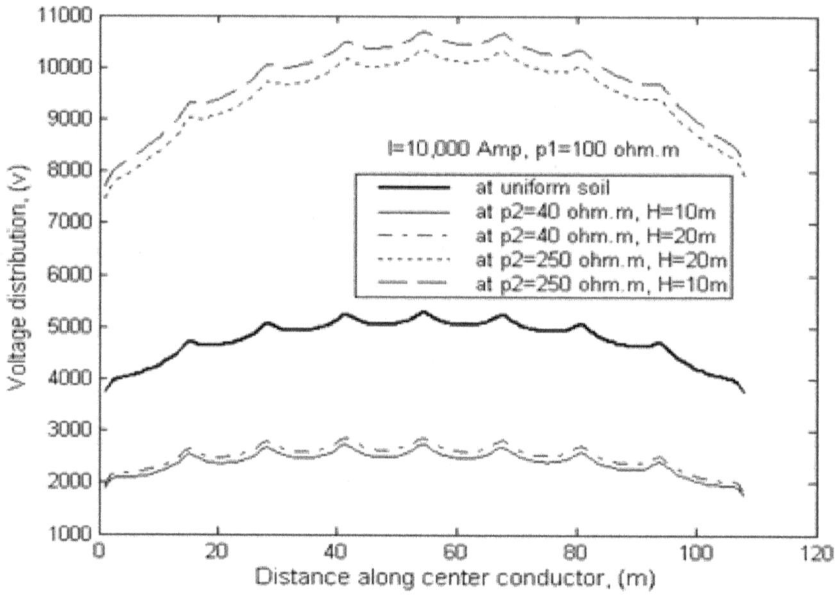

Figure 50. Maximum touch voltage over center conductor for mentioned grid buried in various soil structures as a function of compression ratio

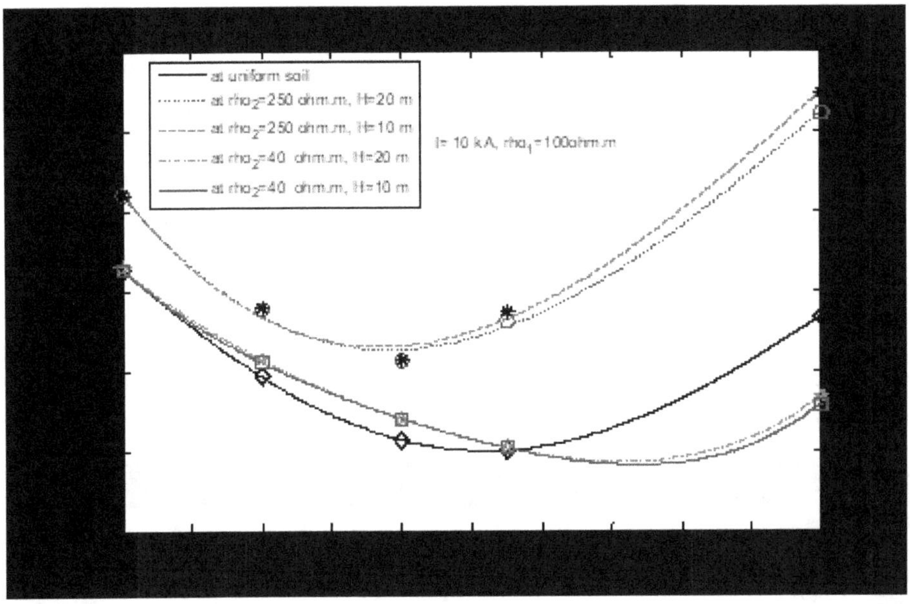

To illustrate the effect of soil non-uniformity on the unequally spaced grounding grid system having different conductors' distribution, the maximum touch voltages along center conductor for the grounding grid module (A) which buried in laying depth of 0.6 meter of the upper layer of double layer soil having the specifications mentioned in Table 4 are plotted in (Figure 51). This figure shows that the compression ratio of grid conductors' arrangement satisfies the lowest touch voltage depends on the soil parameters in which the grid is buried in. From this figure, it is concluded that when $\rho_1<\rho_2$, the compression ratio of lowest touch voltage takes value less than the optimal compression ratio (OCR) which determined by equation 49 for this grid, it's value reaches to 0.4 and this means that the grid conductors will be denser at the edges and having widely conductor spans at the middle. The results will be in the contrary if $\rho_1>\rho_2$ in which the touch voltage reaches its lowest values at the compression ratio of 0.74. This is indication to denser conductors at the grid center. In case of uniform soil $\rho_1=\rho_2$ the obtained results are in agreement with empirical OCR formula given in equation 7.49. The effect of upper layer thickness on OCR can be ignored as shown in Figure 51 and Figure 52.

7.9.4 Effect of Grounding Rods on Voltage Profile of Grounding Grid

Each one of grounding grid modules A, B and C is provided with copper rods, each one has 3 meters length and 25 mm diameter. Thirty two rods are distributed on the grid perimeter nodes in some calculations and eighty one rods are distributed, one at each node of the grounding grid in other calculations. Figures 51, 52 and 53 how the vertical rods effect on the earth surface potential along the center conductor of the grounding grids A, B and C. Slightly decrease of the earth surface potential is noticed when the grids are provided with vertical rods. The range of this decrease reaches to 6.25% - 6.6% when installing rod at each node on the grid perimeter.

Figure 51. Effect of vertical rods on the earth surface potential of square grounding grid (A), short circuit current 10 kA, soil resistivity 100 Ω.m

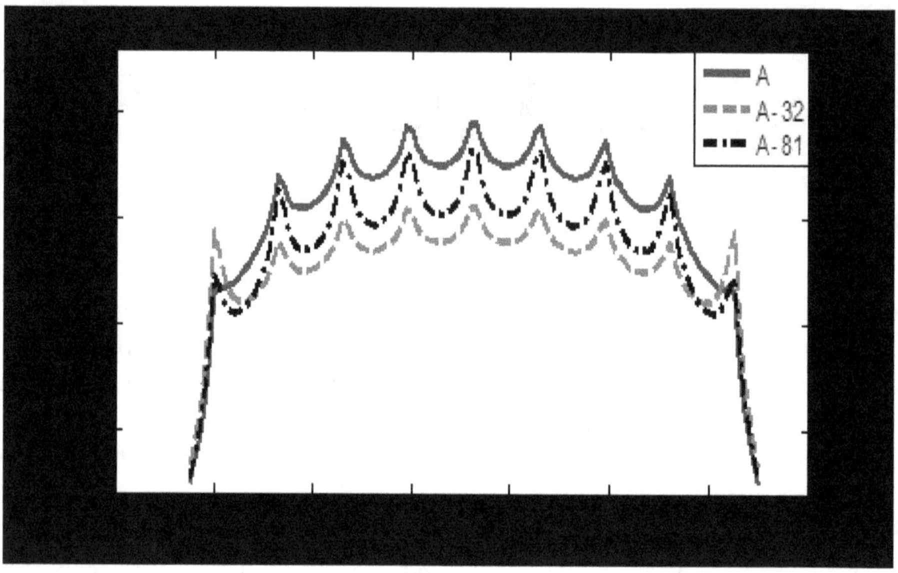

Figure 52. Effect of vertical rods on the earth surface potential of square grounding grid (B), short circuit current 10 kA, and soil resistivity equals100 Ω.m

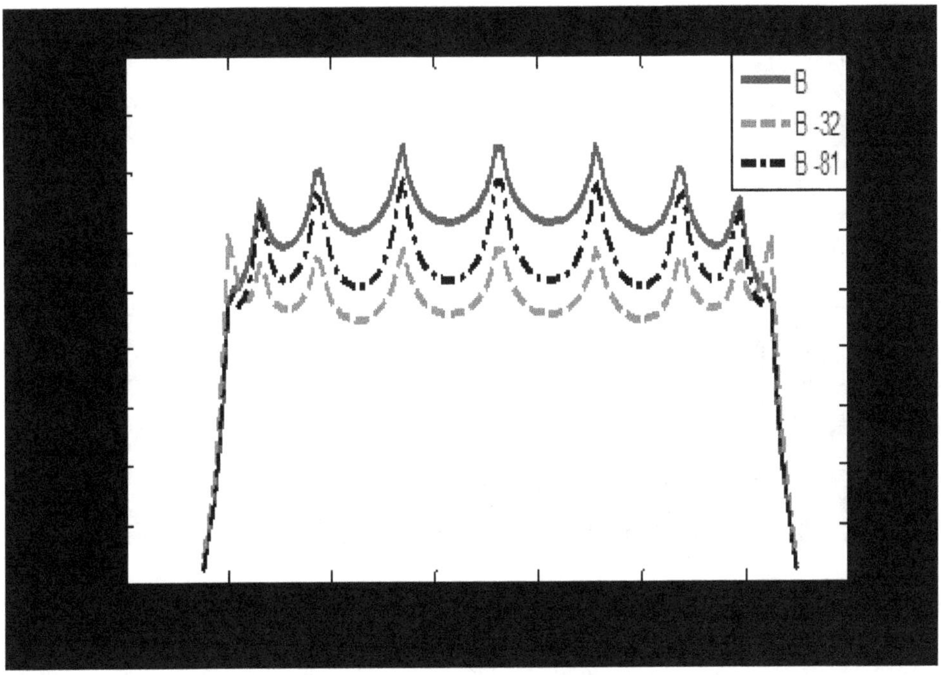

Figure 53. Effect of vertical rods on the earth surface potential of square grounding grid (C), short circuit current 10 kA, soil resistivity 100 Ω.m

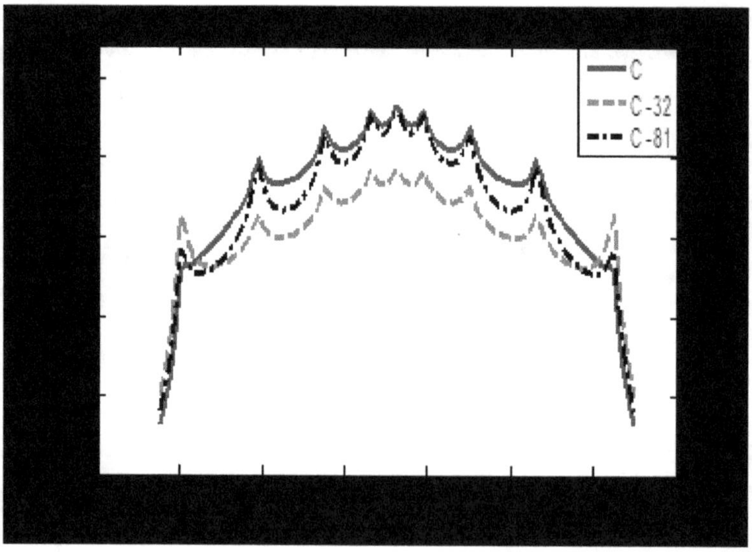

It is noticed also that as given in Table 11 the increase in grounding system length by installing more rods leads to decreasing in grounding resistance and step potential. As well known the grounding system resistance, step and touch voltages are inversely proportional with the grounding system length according to IEEE 80 (2000).

Table 11. Total conductor length, ground resistance, ground potential raise, touch voltage and step voltage of different square grounding grid cases (center profile)

Case	L_t (m)	R_g (Ohm)	GPR (kV)	V_t max % of GPR	V_s max, (kV)
A	1890	0.6578	6.578	37.03	4.76
B	1890	0.6367	6.367	34.75	4.89
C	1890	0.6600	6.600	35.30	4.94
A-32	1986	0.6287	6.287	30.22	5.63
B -32	1986	0.6142	6.142	30.18	5.77
C-32	1986	0.6308	6.308	26.58	5.45
A-81	2133	0.6036	6.036	29.92	5
B -81	2133	0.6118	6.118	31.93	4.80
C-81	2133	0.6182	6.182	28.69	4.911

Where (A), (B), and (C) are square grounding grids without vertical rods; (A-32), (B-32), and (C-32) are square grounding grids with 32 vertical rods; and (A-81), (B-81), and (C-81) are square grounding grids with 81 vertical rods.

7.9.5 Verifications of Calculated Results

A. Experimental Set-Up

For the verification of the calculated results an experimental model contains two identical tanks; each one has 1.5 m × 1.5 m × 0.5 m dimensions, filled with tap water. One is installed above the other, the top ends of the lower tank is coated with plastic to obtain electrical isolation between the metallic bodies of the two tanks. Only the electrical contact between the two mediums (representing the soil two layers) occurs through the metallic base of the upper tank. The lower tank is connected to earth. The restively of the upper and lower water layers can be controlled by adding an amount of sodium chloride.

A 220 volt alternating source provided with variac is connected with the grounding grid model which is suspended in the upper tank. Capacitor (10 Farad) is connected between the variac and the grid model to prevent any flow of D.C current due to the use of dissimilar materials. Two voltmeters are used; one for measuring the input voltage to the model and the other is connected to a probe for measuring the voltage distribution on the grid surface. One ammeter is used to measure the input grid current simulates the short circuit current, as shown in (Figure 55).

B. Grids Simulation

To simulate the grounding grids A, B and C scale factor of 1/100 is chosen, the dimensions of each grid are 105 ×105 cm. In opinion of several researchers, the effects of grid conductor diameter on grid resistance and the earth surface potentials are negligible. The simulated grid conductors are made from copper having 2.5 mm. The scale factor of the grid laying depth and the upper layer thickness is 1/100.

C. Water Resistivity Measurement

The two tanks are filled with water, to control the water resistivity of the two tanks (simulate the two soil layers) an amount of NaCL salt is used, and hence changing the reflection factor. The water conductivity is measured by means of DC probe.

D. Simulation of Short Circuit Current

The input voltage to the grid simulation is controlled by variac, until the current reaches 5 A with scaling factor 1: 2000 of short circuit current 10 kA.

E. Experimental Results

The step potential V is measured using probe as shown in (Figure 7.43).

Table 12. Earth surface potentials in volts and as percentages of GPR along center conductor for equally and unequally spaced grid of (A=105×105 m², N=64)

Equally spaces grid conductors									
x, (m)	**-52.5**	**-39.38**	**-26.25**	**-13.13**	**0**	**13.1**	**26.3**	**39.4**	**52.5**
V, (volt)calculated	4003	4732	5076	5249	5297	5249	5076	4732	4003
V, (volt)measured	4010	4800	5100	5270	5300	5270	5100	4800	4010
Unequally spaces having denser conductors at the grid edges									
x, (m)	-52.4	-46.8	-35.1	-19.1	0.0	19.1	35.1	46.8	52.4
V, (volt)calculated	4080	4431	4629	4762	4799	4762	4629	4431	4080
V, (volt)measured	4100	4350	4650	4780	4800	4780	4650	4350	4100
Unequally spaces having denser conductors at the grid middle									
x, (m)	-52.4	-33.3	-17.3	-5.6	0.0	5.6	17.3	33.3	52.4
V, (volt)calculated	4156	5265	5785	6138	6197	6138	5785	5265	4156
V, (volt)measured	4160	5270	5780	6150	6200	6150	5780	5270	4160

The measurements are done on grid model 105×105 m², N= 64 meshes, for equal conductors distribution, unequal having denser conductors at the grid edges and unequal spaces having denser conductors at the grid middle. The resistivity of each layer is 100 Ω.m, (the soil is considered as uniform soil). The measured step potential V, is multiplied by the scaling factor of current and compared with the calculated values. The results are given in Table 12. From this table it is noticed that the measured values are slightly high compared with the calculated values, the reason may be is due to the boundary effects of the tank walls, as each grid model size is higher than 1/5 of the upper tank size. Similar results are obtained in Table 13 for equally spaces grid conductors for double layers simulation. The soil resistivity of upper and lower layers and the upper layer thickness are changed as given in table 7.4. From this table it is noticed that the lower layer soil resistivity has significant change in the increase of step potential. On the contrary, the change in the thickness of the upper layer has little change in step potential.

Table 13. Earth surface potentials in volts along center conductor for equally spaced grid of A=105×105 m², N=64 in double layer soil

upper and lower layers resistivity	upper layer thickness	Equally spaces grid conductors									
Ohm. Meter	**meter**	**x, (m)**	**-52.5**	**-39.38**	**-26.25**	**-13.13**	**0**	**13.1**	**26.3**	**39.4**	**52.5**
100, 250	10	V, (volt)measured	7800	8200	9000	10000	10900	10000	9000	8200	7800
100, 250	20	V, (volt)measured	7500	8000	8600	9800	10600	0.99	0.96	0.89	0.76
100, 40	10	V, (volt)calculated	2000	2000	2100	2150	2300	2150	2100	2000	2000
100.40	20	V, (volt)measured	2010	2012	2112	2155	2310	2155	2112	2012	2010

Figure 54. Experimental setup

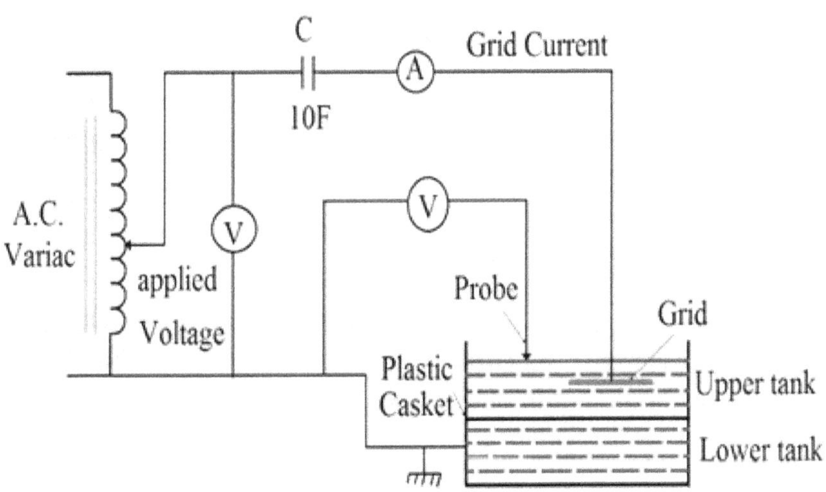

REFERENCES

ANSI/IEEE Std. 80-1986, Guide for Safety in AC substations Grounding, IEEE-John Wiley & Sons, New York August 1986.

Bendito, E., Carmona, A., Encinas, A., M., Jimenez, M., J., (2000), the external charges method in grounding grid design. The commission interministerial de ciencia technology, Spanish research council

Bharat Sanchar Nigam Limited, A Govt. (2005), Measurement Earth Potential Rise (EPR) Zone, No. EPR/A-002, CGM and Technical and Development Circle, 18-03-2005.

Dalziel, C. F., & Massogilia, F. P. (1956). Let-go currents and voltages. *Transactions of the American Institute of Electrical Engineers*, 75, 49–56.

Dawalibi, F., & Mukhedkar, D. (1975). Optimum design of substation grounding in a two layer earth structure, Part I-Analytical study. *IEEE Trans.*, PAS-94(2), 252–261.

El-Khodragy, T. M. (2007), Optimum Design of Earthing System of Substations by Using Artificial Neural Network, Ph. D. thesis, Faculty of Engineering, Cairo University.

El-Sayed, M. (2015). Achievement of the best design for unequally spaced grounding grids. *Ain Shams Engineering Journal*, 6(1), 171–179. 10.1016/j.asej.2014.09.013

Engineering Standards Branch, Telecommunications Standards Division (ESBTSD) (1994), Electrical Protection Grounding Fundamentals, United States department of agriculture, rural electrification administration

Heppe, R. J. (1979). Computation on potential at surface above an energized grid or other electrode, allowing for non-uniform current distribution. *IEEE Transactions on Power Apparatus and Systems*, PAS-98(6), 1978–1989. 10.1109/TPAS.1979.319377

Hung, L., Chen, X., & Yan, H. (1995). Study of Unequally Spaced Grounding Grid. *IEEE Transactions on Power Delivery*, 10(2).

Ibrahim, E. A. (1993), Grounding of the Nuclear Power Plants Using Grounding Grids, Ph. D. thesis, Faculty of Engineering, Cairo University.

IEC60 479-1(1984).Effect of current on human beings and livestock

IEEE 80, (2013), Guide for Safety in AC Substation Grounding, IEEE

IEEE std 142 (1991). IEEE Recommended Practice for Grounding of Industrial and Commercial Power System.

IEEE Std. 80-1976, Guide for Safety in AC substations Grounding, IEEE, New York, June 1976.

Koch, W. (1950), Erdungmassnahmen fur Hoschtspannungsanlagen mit Geerdetem Stempunkt, Elektrotechnische Zeitschrifl, 71(4), 89-91.

No, A. I. E. E. 80 (1961), Guide for Safety in A-Current Substation Grounding, AIEE, New York

Phithakwong, B., Kraisnachinda, N., Banjongjit, S., Chompoo-Inwai, C., & Kando, M. (2000). New Techniques the Computer-Aided Design for Substation Grounding, Power Engineering Society Winter Meeting, 3, 2011-2015.

Puttarach, A. N., Chakpitak, T., Kasirawat, Pongsriwat, C., (2007), Ground Layer Depth and the Effect of GPR, IEEE PES Power Africa 2007 Conference and Exposition, Schon. J.H (1998) Physical Properties of Rocks: Fundamentals and Principles of Petro-physics. Pergamon, Handbook of Geophysical Exploration, 18.

Sverak, J. G. (1984). Simplified analysis of electrical gradients above a ground grid. *IEEE Trans.*, PAS-103(1), 7–25.

Thabet, A. A. (2002). Grounding Systems of Electric Substations in Non-uniform Earth Structure with New Analysis, M. Sc. Thesis, High Institute of Energy, Aswan.

Thapar, B., Gerez, V., Balakrishnan, K., & Blank, D. A. (1991). Simplified Equations for Mesh and Step Voltages in AC Substation IEEE Transaction on Power Delivery, 6(2), 601–607.

Yves, Jaues, Fortin, Lehoux, & Simard. (2005). Earth Potential Rise Influence near HV Substations in Rural Areas, CIRED (Turin).

Zhongdong, Yao, Wen, & Xu. (2005), The Optimum Design of Grounding Grid of Large Substation, Proceedings of the XIVth International Symposium on High Voltage Engineering, Tsinghua University.

Chapter 8
Cathodic Protection of Grounding Grids

Osama El-Sayed Gouda
Faculty of Engineering, Cairo University, Egypt

ABSTRACT

This chapter contains the corrosion theory of grounding electrodes and basic electrochemistry in corrosion reactions. It contains also the forms of substation grounding grid corrosion (uniform corrosion, pitting corrosion, galvanic corrosion, microbial influenced corrosion, survey on corrosion rate of substation grounding grid, copper and steel corrosion rates, corrosion protection methods [coating, cathodic protection (CP)]). The chapter contains also the methods of applying cathodic protection in grounding grids, anode selection, anode spacing, and impressed current in the grounding grid cathodic protection. Finally, it contains the required information for design grounding system cathodic protection and sacrificial anode (galvanic) cathodic protection system design steps.

8.1 CORROSION THEORY AND BASIC ELECTROCHEMISTRY IN CORROSION REACTIONS

In this chapter, corrosion of substation grounding grid or any surround metallic structure due to current discharge buried in soil is fully introduced. Survey on calculating rate of corrosion and factors affecting the rate of corrosion are also introduced in this chapter.

Corrosion is defined as destruction or deterioration of material because of reaction with its environment. Some insist that the definition should be restricted to metals but often corrosion engineers must consider both metal and nonmetal for solution of a given problem as given by Mars G. Fontana (1987). It is also considered the

DOI: 10.4018/979-8-3693-2969-6.ch008

exothermic chemical transformation of a metal or metal alloy to a non-reactive covalent compound such as an oxide or silicate that is often similar or even identical to the mineral from which the metals were extracted as given by A. P. Howard, (2004)

Most of the corrosion in soil is the result of an electrochemical reaction. Each galvanic corrosion cell comprises as shown in (Figure 1):

(i) Anode and cathode areas on metal surface,
(ii) Soil electrolyte to enable occurrence of cathodic and anodic processes and current flow between anode and cathode and,
(iii) Conducting path for the flow of current from cathode to anode.

In the absence of any of these components, the corrosion cell would stop to operate **as reported by** Zhou Mi, Wang Jianguo, Liu Yang, Xiang Nianwen, Sun Zhen, Chen Junjie, Fang Chunhua (2008)

Figure 1. Basic galvanic corrosion cell

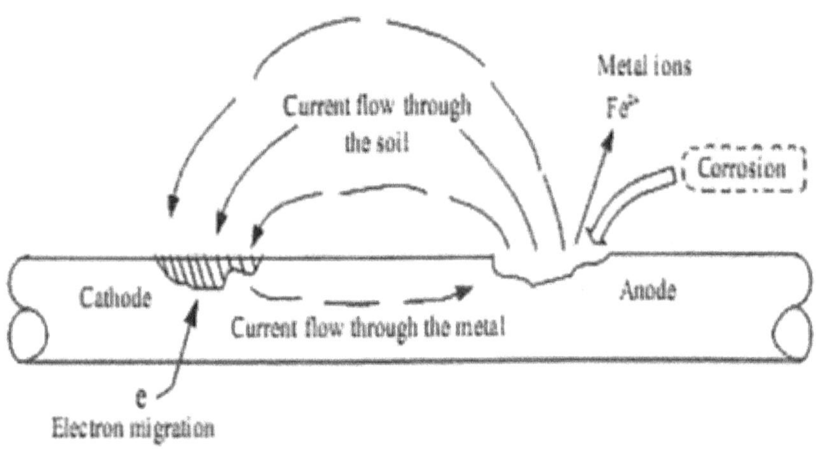

8.2 FORMS OF SUBSTATION GROUNDING GRID CORROSION

There are a number of corrosion forms of substation grounding grid. These forms can be summarized as follows:

8.2.1 Uniform Corrosion

Uniform corrosion refers to corrosion damage occurring evenly over the metal/alloy surface as shown in Figure 2, and the rate of corrosion is nearly the same over the entire surface. Although uniform corrosion is the most important form of corrosion, it is relatively easily measured and predicted, making disastrous failures relatively rare as reported by M. Tullmin and P. R. Roberge, (1995)

Figure 2. Uniform corrosion

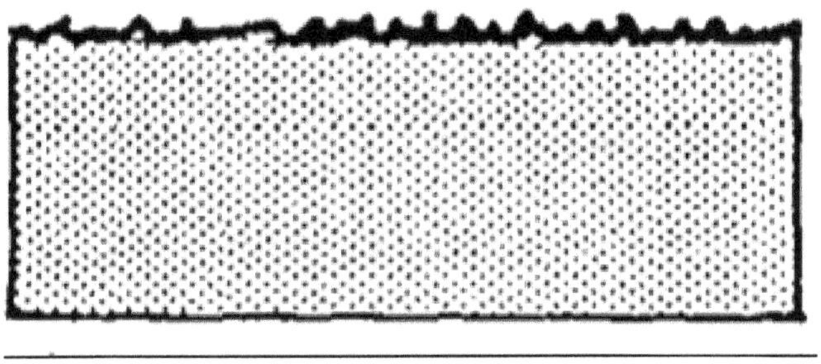

8.2.2 Pitting Corrosion

Pitting corrosion is a form of localized corrosion that produces pits (holes) in the metal/alloy as shown in (Figure 8.3). It occurs when discrete areas of a material undergo rapid attack while the vast majority of the surface remains virtually unaffected. This is in sharp contrast to uniform corrosion in which all parts of the exposed surface recede at approximately the same rate. Pitting of a given material depends strongly upon the presence of an aggressive species in the environment as reported by Zhou Mi, Wang Jianguo, Liu Yang, Xiang Nianwen, Sun Zhen, Chen Junjie, Fang Chunhua (2008).

Figure 3. Pitting corrosion

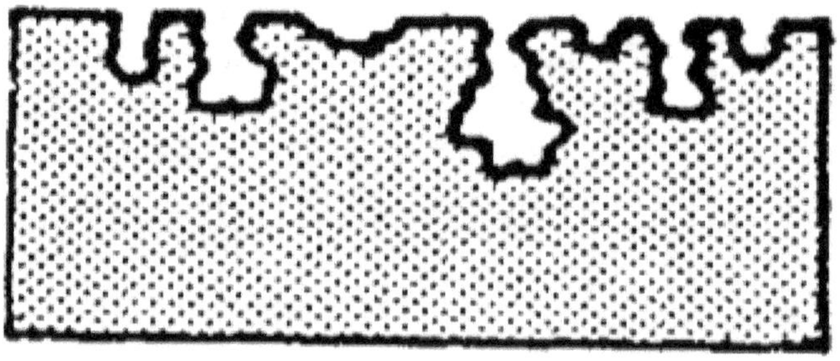

8.2.3 Galvanic Corrosion

When a difference in electric potential exists between two dissimilar metals, as an example, copper and iron, connected externally and buried in soil as shown in Figure 4. The electrolyte (soil) allows some current to flow and causes corrosion in steel. Galvanic corrosion tends to be particularly severe if the anodic material's surface area is small compared to that of the cathode. In this case the galvanic current is concentrated on a small anodic area; it results in rapid corrosive penetration of the anode material as reported by Zhou Mi, Wang Jianguo, Liu Yang, Xiang Nianwen, Sun Zhen, Chen Junjie, Fang Chunhua (2008).

8.2.4 Microbial Influenced Corrosion

In contaminated soils, another form of corrosion called microbial influenced corrosion (MIC) can occur. The most common form is where a range of bacteria called sulphate reducing bacteria (SRB) in anaerobic environments can react to produce hydrogen sulphide (HS). The acidic conditions generated can lead to rapid, localized corrosion.

Pictures of different corrosion forms in substation grounding grid are presented in Figure 5**as** reported by Zhou Mi, Wang Jianguo, Liu Yang, Xiang Nianwen, Sun Zhen, Chen Junjie, Fang Chunhua (2008).

Figure 4. Galvanic corrosion

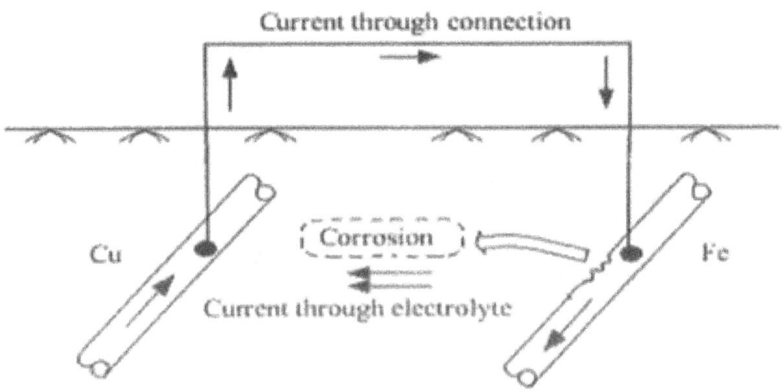

8.3 SURVEY ON CORROSION RATE OF SUBSTATION GROUNDING GRID

Two ways are found to calculate the rate of corrosion. In the following sections, full details of them are introduced.

8.3.1 Steel Corrosion Rate

8.3.1.1 Experimental Formula

a) Formula for calculating expected corrosion rate:

This relationship is obtained from the National Bureau of Standards Data to calculate the corrosion rate of steel (Bessemer Steel). This data was produced from actual experiments in 44 different soils conducted over a period of 12 years. As reported by as reported by P.K. Sen, Keith Malmeda, John P. Nelson (2002)

Figure 5. Corrosion forms of (a) uniform corrosion, (b) pitting corrosion and (c) microbial influenced corrosion in substation grounding grid

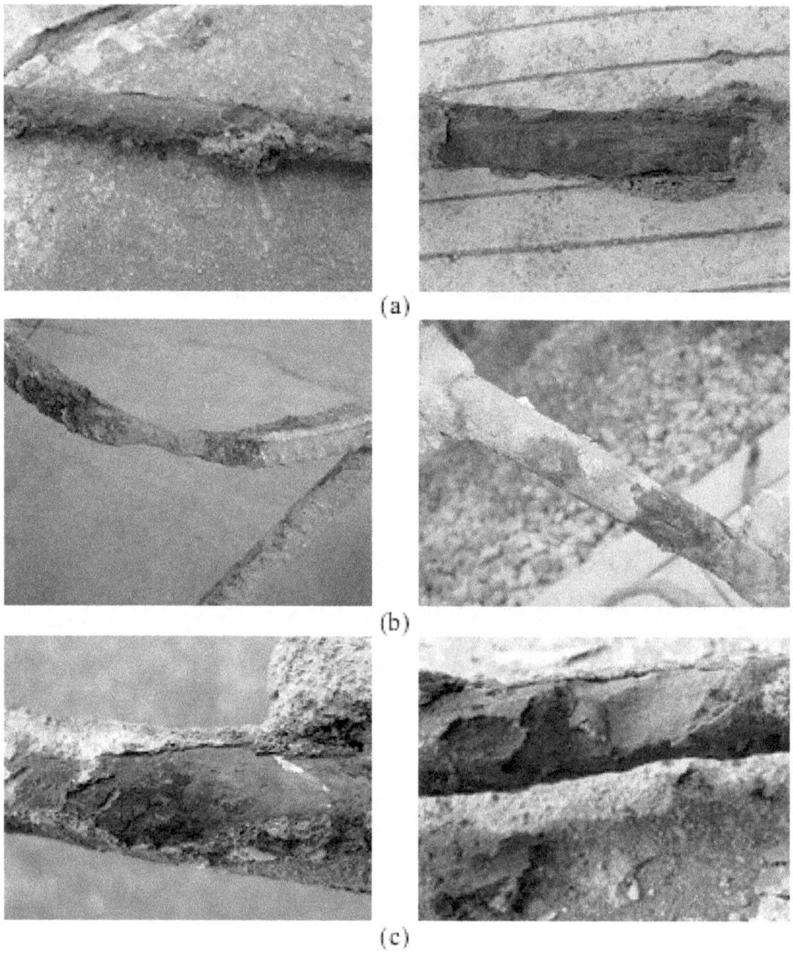

These tests used 1.5-inch and 3-inch diameter samples. The 3-inch samples corroded 13% more than the 1.5-inch sample with an error of 10%. It is found that:

$$Y = F(\rho, x_1, x_2, x_3) \tag{8.1}$$

Where:

Y= Corrosion rate (mils/year), ρ = Resistivity (ohm.cm), x_1 = pH value, x_2= Moisture (%), and x_3= Aeration (%)

Finally, the following equation is obtained:

$$Y = 3.36 - 9.63(10^{-5})(\rho)+0.29(x_1)+0.034(x_2)+0.012(x_3) \quad (8.2)$$

Equation 8.2 is obtained from experimental data and is limited by extreme corrosion conditions such as, extremely high resistivity (> 10,000 ohm.cm) or extremely low aeration quantities (< 3%). Corrosion is reduced by these conditions to almost zero. This equation is applicable for all steel rods up to 3-inch diameter and is applicable for the first 12 years. It is concluded also experimentally that the average corrosion rate in the following 12 years reduces to half that value in the first 12 years. The corrosion rate can be estimated from Table 1 for several material corrosion rate based on equation 8.2 as reported by P.K. Sen, Keith Malmeda, John P. Nelson (2002)

Table 1. Average corrosion rate of several materials (P.K. Sen, Keith Malmeda, John P. Nelson (2002))

Materials	Corrosion rate (mils/years)
Open earth steel	5.90
Wrought iron	5.00
Bessemer steel	5.30
Copper	1.25
Lead	3.00
Zinc	5.00

B) Formula for accounting the weight loss after exposure to a corrosive environment conditions

21 spot-checks by excavation were made on substations of different voltage levels distributed over foshan area, whose operation time varies from 1 to 30 years. Corrosion data computation was done on 70 samples from main grounding conductors (buried in soil), grounding conductors in cable trench and grounding down lead. The average and maximum corrosion rates are calculated as reported by Zhou Mi, Wang Jianguo, Liu Yang, Xiang Nianwen, Sun Zhen, Chen Junjie, Fang Chunhua (2008) as follows:

- The average corrosion rate (mm/y) =

$$10.\Delta W/(d_m.A.t) \quad (8.3)$$

Where: ΔW= weight loss in grams, d_m = metal density (g/cm), A = area of metal (cm), t = time of exposure in corrosive environment (years).

- Thus, maximum corrosion rate (mm/year) =

$$H/t \tag{8.4}$$

Where: H = depth of the deepest hole in mm, and t = time of exposure in corrosive environment (years).

8.3.1.2 Laboratory Formula:

Faraday's first law, that given in (8.5), is also used for conversion of corrosion rates (A/cm^2) to penetration rates (millimeter per year) as given by Xuan Fan, Jianguo Wang, Mi Zhou, Songbo Huang, Yan Wang, Peng Dou. Chengdu, (2011)

$$\Delta m = \frac{A_t . I . s}{n . F_r} \tag{8.5}$$

Where: Δm = mass of metal corroded, A_t = atomic weight of the metal, I = current in amperes, s = time in seconds, n = number of equivalents transferred per mole of metal, F_r = Faraday constant 96,500.

8.3.2. Copper Corrosion Rate

Unlike steel, it is not possible to estimate the copper corrosion rates by oxygen and sulfur mass balance calculation since other species takes part in the copper corrosion process as well. Water molecules as moisture or in aqueous corrosion as hydroxide ions are probably the most frequent species that will react and form copper corrosion products in such a complex environment as expected in are positron with different trace gases and salts as reported by Peter Szakalos, Seshadri Seetharaman (2012)

In grounding grid design for AC substation, copper is used as the primary grid material. Steel-grounding system is widely used in China and readily accepted in many other countries worldwide, where copper is very expensive. However, it is still common in this kind of grounding system for grounding faults problems to occur due to corrosion, which might result in plenty of economic loss as reported by Rong Zheng, Jinliang He, Jun Hu, Guojun Lu and Bin Luo, (2002).

Reliable performance of steel grounding systems can be insured by selecting a conductor cross-section area as explained by V. R. Lawson (1988).

8.4 FACTORS AFFECTING CORROSION RATE

Different factors include factors connected with the metal, with the atmosphere, and with the corrosion product are introduced in the following subsections as reported by B.S. Chauhan (2008)

8.4.1 Factors Connected With the Metal

- Position of the metal in the E.M.F Series
- Purity of the metal
- Surface of the metal
- Stress corrosion
- Anode to cathode area ratio stress
- Physical state of a metal

8.4.2 Factors Connected With the Atmosphere

- Nature of the atmosphere
- Temperature of the atmosphere
- PH of the atmosphere
- Amount of moisture in the atmosphere
- Amount of oxygen in the atmosphere
- Amount of chemical fumes in the atmosphere, etc

8.4.3 Factors Connected With the Corrosion Product

- In some cases, the corroded product sticks to the surface and absorbs more moisture. This induces further corrosion.
- In some cases, the corroded product acts as the protective coating which prevents further corrosion.
- In some other cases the corroded product falls out of position exposing the fresh metal surface for further corrosion.

8.5. CORROSION PROTECTION METHODS

In this section, description of cathodic protection will be demonstrated. Forms, types of anode material, selection of their material, their lifetime, comparison between cathodic protection types, limitations that must be kept in mind and designing steps are also discussed in detail. (Figure 6) gives picture of galvanized rod removed after 10 years. (Figure 7) gives picture of copper-bonded ground rod removed after 10 years (Pawnee Site). (Figure 8) gives picture of copper-bonded rod after 10 years- Electrode type H. Finally. In (Figure 9) Galvanized rod picture after 10 years -Electrode type I is presented.

Some of the most common corrosion prevention measures that are used to minimize corrosion of grounding material and/or steel piping are as reported by P.K. Sen, Keith Malmeda, John P. Nelson (2002):

- Using of tin coated bare copper cable or hot galvanized or corrosion-resistant steel as a grounding conductor to minimize the potential difference between the steel structures and the grounding system of the substation.
- Cathodic protection that achieved by applying a current to the structure to be protected from some outside source. When enough current is applied, the whole structure will be at one potential; thus, anode and cathode sites will not exist. Cathodic protection is commonly used on many types of structures, such as pipelines, underground storage tanks, locks, and ship hulls. Cathodic protection can be designed for 30 years of life expectancy, and it can be used to protect the grounding system of electrical substation.
- Using of insulated copper ground cable in areas near pipes, which is a method to reduce corrosion by minimizing the difference in potential between anode and cathode.
- Electrical isolation of piping from other plant structures and the grounding system.

The first two methods are the most commonly adopted in practical designs.

Figure 6. Galvanized rod removed after 10 years (Pawnee site)

Figure 7. Copper-bonded ground rod removed after 10 years (Pawnee site)

Figure 8. Copper-bonded rod after 10 years- Electrode type H

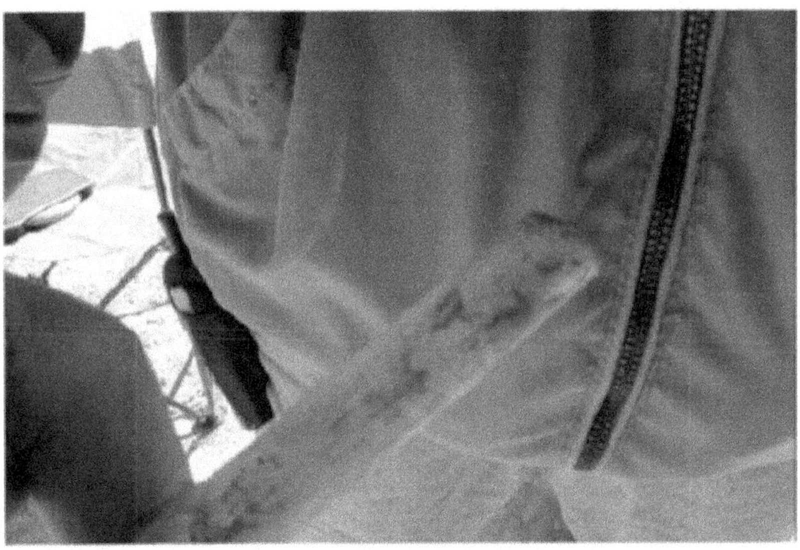

Figure 9. Galvanized rod after 10 years -Electrode type I

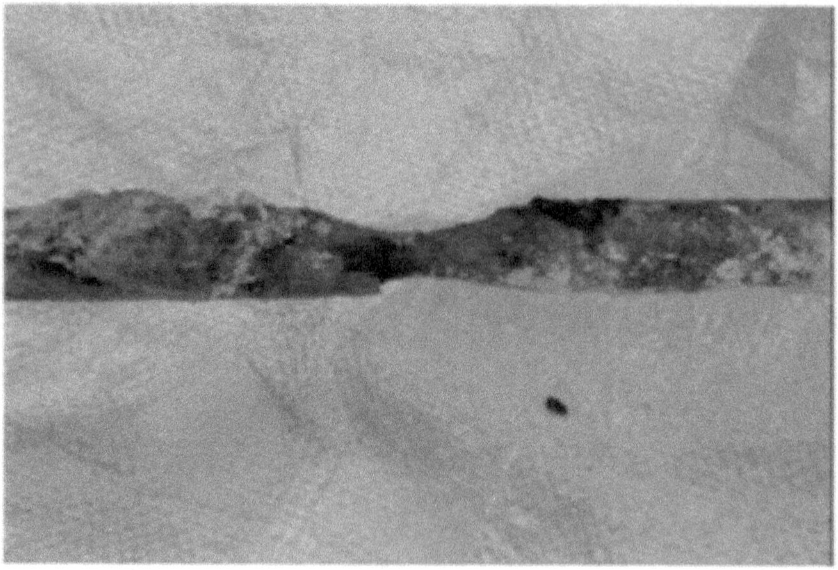

8.6. COATING

It has been used to control corrosion rate. It initially protects the underlying metal. When the continuity of the coating is destroyed, it is observed that the corrosion rate of the base metal is normal or above normal depending on the type of coating and the underlying metal. Zinc-coated (galvanized) steel is the most used type of coating. However, any coating as a sole means of protection is not very effective 5 as reported by P.K. Sen, Keith Malmeda, John P. Nelson (2002). This type of protection to control corrosion can be used in protection of grounding system of substaon by using galvanized steel.

External coatings must be properly selected and applied, and the coated system must be carefully handled and installed to fulfill these functions. Various types of external coatings can accomplish the desired functions. Desirable characteristics of external coatings include effective electrical insulator; effective moisture barrier; and applied to a pipe by a method that does not adversely affect the properties of system as reported by NACE Standard RP0169 (2002).

8.7 CATHODIC PROTECTION (CP)

One definition of cathodic protection is a technique to reduce the corrosion rate of a metal surface by making it the cathode of an electrochemical cell. It has inherent personnel protection by keeping the grounding system protected from corrosion to maintain system performance.

CP systems are typically designed to drive the potential to about 1 V negative. Potentials more negative than 2.5 V can be damaging to coatings. Equipment properly connected to a CP system has a very low resistance path to ground (< 2 Ω). This provides an adequate path for dissipation of current in a fault condition as given by Marcus O. Durham& Robert A., (2005).

In the case of an impressed current system, a current is impressed on the structure by means of a power supply, referred to as a rectifier, and an anode buried in the ground. In the case of a sacrificial anode system, the galvanic relationship between a sacrificial anode material, such as zinc or magnesium, and the pipe steel is used to supply the required cathodic protection current as reported by Beavers, John A., (2001).

8.7.1. Methods of Applying Cathodic Protection

Cathodic protection may be achieved in either of two ways: sacrificial (or galvanic) anode cathodic protection (SACP) and impressed current cathodic protection (ICCP). The main difference between the two methods is that ICCP uses an external power source with inert anodes and SACP uses the naturally occurring electrochemical potential difference between different metallic elements to provide protection.

8.7.1.1. Sacrificial or Galvanic Anodes

They are applicable and effective where current requirements are low (less than 100 mA), and the structures to be protected are well-coated in a low resistivity soil. The anode consists of a metal, which is electro-negative to the protected structure. These anodes are self-energized and are connected directly to the protected structure as shown in Figure 10. Sacrificial anodes, usually distributed along or around the protected metal parts, consume themselves in protecting the other metal. The consumption rate depends on the magnitude of current generated as villas the material of which the anode is made as reported by Bianchetti, Ronald L., (2001).

Figure 10. Sacrificial or galvanic anode material

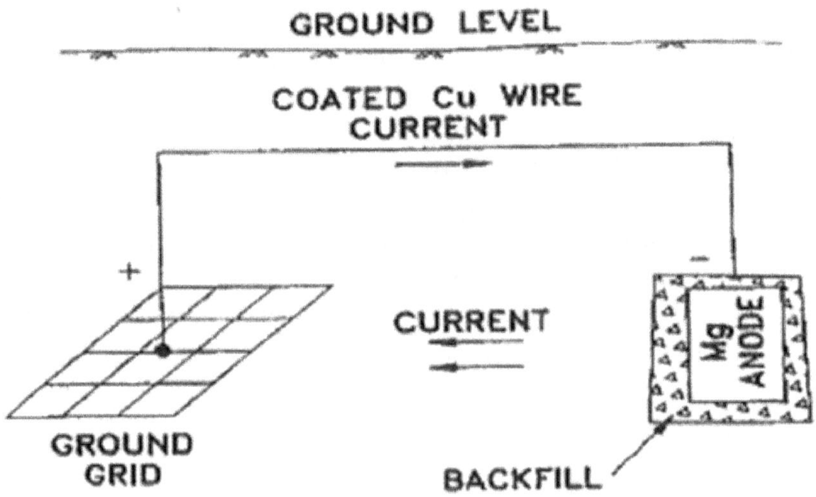

8.7.1.2. The Main Advantages of Sacrificial Anodes

- Economically feasible when installed with the structure.
- Very little operation or maintenance requirements (very small chance of premature failure or breakdown).
- Extremely small possibility of overprotection (which may cause coating damage or hydrogen embitterment).
- Small likelihood of stray current causing interference damage to other metallic (foreign) structures.

8.7.1.3. The Main Disadvantages of Sacrificial Anode

- Small driving voltage available (limited potential difference).
- Extremely small current available in higher resistivity electrolytes.
- Not economically feasible to install or replace anodes.
- Extensive existing structures.

8.7.1.4. Galvanic Anodes, Anode Selection and Anode Backfill

The use of galvanic anodes for cathodic protection (CP) is a simple application of the dissimilar metal corrosion cell, where a steel grounding system is electrically connected to a metal higher in the electromotive force series and both are in a common conductive electrolyte such as the soil, the more active metal is corroded and discharges current in the process. Magnesium and zinc are such metals. If the amount of current needed for a given CP application is known, anode systems can be designed using sufficient anode material to produce the desired current output continuously over a desired number of years as reported by Bianchetti, Ronald L., (2001).

Galvanic anodes consume themselves in protecting the other metal. The two galvanic anode metals commonly used for buried any metallic structure like grounding system for electrical substations are magnesium and zinc (A_t = 24.32 for Mg and 65.38 for Zn, n= 2 for Mg and 2 for Zn).

The consumption rate depends on the magnitude of current generated as well as the material of which the anode is made and is given by Farady's Second Law of Electrolysis as given in equation 8.5 as reported by P.K. Sen, Keith Malmeda, John P. Nelson (2002).

i. **Magnesium anode**

Magnesium is the most widely used material for galvanic anodes. Typical characteristics are given in Table 2. Magnesium anodes are available in various shapes and weights from the manufacturers. Some of the sizes available, suitable for use in soil, are listed in Tables 3 and 4 as reported by Bianchetti, Ronald L., (2001).

Table 2. Characteristics of Magnesium Anodes (Bianchetti, Ronald L., (2001))

Specific Gravity	**1.94**
Pounds per Cubic Foot	121
Theoretical Amp Hours per Pound	1000
Theoretical Pounds per Amp per Year	8.7
Current Efficiency - Percent	50
Actual Amp Hours per Pound	500
Actual Pounds per Amp per Year	17.4
Solution Potential - Volts to CSE	
Standard H -1 Alloy	-1.50 to 1.55

continued on following page

Table 2. Continued

Specific Gravity	**1.94**
High Potential Alloy	-1.75 to 1.77
Driving Potential to Pipe Line	
Polarized to -0.90 Volt to $CuSO_4$	
Standard Alloy - Volts	0.55
High Potential Alloy - Volts	0.80

Table 3. Standard alloy chemical composition (Bianchetti, Ronald L., 2001)

Weight content %			
Element	**Grade A**	**Grade B**	**Grade C**
Al	5.3 - 6.7	5.3 - 6.7	5.0 - 7.0
Mn	0.15 min	0.15 min	0.15 min
Zn	5.3 - 6.7	5.3 - 6.7	5.3 - 6.7
Si	0.10 max	0.30 max	0.30 max
Cu	0.02 max	0.05 max	0.10 max
Ni	0.002 max	0.003 max	0.003 max
Fe	0.003 max	0.003 max	0.003 max
Other	0.30 max	0.30 max	0.30 max
Magnesium	Remainder	Remainder	Remainder

Table 4. High potential alloy chemical composition (Bianchetti, Ronald L., 2001)

Element	**Weight Content %**
Al	0.010
Mn	0.50 to 1.30
Cu	0.02 max
Ni	0.001 max
Fe	0.03 max
Other	0.05 each or 0.3 max total
Magnesium	Remainder

ii. Zinc anode

Zinc anodes are available in a number of sizes for use in earth anode beds. Typical characteristics of zinc used as an anode material are given in Table 5. These are long slender shapes to achieve low resistances to earth and practical current output at the usual low driving voltage between anode and protected structure. Some available sizes as standard commercial items are shown in Table 6. Nonstandard sizes may be obtained on special order as reported by Bianchetti, Ronald L., (2001).

For reliable operation in earth installations, both zinc and magnesium anodes are used with a chemical backfill to surround the anode completely. Typical data on chemical backfill are given in Table 7.

Table 5. Characteristics of zinc anodes (Bianchetti, Ronald L., 2001)

Specific Gravity	**7**
Pounds per Cubic Foot	440
Theoretical Amp Hours per Pound	372
Theoretical Pounds per Amp per Year	23.5
Current Efficiency - Percent	90
Actual Amp Hours per Pound	335
Actual Pounds per Amp per Year	26
Solution Potential - Volts to CSE	
Driving Potential to Pipeline	-1.1

Table 6. Standard H-1 chemical composition (Bianchetti, Ronald L., 2001)

Element	**Weight content %**	
	MLA – A – 18001 (ASTM B – 418 Type I)	**ASTM B-418 Type II**
Al	0.1 – 0.5	0.05 max
Cd	0.02 – 0.07	0.003 max
Fe	0.005 max	0.0014 max
Pb	0.006 max	0.003 max
Cu	0.005 max	0.002 max
Zinc	Remainder	Remainder

With the anode surrounded with a uniform material of known composition, anode current is more efficient. If the soil contacts the anode directly, variations in soil composition can set up local corrosion on the anode resulting in non-uniform consumption of the anode. By isolating the anode material from the native soil, the backfill material greatly reduces the possibility of adverse effects on anode performance. In the presence of phosphates, carbonates and bicarbonates, zinc anodes can develop passive films and cease to produce useful amounts of current. Carbonates and bicarbonates will influence magnesium in the same way. Chlorides tend to increase self-corrosion of magnesium and reduce its current efficiency as reported by Bianchetti, Ronald L., (2001). Backfill serves some purposes such as providing uniform soil around the anode in order to assure uniform corrosion of the anode, minimizing the resistance between the anode and the earth, retaining moisture which aids in minimizing their distance, and aiding in depolarizing any film build up, which would reduce the flow of the galvanic currents as reported by J.P., Nelson, Holm, William K., (1985).

Table 7. Backfill data for magnesium and zinc anodes

Gypsum ($CaSo_4$) %					
Backfill	**Hydrated**	**Molding plaster (plaster of paris)**	**Bentonite Clay %**	**Sodium Sulfate %**	**Approximate soil resistivity in ohm.cm**
(A)	50	--	50	--	250
(B)	70	--	20	5	50

- Backfill (A) is commonly used with zinc anodes.
- Backfill (B) with low resistivity is useful in high soil resistivity areas to reduce the anode resistance to earth.

After considering the available materials, one must make the suitable selection. The criterion for selection is as one would expect an analysis of performance versus cost, the performance of an anode is measured by the following criteria:

i. Anode life:

Life is a function of three factors: weight, current output, and efficiency. Longer life is achieved through heavier weight, lower current output and high efficiency. The following expressions are used to calculate the estimated anode life time as reported by James B. Bushman PE, (2001).

: Anode life time =

$$\frac{F_c}{h_y} \times \frac{W_A \cdot \eta_A \cdot X}{I_A} \tag{8.6}$$

Where: F_c is Faradays consumption rate (amperes/hours/pound), h_y is no. of hours per year, W_A is anode weight (Ibs), η_A is anode efficiency, X is the utilization factor, and I_A is anode current in Amperes. The utilization factor may be taken as 85% meaning that when the anode is 85% consumed, it will require replacement because there is insufficient anode material remaining to maintain a reasonable percentage of its original current output as reported by James B. Bushman PE., (2001).

1. *Magnesium lifetime as given by Bianchetti, Ronald L., (2001):*

L_m = 0.116 × Anode weight (pound) × Efficiency × Utilization Factor / Design current (Amperes) (8.7)

It can be reduced to the simpler form as follows:

$$L_m = \frac{48.5\ W}{I_m} \tag{8.8}$$

Where: L_m is the magnesium anode life in years, W is the anode metal weight in pounds, and I_m is the current output in mA as given by James B .Bushman PE, (2001).

2. Zinc life time as given by Bianchetti, Ronald L., (2001):

L_Z =0.0424 × Anode weight (pound) × Efficiency × Utilization Factor / Design current (Amperes) (8.9)

It can be reduced to the simpler form as follows:

$$L_Z = \frac{32.5.W}{I_m} \tag{8.10}$$

Where: L_Z is the zinc anode life in years, W is the anode metal weight in pounds and I_m is the current output in mA as given by James B .Bushman PE.,(2001).

ii. Anode current:

The current output of anode can be calculated in accordance with ohms law as follows

$$I_A = \frac{E_{A-}E_P}{R_A + R_{AP} + R_{Pg} + R_C + R_W} \tag{8.11}$$

Where: I_A is the anode output current in amperes, EA is the potential of anode, E_p :is the backfill potential, $R_A$is the anode film resistance, $R_{AP}$is the backfill resistance, R_{Pg} is the backfill to earth resistance, $R_C$is the cathode film resistance, and $R_W$is the resistance of connecting wire. Correction factors are then applied to adjust the result of various shapes and structures potentials:

iii. Output current for bare structure as given by James B. Bushman PE.,(2001).

For Magnesium:

$$I_{mb} = 150,000.\frac{F.Y}{\rho} \quad (8.12)$$

For Zinc:

$$I_{Zb} = 50,000.\frac{F.Y}{\rho}. \quad (8.13)$$

iv. Output current for coated structure [17]:

For Magnesium:

$$I_{mc} = 110,000.\frac{F.Y}{\rho} \quad (8.14)$$

For Zinc:

$$I_{Zc} = 40,000.\frac{F.Y}{\rho} \quad (8.15)$$

Where: I_{mb} is the current output for magnesium anode on bare structure, mA

- I_{Zb} is the current output for zinc anode on bare structure, mA
- I_{mc} is the current output for magnesium anode on coated structure, mA
- I_{Zc} is the current output for zinc anode on coated structure, mA
- Y is the factor of driving voltage as given in Table 8.
- F is the factor from anode shape as given in Table 9.
- ρ is the soil resistivity, ohm.cm

Table 8. Driving voltage correction (Y) (James B. Bushman PE., 2001)

Structure potential (Cu-Cuso4)	Std. Mag.	High potential Mag.	Zinc
-0.7	1.21	2.14	1.60
-0.8	1.07	1.36	1.20
-0.85	1.00	1.29	1.00
-0.90	0.93	1.21	0.80
4.00	0.79	1.07	0.40
-1.10	0.64	0.93	n/a
-1.20	0.50	0.79	n/a

Anodes may be connected in parallel in order to achieve a higher total current output at a given location, the closer together the anode are spaced the more the current output is restricted. To determine the approximate current output of a multiple anode ground bed, the single anode current previously calculated is multiplied by the appropriate adjusting factor that found in Table 10 as given by James B. Bushman PE., (2001).

Table 9. Anode shape correction (F) (James B. Bushman PE., 2001)

Alloy	Weight (Ibs)	Package dimension	Anode factor (F)
Mg	3	3" × 3" × 4.5"	0.53
Mg	5	3" × 3" × 7.5"	0.60
Mg	9	3" × 3" × 13.5"	0.71
Mg	9	2.75" × 2.75" × 26"	1.01
Mg	10	1.5" × 1.5" × 72" ingot 4" × 78" package	1.71
Mg	15	1.6" dia. × 10" extrusion 6" × 10" backfill	2.61
Mg	17	4"×4"×17"	1.00
Mg	18	2" × 2" × 72" ingot 5" × 78" package	1.81
Mg	20	2.5" × 2.5" × 60" ingot 5" × 66" package	1.60
Mg	20	1.3" dia. × 20" extrusion 6" × 20" backfill	4.28
Mg	25	2" dia. × 10" extrusion 8" × 10" backfill	2.81
Mg	32	5" × 5" × 21"	1.06
Mg	40	3.75" × 3.75" × 60" ingot 6.5" × 66" package	1.72

Alloy	Weight (Ibs)	Package dimension	Anode factor (F)
Mg	42	3" × 3" × 72" ingot 6" × 78" package	1.90
Mg	50	8" dia. × 16"	1.09
Mg	50	5" × 5" × 31"	1.29
Zn	10	1.4" × 1.4" × 36" ingot 5" × 42"package	1.68
Zn	30	2" × 2" × 30" ingot 5" × 36" package	1.44
Zn	30	1.4" × 1.4" × 72" ingot 5" × 78" package	1.81
Zn	36	2" × 2" × 60" ingot 6.5" × 66" package	1.72

Table 10. Multiple Anode adjusting factor (James B. Bushman PE., 2001)

No. of anodes Bank	Anode spacing in feet				
	5'	10'	15'	20'	25'
2	1.84	1.92	1.95	1.97	2.03
3	2.46	2.71	2.80	2.85	3.02
4	3.04	3.46	3.63	3.71	4.01
5	3.59	4.19	4.43	4.56	4.98
6	4.13	4.90	5.22	5.31	5.96
7	4.65	5.60	6.00	6.23	6.91
8	5.15	6.28	6.77	7.04	7.85
9	5.67	6.96	7.54	7.88	8.82
10	6.16	7.64	8.30	8.68	9.75
11	6.76	8.41	9.14	9.56	10.75
12	7.30	9.12	9.93	10.40	11.71
13	7.83	9.83	10.72	11.23	12.68
14	8.37	10.54	11.51	12.07	13.64
15	8.91	11.25	12.30	12.91	14.61
16	9.44	11.96	13.09	13.75	15.57
17	9.98	12.68	13.89	14.58	16.54
18	10.51	13.39	14.68	15.45	17.50
19	11.05	14.10	15.47	16.26	18.47
20	11.59	14.81	16.26	17.10	19.43

For a more exact calculation an adjusting factor may be determined from Equation (8.16) which is given by James B. Bushman PE., (2001) as follows:

$$MA_{ADJ} = \frac{N_P}{1 + \frac{2.L(In0.656.N_P)}{S\left[\left(In\frac{8L_a}{d_a}\right) - 1\right]}} \tag{8.16}$$

Where: MA_{ADJ} is the multiple anode adjusting factor, N_P is the No. of anodes in parallel, L_a is the length of anode in feet, S is the spacing center-to-center in feet, d_a is the diameter of anode in feet.

Impressed-current systems employ inert (zero or low dissolution) anodes and use an external source of DC power (rectified AC) to impress a current from an external anode onto the cathode surface as reported by R. L. Kean, K.G. Davies (1981).

An electrolytic cell is developed with the protected structure and the ground rod of the rectifier. The ground bed (anode) consists of a number of parallel graphite, carbon, duriron, or junk iron anodes, usually distributed along the protected structure and placed deep in the soil. The structure, connected to the negative of the rectifier, receives the current from the soil and is protected. The protective current from the rectifier is usually sufficient to overcome the galvanic currents leaving the anodic areas of the structure. The general arrangement is shown in Figure 11 as reported by P.K. Sen, Keith Malmeda, John P. Nelson (2002).

Figure 11. Impressed current or Rectifier type

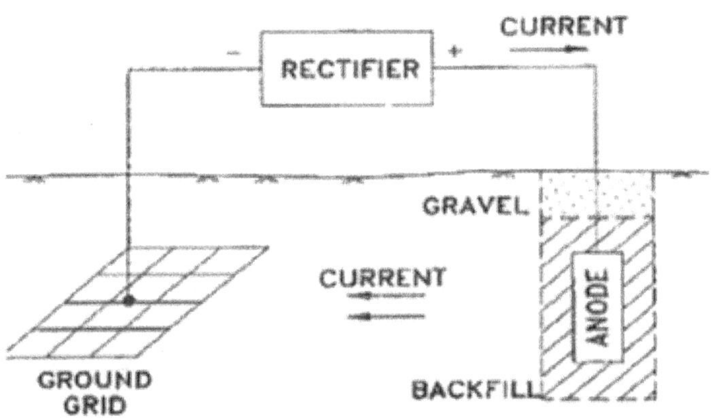

8.7.2. The Main Advantages of Impressed Current

- Unlimited current output capacity: the amount of current that can be designed into an impressed current system can be from a few amperes to as much as several hundred amperes. The amount of current available will be a function of the number of provided anodes, the rectifier voltage and amperage capacity and the soil (electrolyte) resistivity in which the anodes are installed.
- Adjustable output capacity: the output of the rectifier power supply is easily adjusted to accommodate either change in circuit resistance or current requirement. They can be provided with automatic control circuitry to eliminate the need for manual adjustment.
- Lower cost per ampere of cathodic protection current. Galvanic anode systems are significantly more expensive where amperes rather than mill-amperes of total current are required.

8.7.3. The Main Disadvantages of Impressed Current

They are more costly if only a few mill- amperes of current are required to protect a small or very well-coated structure. Impressed current systems typically have a base cost of several thousand dollars. If only a few galvanic anodes are required for protection of a specific structure, this will often be the more economic choice.

- Impressed current systems are inherently more prone to failure. Switches can be turned off and fuses can blow. Thus, the system needs to be monitored more frequently and some repair may be required once every several years.
- Impressed current systems may create stray current corrosion on other nearby structures. This is an inherent potential problem with any impressed current system. This can sometimes be minimized through the use of distributed anode designs where the anodes are placed nearby the protected structures.

8.7.4. Impressed Current Anodes

Materials currently popular for use as anode material include graphite, high silicon cast iron, mixed metal oxide, platinum, and steel.

Table 11 gives a comparison between the two types of cathodic protection as shown below as reported by Pierre R. Roberge (2014). The comparison includes the general uses, the benefits, and features.

8.8 REQUIRED INFORMATION FOR DESIGN OF GROUNING GRID CTHODIC PROTECTION

Before deciding which type, galvanic or impressed current cathodic protection system will be used and before the system is designed, certain preliminary data must be gathered as reported by Headquarters Department of the army Washington (2005).

Table 11. Comparison between the two types of cathodic protection (Pierre R. Roberge, 2014)

Sacrificial anodes (galvanic)	Impressed current (ICCP)
Uses	
§ Generally used for protection of well coated areas where protective current requirements and soil or water resistivity are low. § Where the surface area of a protected structure is relatively small due to economic restrictions.	§ For structures where protective current requirements and life requirements are high. § Can be used over a wider range of soil and water resistivity. § For protection of large uncoated areas, where relatively few anodes are required.
Benefits and Features	
§ No independent source of electric power required. § Limited effects on neighboring structures. § Extremely simple to install. May be directly fixed to the structure. § Simple additions can be made until the desired effect is achieved. § Anode connections are also protected. § Self-adjusting but the output generally cannot be controlled § Correct material selection ensures no over protection, thus avoiding metal embitterment and coating damage § No possibility of plant damage due to incorrect connections (i.e., reversed polarity). § Straight forward to install, operate and maintain.	§ Requires external power source restrictions. § Can be applied to a wide range of structures in various states of coating condition. § May be adjusted manually or automatically to cater for changing conditions. § May be remotely adjusted, monitored and connected to Plant Alarm System. § Anodes are very compact, thus drag and water flow restriction are negligible. § Requires a small number of anodes compared to a galvanic system. Needs careful design and operation to ensure ongoing protection § Can affect other structures if not properly monitored § Installation needs to ensure all connections have a high integrity of insulation and that damage does not occur due to reversed polarity.

a. Physical Dimensions of Structure to Be Protected

One important element in designing a cathodic protection system is the structure's physical dimensions (for example, length, width, height). This data are used to calculate the surface area to be protected (A) as reported by Headquarters Department of the army Washington (2005).

b. Electrical Isolation

If a structure is to be protected by the cathodic system, it must be electrically connected to the anode. Sometimes parts of a structure or system are electrically isolated from each other by insulators as reported by Headquarters Department of the army Washington (2005).

c. Short Circuits

All short circuits must be eliminated from existing and new cathodic protection systems. A short circuit can occur when one protected system contacts another, causing interference with the cathodic protection system. When updating existing systems, eliminating short circuits would be a necessary first step as reported by Headquarters Department of the army Washington (2005).

d. Corrosion History of Structures in the Area

Studying the corrosion history in the area can prove very helpful when designing a cathodic protection system. The study should reinforce predictions for corrosivity of a given structure and its environment; in addition, it may reveal abnormal conditions not otherwise suspected as reported by Headquarters Department of the army Washington (2005).

e. Electrolyte Resistivity Survey

A structure's corrosion rate is proportional to the electrolyte resistivity. Without cathodic protection, as electrolyte resistivity decreases, more current is allowed to flow from the structure into the electrolyte; thus, the structure corrodes more rapidly. As electrolyte resistivity increases, the corrosion rate decreases as shown in Table 12. Resistivity can be measured either in a laboratory or at the site with the proper instruments. The resistivity data will be used to calculate the sizes of anodes and rectifiers required in designing the cathodic protection system as reported by Headquarters Department of the army Washington (2005).

Table 12. Soil resistivity and soil

Soil resistivity range (ohm.cm)	Corrosistivity
0 to 2000	Severe
2000 to 10,000	Moderate to severe

continued on following page

Table 12. Continued

Soil resistivity range (ohm.cm)	Corrosistivity
10,000 to 30,000	Mild
Above 30,000	Not likely

f. Electrolyte pH Survey

Corrosion is also proportional to electrolyte pH. The value of soil pH affects the corrosion process greatly. The more acidic the soil is, the higher the corrosion rate. The value ranges generally from 5 to 10 in soil; a value of 7 indicates neutrality (lower values, acidity; and higher values, alkalinity). The general relationship between the pH values and the corresponding corrosion is shown in Table 13 as reported by P., K., Sen, Keith, Malmeda, John P., Nelson, (2002).

Table 13. pH values and corrosion

Soil Characteristics	pH Values	Corrosion Rate
Extremely Acid	Below 4.5	Highest Corrosion
Very Strongly Acid	4.5 - 5.0	
Strongly Acid	5.1 – 5.5	
Medium Acid	5.6 – 6.0	
Slightly Acid	6.1 – 6.5	
Mildly Alkaline	7.4 – 7.8	Higher Corrosion
Moderately Alkaline	7.9 – 8.4	
Strongly Alkaline	8.5 – 9.0	
Very Strongly Alkaline	9.1-Higher	
Neutral	6.6 – 7.3	Least Corrosion

g. Structure Versus Electrolyte Potential Survey

For existing structures, the potential between the structure and the electrolyte will give a direct indication of corrosivity. According to NACE Standard number RP-01, the potential requirement for cathodic protection is a negative (cathodic) potential of at least 0.85 volt as measured between the structure and a saturated copper-copper sulfate reference electrode in contact with the electrolyte. A potential which is less than negative -0.85 volt would probably be corrosive, with corrosivity increasing as the negative value decreases (becomes more positive) as reported by Headquarters Department of the army Washington (2005).

h. Current Requirement

A critical part of design calculations for cathodic protection systems on existing structures is the amount of current required per square foot (called current density) to change the structure's potential to -0.85 volt. The current density required to shift the potential indicates the structure's surface condition.

The amount of current required for complete cathodic protection can be determined by three ways:

i. An actual test on existing structures using a temporary cathodic protection setup.
ii. A theoretical calculation based on coating efficiency. The efficiency of the coating as supplied will have a direct effect on the total current requirement, as given in equation 12 as reported by Headquarters Department of the army Washington (2005).

$$I_T = (A_s)(I_r)(1.0 - CE) \tag{8.17}$$

Where: $I$is the total protective current, $A_s$is the total structure surface area in square feet, (I_r) is the required current density as given in Table 8.14, and $CE$is the coating efficiency.

iii. An estimate of current requirements using tables based on field experience.

j. Coating Resistance

A coating's resistance decreases greatly with age and directly affects structure to-electrolyte resistance for design calculations. The coating manufacturers supply coating resistance values are given in reference as reported by Headquarters Department of the army Washington (2005).

Table 14. Typical current density requirements for cathodic protection of uncoated steel (Headquarters Department of the army Washington, 2005)

Environment	**(I_r)(mA/sq.ft) AFM 88 – 9[22]**	**(I_r)(mA/sq.ft) I.S Gerrard[23]**
Neutral soil	0.4 to 1.5	0.4 to 1.5
Well aerated neutral soil	2 to 3	2 to 3
Wet soil	1 t 6	2.5 t 6
Highly acidic soil	3 to 15	5 to 15

Environment	(I_r)(mA/sq.ft) AFM 88 – 9[22]	(I_r)(mA/sq.ft) I.S Gerrard[23]
Soil supporting active sulfate reducing bacteria	6 to 42	Up to 42
Head soil	3 to 25	5 to 25
Stationary freshwater	1 to 6	5
Moving freshwater containing dissolved	5 to 15	5 to 15
Seawater	3 to 10	5 to 25

k. Protective Current Required

By knowing the physical dimensions of the structure to be protected, the surface area can be calculated. The product of the surface area multiplied by current density obtained previously in Table 14 gives the total current required as reported by Headquarters Department of the army Washington (2005).

l. The Need for Cathodic Protection

For existing structures, the current requirement survey will verify the need for a cathodic protection system. For new systems, standard practice is to assume a current density of at least 2 mill amperes per square foot of bare area will be needed to protect the structure. However, local corrosion history may demand a different current density. In addition, cathodic protection is mandatory for underground gas distribution lines and for water storage tanks with a 250,000-gallon capacity or greater. Cathodic protection is also required for underground piping systems located within 10 feet of steel reinforced concrete because galvanic corrosion will occur between the steel rebar and the pipeline. It can also be used for protecting the underground metallic structure of grounding system as reported by Headquarters Department of the army Washington (2005).

8.9 DETERMINING TYPE AND DESIGN OF CATHODIC PROTECTION SYSTEM

Before the design work starts, a design basis shall be prepared and accepted by the customer/ management. The design basis shall include the following information as reported by Johnsen, Roy (2006)

- Which standard or recommended practice to base the design upon.

· Type of protection; sacrificial anodes or impressed current.
· Combined protection with coating.
· Current densities to be used (if not directly in accordance with the standard).
· Protection potentials.
· If coating, type of coating and degradation rate.
· Documentation to base the design upon.
· Is the structure in electrical contact with other structures permanently or from time to time?

If the soil resistivity is low (less than 5000 ohm-centimeters) and the current density requirement is low (less than 1 mill ampere per square foot), a galvanic system can be used. However, if the soil resistivity and/or current density requirement exceed the above values, an impressed current system should be used as reported by Headquarters Department of the army Washington (2005).

8.9.1 Sacrificial Anode (Galvanic) Cathodic Protection System Design Steps

- **Step 1:** Review soil resistivity (p) in Ω. cm.
- **Step 2:** Select anode material

If p < 2000, Zinc Anode is selected
If ρ > 2000, Magnesium Anode is selected
Each anode specification will include anode wei ght, anode dimensions, and package dimensions (anode plus backfill). Table 8.15 shows size and weight of high potential magnesium alloy anodes.

- **Step 3:** Calculate the net driving potential for anodes (E) Volt.
- **Step 4:** Calculate the number of anodes needed to meet grounded resistance limitations. The total resistance (R_T) of the galvanic circuit is given by the following equations:

$$R_T = R_A + R_{Wa} + R \tag{18}$$

$$R_T = E/I \tag{19}$$

$$R_c = R_{av}/A_s \tag{20}$$

Where: R_A is the anode-to-electrolyte resistance, R is the structure-to-electrolyte resistance, R_{Wa} is the anode lead wire resistance (negligible), R_c is the coating resistance in ohms per square feet, R_{av} is the average coating resistance (specified by the supplier) Ω/sq.ft, and A_s: is the structure's surface area in square feet.

$$N = \frac{0.0052.\rho}{R_A.L_b}\left[1n\frac{8L_b}{d_b} - 1\right] \quad (8.21)$$

Where: N is the number of anodes, d_b is the diameter of the backfill column in feet (specified by supplier), and L_b is the length of the backfill column in feet (specified by supplier).

Step 5: Calculate the number of anodes for systems by the following equation

$$N = \frac{L_f.I}{49.3 \times W_a} \quad (8.22)$$

Where: W_a is the weight of one anode (in pound) and L_f is the expected lifetime in years.

Step 6: Select number of anodes to be used. It is the greater value of the two equations.

Step 7: Select ground bed layout. When the required number of anodes has been calculated, the area to be protected by one anode in square feet is calculated by the following equation as reported by Headquarters Department of the army Washington (2005).

$$A_a = \frac{A_s}{N} \quad (8.23)$$

Step 8: Calculate the life-cycle cost for proposed design according to NACE Standard RP-02. The design process should be done for several different anode choices to find the one with minimal life-cycle cost.

Step 9: Prepare plans and specifications. When the design procedure has been done for several different anodes and the final anode has been chosen, plans and specifications can be completed as reported by Headquarters Department of the army Washington (2005).

8.9.2. Impressed Current Cathodic Protection System Design Steps

The steps of designing ICCP are discussed below in full details.

- **Step 1:** Review soil resistivity ρ in Ω. cm.
- **Step 2:** Review current requirement test, it can be obtained from Table 14
- **Step 3:** The used anodes are most often made of high silicon chromium-bearing cast-iron (HSCBCI). When impressed current-type cathodic protection systems are used to mitigate corrosion on an underground steel structure, Table 16 gives (HSCBCI) anode sizes and specifications.

Table 15. Weights and dimensions of selected high potential magnesium-alloy anodes for use on soil or water

Weight (Ib)	Size (in.)	Packaged weight (Ib)	Packaged size (in.)
3	3.75 × 3.75 × 5	12	6×10
5	3.75 × 3.75 × 7.5	17	6×12
9	2.75 × 2.75 × 26	35	6×31
9	3.75 × 3.75 × 13.25	27	6×17
12	3.75 × 3.75 × 18	36	6 ×23
14	2.75 × 2.75 × 4	50	6 ×46
14	13.75 × 3.75 × 21	42	6.5 × 26
17	2.75 × 2.75 × 50	60	6×55
17	3.75 × 3.75 × 26	45	6.5 × 29
20	2.5 × 2.5 × 59.25	70	5 ×66
24	4.5×4.5 ×23	60	7×30
32	5.5 × 5.5 × 21	74	8×28
40	3.75 × 3.75 ×59.25	105	6.5 × 66
48	5.5 × 5.5 × 30	100	8×38
48	8×16	100	12×25
60	4.5 ×4.5 ×60	---	---

Table 16. Weights and dimensions of selected circular high silicon chromium bearing cast iron

Anode weight (lb)	Anode dimensions (in.)	Anodes Surface size (in.)	Package area (sq.ft)
12	1×60	1.4	10×84
44	2×60	2.6	10×84
60	2×60	2.8	10×84
110	3×60	4.0	10×84

Step 4: Calculate the number of anodes needed to satisfy manufacture's current density limitations. Impressed current anodes are supplied with a recommended maximum current density.

$$N = \frac{I}{A_1 . I_1} \quad (8.24)$$

Where: A_1 is the anode surface area per square feet per anode, I_1 is the recommended maximum current density output in mA.

Step 5: Calculate number of anodes needed to meet design life requirement as reported by Headquarters Department of the army Washington (2005).

$$N = \frac{L_f . I}{1000 \ W_a} \quad (8.25)$$

Where: W_a is the weight of one anode (in pound), L_f is the expected lifetime in years.

Step 6: Calculate the number of anodes needed to meet maximum anode ground bed resistance requirements as reported by Headquarters Department of the army Washington (2005).

$$R_a = \frac{\rho . K}{N . L_b} \bullet \frac{\rho . P}{S_b} \quad (8.26)$$

Where: K is the anode shape factor it can be obtained from Table (17), S_b is the center-to-center spacing between anode backfill columns in feet, L_b is the length of the anode backfill column in feet, and P is the paralleling factor that can be obtained from Table (18).

- **Step 7:** Select number of anodes to be used. The highest number calculated by equation will be the number of anodes used.
- **Step 8:** Select area for placement of anode bed. The area with the lowest soil resistivity will be chosen to minimize anode-to-electrolyte resistance.
- **Step 9:** Determine total circuit resistance. The total circuit resistance will be used to calculate the rectifier size needed.
- Calculate anode ground bed resistance using Equation (8.26) as reported by Headquarters Department of the army Washington (2005).

Table 17. Shape function (K) for impressed current cathodic protection anodes where L_a is effective anode length and d_a is anode / backfill diameter (W.T. Bryan, 1975)

L/d	K	L/d	K
5	0.0140	20	0.0213
6	0.0150	25	0.0224
7	0.0158	30	0.0234
8	0.0165	35	0.0242
9	0.0171	40	0.0249
10	0.0177	45	0.0255
12	0.0186	50	0.0261
14	0.0194	55	0.0266
16	0.0201	60	0.0270
28	0.0207	20	0.0213

Table 18. Anode paralleling factor (P) of various numbers of anodes (N) installed in parallel (W.T. Bryan, 1975)

N	P	N	P
2	0.00261	14	0.00168
3	0.00289	16	0.00155
4	0.00283	18	0.00145
5	0.00268	20	0.00135
6	0.00252	22	0.00128
7	0.00237	24	0.00121
8	0.00224	26	0.00168
9	0.00212	28	0.00155
10	0.00201	30	0.00145
12	0.00182	----	0.00135

(a) Calculate ground bed header cable resistance. The cable is typically supplied with a specified resistance in ohms per 100 feet. The wire resistance then is calculated as reported by Headquarters Department of the army Washington (2005)

$$R_W = \frac{R_m . L_w}{100ft} \quad (8.27)$$

Where: R_m is the wire resistance specified by the manufacturer, L_w is the structure's length in feet.

(b) Calculate structure-to-electrolyte resistance:

$$R_C = \frac{R}{N_c} \quad (8.28)$$

Where: R is the structure-to-electrolyte resistance, R_C is the coating resistance in ohms per square feet, N_c is the coated grounding system or pipe area in square feet.

(c) Calculate total circuit resistance as reported by Headquarters Department of the army Washington (2005):

$$R_T = R_A + R_W + R_C \quad (8.29)$$

- **Step 10:** Calculation of rectifier voltage. Equation 8.30) is used to determine voltage output (V_{rec}) of the rectifier as reported by Headquarters Department of the army Washington (2005)

$$V_{rec} = I * R_T * 150\% \quad (8.30)$$

Where: V_{rec} is the voltage output of rectifier, I is the total protection current in amperes, and R_T is the total circuit resistance.

- **Step 11:** Selection of a rectifier. A rectifier must be chosen based on the results of Equation 30). Many rectifiers are available commercially; one that satisfies the minimum requirements of (I) and (V_{rec}) in Equation (8.25) should be chosen.
- **Step 12:** Calculation of system cost. As with the galvanic cathodic protection system, the choice of anode for design calculation is arbitrary. When several anodes have been used in the design calculations, an economic analysis should be done as recommended in NACE Standard RP-02.
- **Step 13:** Prepare plans and specifications as reported by Headquarters Department of the army Washington (2005)

REFERENCES

Air Force Manual AFM 88-9. (1962). Corrosion Control.

Beavers, J. A. (2001). Cathodic Protection - How It Works, Control of Pipeline Corrosion, NACE International.

Bianchetti, R. L. (2001). Cathodic Protection with Galvanic Anodes, Control of Pipeline Corrosion, NACE International.

Bryan, W. T. (1970), Designing Impressed Current Cathodic Protection Systems with Durco Anodes. The Duriron Company.

Chauhan, B. S. (2008). *Engineering Chemistry*. Laxmi Publications Pvt Limited.

Gerrard, J. S. (1976). Practical Applications of Cathodic Protection, Corrosion, Vol. 2 (L.L. Shreir, Ed.), Newnes Butter worths.

Headquarters Department of the Army. (2003), Operation and Maintenance Cathodic Protection Systems, Unified Facilities Criteria (UFC), UFC 3-570-06, January 31, 2003

Headquarters Department of the Army, Washington, (2005), Electrical Design Cathodic Protection, Unified Facilities Criteria (UFC), UFC 3-570-02A.

Howard, A. P. (2004). *Introduction to Corrosion and Galvanizing of Helix Foundations*. Presented at Helical Foundations and Tiebacks Committee, Deep Foundation Institute.

IEEE Guide for Safety in AC Substation 80 (1986), IEEE Std. No. 80-1986.

IEEE Guide for Safety in AC Substation 80(2000), IEEE Std. No. 80-2000.

James, B., & Bushman, P. E. (2001). *Galvanic Anode System Design, Bushman, Associates,Inc. Corrosion Consultants*.

James, P., & Busman, P. E. (2001). *Impressed Current Cathodic Protection Design, Bushman. Associates,Inc. Corrosion Consultants*.

Johnsen, R. (2006). Cathodic Protection, Institutt Produktutvikling og materialer (NTNU), DocumBase.com.

J.P. (1985). A Cathodically Protected Electrical Substation Ground. *IEEE Transactions on Industry Applications*, IA-21(2), 357–361. 10.1109/TIA.1985.349656

Kean, R. L., & Davies, K. G. (1981), Cathodic Protection, updated www.npl.co.uk/upload/pdf/cathodic_ororection.pdf

Lawson, V. R. (1988) Problems and detection of line anchor and substation ground grid corrosion, IEEE Transactions on Industry Applications, 24, 25-32.

Marcus, O. & Robert, A. (2005), Consequences and standards from using CP systems to prevent corrosion, IEEE Industry Applications Magazine, 41-47.

Mars, G. (1987). *Fontana.* Corrosion Engineering, Mc-Graw Hill Books Company.

Mi, Z., Wang, J., Yang, L., Xiang, N., Zhen, S., Chen, J., & Fang, C. (2008) Causes, Forms and Remedies of Substation Grounding Grid Corrosion. *International Conference on High Voltage Engineering and Application*, 1-2-3-4. 10.1109/ICHVE.2008.4773904

Pierre, R. (2017).Cathodic Protection Part 1 – Code of Practice for Land and Marine Applications, Cathodic Protection Co. Limited, www.cathodic.co.uk/files/1246442149cathodic%20protection%20overview.pdf

Sen, Malmeda, & Nelson. (2002). Steel Grounding Design Gide and Application Notes, IEEE Rural Electric Power Conference, 6-7.

NACE Standard, RP0169-2002(2002), Control of External Corrosion on Underground or Submerged Metallic Piping Systems Rigoberto Castro Ignacio.

Szakalos, P., & Seetharaman, S. (2012). Corrosion of copper canister, Technical Note 2012:17. Germany, www.stralsakerhetsmyndigheten.se

Tullmin, M., & Roberge, P. R. (1995). Tutorial corrosion of metallic. *IEEE Transactions on Materials Reliability*, 44, 271–278. 10.1109/24.387383

Zheng, R., He, J., Hu, J., Lu, G., & Luo, B. (2002).The theory and implementation of corrosion diagnosis for grounding system, *Industry Applications Conference*, 2, 1120-1126. 10.1109/IAS.2002.1042700

Compilation of References

Air Force Manual AFM 88-9. (1962). Corrosion Control.

Ala, G., Viola, F., Buccheri, P. L., & Romano, P. (2008). Finite difference time domain simulation of earth electrodes soil ionisation under lightning surge condition. *IET Science, Measurement & Technology*, 2(3), 134–145. 10.1049/iet-smt:20070001

Ala, G., Viola, F., Buccheri, P. L., & Romano, P. (2008, May). Finite difference time domain simulation of Earth electrodes soil ionization under lightning surge condition, IET Sci. *Measurement Techniques*, 2(3), 134–145.

Alipio, R., & Visacor, S. (2012), How the Frequency Dependence of Soil Parameters Affect the Lightning Response of Grounding Electrodes, *International Conference on Lightning Protection*, 1-4. 10.1109/ICLP.2012.6344354

Alipio, R., & Visacro, S. (2013). Frequency Dependence of Soil Parameters: Effect on the Lightning Response of Grounding Electrodes. *IEEE Transactions on Electromagnetic Compatibility*, 55(1), 132–139. 10.1109/TEMC.2012.2210227

Andre, M., & Mattos, F. (2005). Grounding grids transient simulation. *IEEE Transactions on Power Delivery*, 20(2), 1370–1379. 10.1109/TPWRD.2004.839177

Andrews, C. (2003), Electrical aspects of lightning strike to humans. In The Lightning Flash. IEEE Press.

Andrews, C. J. (2007). Lightning caused injuries in humans. *Journal of Electrostatics*, 65(5-6), 386–394. 10.1016/j.elstat.2006.09.016

ANSI/IEEE C37.5-1979. (n.d.). IEEE Standards, https://standards.ieee.org/ieee/C37.5/6581/

ANSI/IEEE Std. 80-1986, Guide for Safety in AC substations Grounding, IEEE-John Wiley & Sons, New York August 1986.

ATP rule book, (1995), Can/Am EMTP User Group.

Attia, E.-F. (2013). Accelerated Particle Swarm Optimization-based Approach to the Optimal Design of Substation Grounding Grid. *Przegląd Elektrotechniczny*.

Baba, Y., Nagaoka, N., & Ametani, A. (2005). Modeling of thin wires in a lossy medium for FDTD simulations. *IEEE Transactions on Electromagnetic Compatibility*, 47(1), 54–61. 10.1109/TEMC.2004.842115

Beavers, J. A. (2001). Cathodic Protection - How It Works, Control of Pipeline Corrosion, NACE International.

Becerra, M., & Cooray, V. (2009, August). on the interaction of lightning upward connecting positive leaders with humans. *IEEE Transactions on Electromagnetic Compatibility*, 51(4), 1001–1008. 10.1109/TEMC.2009.2033265

Bellaschi, P. L., Armington, R. E., & Snowden, A. E. (1942). Impulse and 60-cycle characteristics of driven grounds-II. *Transactions of the American Institute of Electrical Engineers*, 61(6), 349–363. 10.1109/T-AIEE.1942.5058539

Bendito, E., Carmona, A., Encinas, A., M., Jimenez, M., J., (2000), the external charges method in grounding grid design. The commission interministerial de ciencia technology, Spanish research council

Bharat Sanchar Nigam Limited, A Govt. (2005), Measurement Earth Potential Rise (EPR) Zone, No. EPR/A-002, CGM and Technical and Development Circle, 18-03-2005.

Bianchetti, R. L. (2001). Cathodic Protection with Galvanic Anodes, Control of Pipeline Corrosion, NACE International.

Bronzino, J. D. (2000). *The Biomedical Engineering Handbook* (2nd ed.). CRC Press.

Bryan, W. T. (1970), Designing Impressed Current Cathodic Protection Systems with Durco Anodes. The Duriron Company.

Characterestics of TT, TN and IT systems. (n.d.). Wiki Electrical installation, Presented by Schneider (2016), http://www.electrical-installation.org/enwiki/Characteristics_of_TT,_TN_and_IT_systems

Chauhan, B. S. (2008). *Engineering Chemistry*. Laxmi Publications Pvt Limited.

Chisholm, W. A., & Janischewskyj, W. (1989). Lightning surge response of ground electrodes'. *IEEE Transactions on Power Delivery*, 4(2), 1329–1337. 10.1109/61.25620

CIGRE Working Group 33-01 (Lightning) of Study Committee 33, (1991), Over voltages and insulation coordination, Guide to procedures for estimating the lightning performance of transmission lines, Paris.

CIGRE, (1991), Guide to procedures for estimating the lightning performance of transmission lines,' WG 01 (Lightning), Study Committee 33

Dalziel, C. F., & Massogilia, F. P. (1956). Let-go currents and voltages. *Transactions of the American Institute of Electrical Engineers*, 75, 49–56.

Damir, C., Nicolas, M., & Farhad, R. (2014). A Comparison of Frequency-Dependent Soil Models: Application to the Analysis of Grounding Systems. *IEEE Transactions on Electromagnetic Compatibility*, 56(1), 177–187. 10.1109/TEMC.2013.2271913

Datsios, Z. G., & Mikropoulos, P. N. (2012), Safe grounding systems design for a photovoltaic power station. *8th Mediterranean Conference on Power Generation, Transmission, Distribution and Energy Conversion MEDPOWER*. 10.1049/cp.2012.2041

Dawalibi, F., & Mukhedkar, D. (1975). Optimum design of substation grounding in a two layer earth structure, Part I-Analytical study. *IEEE Trans.*, PAS-94(2), 252–261.

Djamel I., Slaoui F.H., & Georges, S. (2016), Transient Response of Grounding Systems under Impulse Lightning Current. Conf. Electric Power Quality and Supply Reliability (PQ), 71-75.

Dong-Hyuk Ha, Choi Jong-Kee, & Yoon Yong-Beum, (2019), Parametric Analysis of Lightning Induced Voltage on Distribution Lines using EMTP, 33(3), 19-26.

Easy power. (n.d.). Grounding. https://www.easypower.com/resources/article/grounding-analysis-ground-fault-current

Effects of lightning strikes on the human body, (2019). Cardiac Arrest, Brain Injury, Seizures, Spinal Cord Damage, Amnesia, https://www.prevention.com/health/a32851873/struck-by-lightning-effects

El-Khodragy, T. M. (2007), Optimum Design of Earthing System of Substations by Using Artificial Neural Network, Ph. D. thesis, Faculty of Engineering, Cairo University.

El-Sayed, M. (2015). Achievement of the best design for unequally spaced grounding grids. *Ain Shams Engineering Journal*, 6(1), 171–179. 10.1016/j.asej.2014.09.013

Engineering Standards Branch, Telecommunications Standards Division (ESBTSD) (1994), Electrical Protection Grounding Fundamentals, United States department of agriculture, rural electrification administration

Gauss-Seidel Method. (2022), https://math.libretexts.org/Bookshelves/Linear_Algebra/Introduction_to_Matrix_Algebra_ (Kaw)/01%3A_Chapters/1.08%3A_Gauss-Seidel_Method

Gazzana, D. S., Bretas, A. S., Dias, G. A. D., Telló, M., Thomas, D. W. P., & Christopoulos, C. (2014). A study of human safety against lightning considering the grounding system and the evaluation of the associated parameters. *Electric Power Systems Research*, 113, 88–94. 10.1016/j.epsr.2014.03.015

Gerrard, J. S. (1976). Practical Applications of Cathodic Protection, Corrosion, Vol. 2 (L.L. Shreir, Ed.), Newnes Butter worths.

Gouda O., Amer G. & EL-Saied T. (2006). Computation of Potential Profile at a Surface above Energized Unequally Spaced Grounding Grid. The eleventh international middle east power systems conference, MEPCON', 2, 517–22.

Gouda Osama El-Sayed. (2017). *Design Parameters of Electrical Network Grounding Systems, book*. Advances in Computer and Electrical Engineering. IGI Global.

Gouda Osama, E., & El Dein Adel, Z. (2017). Ground potential rise of faulty substations having equal and unequal spacing grounding grids conductors. *IET Generation, Transmission & Distribution*, 11(1), 18–26. 10.1049/iet-gtd.2016.0263

Gouda, O. E., & Ibrahim, H. M. (2010), Earth Surface Potentials and GPR for Grids Having Different Conductors' Distributions Buried in different Layer Soil Structures, MEPCON'10, Cairo University, Egypt, Dec., Paper ID 229.

Gouda, O. E. (2011). *Optimum design of grounding system of high voltage substations*. LAMBERT Academic Publishing.

Gouda, O. E., Mohamed, A. Z. E. D., Al-Harthi, M. M., Omar, S. Y., & Ghoneim, S. S. M. (2022). Performance of grounding electrodes under lightning strokes in uniform and two-layer soils considering soil ionization. *IEEE Access : Practical Innovations, Open Solutions*, 10, 76855–76869. 10.1109/ACCESS.2022.3192394

Grcev, L. (1992), Computation of transient voltages near complex grounding systems caused by lightning currents. *Proc. IEEE Int. Symp. Electromagnetic Compatibility*, 393-400. 10.1109/ISEMC.1992.626123

Grcev, L., & Arnautovski, V. (1998), Comparison between simulation and measurement of frequency dependent and transient characteristics of power transmission line grounding, *Proc. Int. Conf. Lightning Protection*, 524–529.

Grcev, L., & Grceva, S. (2009). on high-frequency circuit of horizontal grounding electrodes. *IEEE Transactions on Electromagnetic Compatibility*, 51(3), 873–875. 10.1109/TEMC.2009.2023330

Grcev, L., & Haznadar, Z. (1988), A novel technique of numerical modeling of impulse current distribution in grounding systems. *Proc. Int. Conf. on Lightning Protection*, 165-169.

Habjanic, A., & Trlep, M. (2006). The simulation of the soil ionization phenomenon around the grounding system by the finite element method. *IEEE Transactions on Magnetics*, 42(4), 867–870. 10.1109/TMAG.2006.871625

He, Zeng, Gao, Tu, Sun, Zou, & Guan. (2005), Optimal Design of Grounding System Considering the Influence of Seasonal Frozen Soil Layer, IEEE Trans. Power Del., 20(1).

Headquarters Department of the Army, Washington, (2005), Electrical Design Cathodic Protection, Unified Facilities Criteria (UFC), UFC 3-570-02A.

Headquarters Department of the Army. (2003), Operation and Maintenance Cathodic Protection Systems, Unified Facilities Criteria (UFC), UFC 3-570-06, January 31, 2003

Heidler F., & Cvetic J. (2002), A Class of Analytical Functions to Study the Lightning Effects Associated with the Current Front. ETEP, 12(2).

Heidler, F., & Cvetic´, J. (2002). A Class of Analytical Functions to Study the Lightning Effects Associated with the Current Front'. *European Transactions on Electrical Power*, 12(2), 141–150. 10.1002/etep.4450120209

He, J., Rong, Z., & Bo, Z. (2013). *Methodology and Technology for Power System Grounding*. John Wiley & Sons.

Heppe, R. J. (1979). Computation on potential at surface above an energized grid or other electrode, allowing for non-uniform current distribution. *IEEE Transactions on Power Apparatus and Systems*, PAS-98(6), 1978–1989. 10.1109/TPAS.1979.319377

Hetita, I. D., Mansour, A., Han, Y., Yang, P., & Zalhaf, A. S. (2022). Experimental and Numerical Analysis of Transient Over voltages of PV Systems When Struck by Lightning, IEEE Transactions on Instrumentation and Measurement.

Howard, A. P. (2004). *Introduction to Corrosion and Galvanizing of Helix Foundations*. Presented at Helical Foundations and Tiebacks Committee, Deep Foundation Institute.

Hung, L., Chen, X., & Yan, H. (1995). Study of Unequally Spaced Grounding Grid. *IEEE Transactions on Power Delivery*, 10(2).

Ibrahim, E. A. (1993), Grounding of the Nuclear Power Plants Using Grounding Grids, Ph. D. thesis, Faculty of Engineering, Cairo University.

IEC/TR 60479-4, (2004), Effects of Current on Human Beings and Livestock – Part 4: Effects of Lightning Strokes on Human Beings and Livestock", Edition 1.0.

IEC/TS 60479-1, (2005), Effects of Current on Human Beings and Livestock – Part 1, General Aspects, Edition 4.0.

IEC60 479-1(1984).Effect of current on human beings and livestock

IEEE 80, (2013), Guide for Safety in AC Substation Grounding, IEEE

IEEE Guide for Safety in AC Substation 80 (1986), IEEE Std. No. 80-1986.

IEEE Guide for Safety in AC Substation 80(2000), IEEE Std. No. 80-2000.

IEEE Guide for safety in AC substation, (2000), IEEE Std 80 -2000.

IEEE Standards 1410, (2004), IEEE Guide for Improving the Lightning Performance of Electric Power Overhead Distribution Lines. https://www.nvent.com/en-eg/erico/solutions/lightning-protection

IEEE std 142 (1991). IEEE Recommended Practice for Grounding of Industrial and Commercial Power System.

IEEE Std 80, (2000), Guide for Safety in Ac Substation Grounding, IEEE, New York.

IEEE Std. 80. (2000). IEEE Guide for Safety in AC Substation Grounding, IEEE Std, 80, 2000, 4th edition.

IEEE Std. 80-1976, Guide for Safety in AC substations Grounding, IEEE, New York, June 1976.

J.P. (1985). A Cathodically Protected Electrical Substation Ground. *IEEE Transactions on Industry Applications*, IA-21(2), 357–361. 10.1109/TIA.1985.349656

James, B., & Bushman, P. E. (2001). *Galvanic Anode System Design, Bushman, Associates,Inc. Corrosion Consultants.*

James, P., & Busman, P. E. (2001). *Impressed Current Cathodic Protection Design, Bushman. Associates,Inc. Corrosion Consultants.*

Johnsen, R. (2006). Cathodic Protection, Institutt Produktutvikling og materialer (NTNU), DocumBase.com.

Kean, R. L., & Davies, K. G. (1981), Cathodic Protection, updated www.npl.co.uk/upload/pdf/cathodic_ororection.pdf

Khodr, H. M., Salloum, G. A., & Miranda, V. (2006). *Grounding System Design in Electrical Substation: An Optimization Approach*. IEEE PES Transmission and Distribution Conference and Exposition Latin America.

Koch, W. (1950), Erdungmassnahmen fur Hoschtspannungsanlagen mit Geerdetem Stempunkt, Elektrotechnische Zeitschrifl, 71(4), 89-91.

Koehler, F., & Swingler, J. (2018), Practical model for tower earthing systems in lightning simulations, Electric Power Systems Research, 158, 306-314.

Koont, D. R. (2015). *Lightning*. Headline Publishing Group.

Kosztaluk, R., Loboda, M., & Mukhedkar, D. (1981). Experimental study of transient ground impedances. *IEEE Power Engineering Review*, 1(11), 44–51. 10.1109/MPER.1981.5511930

Lawson, V. R. (1988) Problems and detection of line anchor and substation ground grid corrosion, IEEE Transactions on Industry Applications, 24, 25-32.

Lee, C.-Y., & Shen, Y.-X. (2009). Optimal Planning of Ground Grid Based on Particle Swam Algorithm, World Academy of Science, Engineering and Technology. *International Journal of Electrical, Computer, Energetic, Electronic and Communication Engineering, Vol*, 3(12).

Leonid, G. (2005). on high-frequency circuit equivalents of a vertical ground rod. *IEEE Transactions on Power Delivery*, 20(2), 1598–1603. 10.1109/TPWRD.2004.838460

Leonid, G. (2009). Impulse efficiency of ground electrodes. *IEEE Transactions on Power Delivery*, 24(1), 441–450. 10.1109/TPWRD.2008.923396

Leonid, G. (2009). Modeling of grounding electrodes under lightning currents. *IEEE Transactions on Electromagnetic Compatibility*, 51(3), 559–571. 10.1109/TEMC.2009.2025771

Leonid, G. (2009). Time and Frequency – Dependent Lightning Surge Characteristics of Grounding Electrodes'. *IEEE Transactions on Power Delivery*, 24(4), 2186–2196. 10.1109/TPWRD.2009.2027511

Liew, A. C., & Darveniza, M. (1974). Dynamic model of impulse characteristics of concentrated earths. *Proceedings of the Institution of Electrical Engineers*, 121(2), 123–135. 10.1049/piee.1974.0022

Liu, Y. Q. (2004). An improved model for soil ionization around grounding system and its application to stratified soil. *Journal of Electrostatics*, 60(2–4), 203–209. 10.1016/j.elstat.2004.01.012

Loewer, Igel, & Norman. (2016), Electrical and Dielectric Properties of Soils in the mHz to GHz Frequency Range, Conference: 11th International Conference on Electromagnetic Wave Interaction with Water and Moist Substances (ISEMA), 449-461.

Majed, A., Keyhan, S., & Mohamad, R. A. (2013). The Effect of Frequency Dependence of Soil Electrical Parameters on the Lightning Performance of Grounding Systems. *IEEE Transactions on Electromagnetic Compatibility*, 55(4), 739–746. 10.1109/TEMC.2012.2222416

Malik, N. H. (1989, February). A review of charge simulation method and its application. *IEEE Transactions on Electrical Insulation*, 24(1), 3–20. 10.1109/14.19861

Manna, T. K., & Chowdhuri, P. (2007). Generalized equation of soil critical electric field EC based on impulse tests and measured soil electrical parameters. *IET Generation, Transmission & Distribution*, 1(5), 811–817. 10.1049/iet-gtd:20060559

Marcus, O. & Robert, A. (2005), Consequences and standards from using CP systems to prevent corrosion, IEEE Industry Applications Magazine, 41-47.

Mars, G. (1987). *Fontana*. Corrosion Engineering, Mc-Graw Hill Books Company.

Messier, M. A. (1980), The propagation of an electromagnetic impulse through soil: Influence of frequency dependent parameters, Mission Res. Corp., Santa Barbara, CA, USA, Tech. Rep. MRC-N-415.

Messier, M. A. (1980), The propagation of an electromagnetic impulse through soil: Influence of frequency dependent parameters', Mission Res. Corp., Santa Barbara, CA, USA, Tech. Rep. MRC-1980.

Messier, M. A. (1985). *Another soil conductivity model, internal rep*. JAYCOR.

Misbah, N. R., Ab Kadir, M. Z. A., & Gomes, C. (April, 2011), Modeling and analysis of different aspect of mechanisms in lightning injury, *Proceedings of the 4th International Conference on Modeling, Simulation and Applied Optimization*

Mi, Z., Wang, J., Yang, L., Xiang, N., Zhen, S., Chen, J., & Fang, C. (2008) Causes, Forms and Remedies of Substation Grounding Grid Corrosion. *International Conference on High Voltage Engineering and Application*, 1-2-3-4. 10.1109/ICHVE.2008.4773904

Mokhtari, M., Zulkurnain, A.-M., & Gharehpetian, G. B. (2016), A critical review on soil ionization modeling for grounding electrodes, Archives of Electrical Engineering, 449-461.

Mokhtari, M., Abdul-Malek, Z., & Gharehpetian, G. B. (2016). A critical review on soil ionization modeling for grounding electrodes. *Archives of Electrical Engineering*, 65(3), 449–461. 10.1515/aee-2016-0033

Mokhtari, M., Malek, Z. A., & Salam, Z. (2015). An improved circuit based model of a grounding electrode by considering the current rate of rise and soil ionization factors'. *IEEE Transactions on Power Delivery*, 30(1), 211–219. 10.1109/TPWRD.2014.2347283

Mousa, A. M. (1994). *The soil ionization gradient associated with discharge of high currents into concentrated electrodes. Power Delivery*. IEEE Transactions.

Mukherjee, P. K., & Roy, C. K. (1978). Computation of Fields in and around Insulators by Fictitious Point Charges. *IEEE Transactions on Electrical Insulation*, EI-13(1), 24–31. 10.1109/TEI.1978.298095

NACE Standard, RP0169-2002(2002), Control of External Corrosion on Underground or Submerged Metallic Piping Systems Rigoberto Castro Ignacio.

Nahman, L., & Paunovic, I. (2006). Resistance to earth of earthing grids buried in multi-layer soil. *Electrical Engineering*, 88(4), 281–287. 10.1007/s00202-004-0282-y

National Electric Code (NEC). (2017), 250.52(A) (5), https://www.electricallicenserenewal.com/Electrical-Continuing-Education-Courses/NEC-Content.php?sectionID=259.1

Nezhad, N. K., Hadi Fallahi, M., & Dozein, M. G. (2013). An Optimal Design of Substation Grounding Grid Considering Economic Aspects Using Particle Swarm Optimization. *Research Journal of Applied Sciences, Engineering and Technology*, 6(12), 2159–2165. 10.19026/rjaset.6.3840

Nixon, K. J. (2006), The Lightning Transient Behavior of a Driven Rod Earth Electrode in Multi-Layer Soil, PhD Thesis, University of the Witwatersrand, Johannesburg.

No, A. I. E. E. 80 (1961), Guide for Safety in A-Current Substation Grounding, AIEE, New York

Nor, . (2012). Comparison between utility sub-station and imitative earthing systems when subjected under lightning response. *International Journal of Electrical Power & Energy Systems*, 43(1), 156–161. 10.1016/j.ijepes.2012.04.042

Oettle, E. E. (1988). A new general estimation curve for predicting the impulse impedance of concentrated earth electrodes'. *IEEE Transactions on Power Delivery*, 3(4), 2020–2029. 10.1109/61.194013

Oettle, E. E. (1988). *The characteristics of electrical breakdown and ionization processes in soil.* Trans. SAIEE.

Omar, K., Sofiane, C., Madjid, T., Madjid, T., Abdelouahab, M., & Noureddine, H. (2017). time-domain modeling of grounding systems impulse response incorporating nonlinear. *IEEE Transactions on Electromagnetic Compatibility*, 60(4), 1–9.

Papalexopoulos, A. D., & Meliopoulos, A. P. (1987). Frequency dependent characteristics of grounding systems. *IEEE Transactions on Power Delivery*, 2(4), 1073–1081. 10.1109/TPWRD.1987.4308223

Pedrosa, A. G., Alípio, R. S., Schroeder, M. A. O., & Afonso, M. M. (2010), Simulation of behavior electrical grounding Comparative analysis for performance in frequency-domain. *Proc. Int. Workshop Appl. Model. Simul.*, 1, 89–92.

Pedrosa, A. G., Shroeder, M. A. O., Afonso, M. M., Alípio, R., Assis, S. C., Oliveira, T. A. S., & Braga, A. R. (2010). Transient response of grounding electrodes for the frequency-dependence of soil parameters, IEEE/PES Transmission and Distribution Conference and Exposition. *Latinskaia Amerika = América Latina*, 839–845.

Permal, N., Osman, M., Kadir, M. Z. A. A., & Ariffin, A. M. (2020). Review of substation grounding system behavior under high frequency and transient faults in uniform soil. *IEEE Access : Practical Innovations, Open Solutions*, 8, 142468–142482. 10.1109/ACCESS.2020.3013657

Phithakwong, B., Kraisnachinda, N., Banjongjit, S., Chompoo-Inwai, C., & Kando, M. (2000). New Techniques the Computer-Aided Design for Substation Grounding, Power Engineering Society Winter Meeting, 3, 2011-2015.

Pierre, R. (2017).Cathodic Protection Part 1 – Code of Practice for Land and Marine Applications, Cathodic Protection Co. Limited, www.cathodic.co.uk/files/1246442149cathodic%20protection%20overview.pdf

Puttarach, A. N., Chakpitak, T., Kasirawat, Pongsriwat, C., (2007), Ground Layer Depth and the Effect of GPR, IEEE PES Power Africa 2007 Conference and Exposition, Schon. J.H (1998) Physical Properties of Rocks: Fundamentals and Principles of Petro-physics. Pergamon, Handbook of Geophysical Exploration, 18.

Rachidi, F., Janischewskyj, W., Hussein, A. M., Nucci, C. A., Guerrieri, S., Kordi, B., & Jen-Shih Chang, . (2001, August). Current and electromagnetic field associated with lightning return strokes to tall towers. *IEEE Transactions on Electromagnetic Compatibility*, 43(3), 356–367. 10.1109/15.942607

Ramamoorty, M., Narayanan, M. M. B., Parameswaran, S., & Mukhedkar, D. (1989). Parameswaran S., Transient performance of grounding grids. *IEEE Transactions on Power Delivery*, 4(4), 2053–2060. 10.1109/61.35630

Rochereau, H. (1988). *Response of earth electrodes when fast fronted currents are flowing out," Response of Earth Electrodes When Fast Fronted Currents are Flowing Out, ser. EDF, B*. Bulletin de la Direction des Études et Recherches.

Schwarz, S. J. (1954), Analytical expression for resistance of grounding systems, AIEE Transactions on Power Apparatus and Systems, 73(13), 1011–1016.

Scott, J. H. (1966, May), Electrical and magnetic properties of rock and soil, Theoretical Notes, Note 18, U.S. Geological Survey. Available: https://www.ece.unm.edu/summa/notes/Theoretical.html

Scott, J.H., Carroll, R.D., & Cunningham, D.R., (1964), Dielectric Constant and Electrical Conductivity of Moist Rock from Laboratory Measurements, US Dept. of interior geological survey technical letter, Special projects-12.

Sen, Malmeda, & Nelson. (2002). Steel Grounding Design Gide and Application Notes, IEEE Rural Electric Power Conference, 6-7.

Sherif, G. S. (2007). *Optimization of Grounding Grids Design with Evolutionary Strategies, Ph.* D., Univeristät Duisburg-Essen.

Smith K. S., Longmire C., L., (1975), A universal impedance for soils, Defense Nuclear Agency, Alexandria, VA, USA, Topical Report for Period Jul.

Sreedhar, S., & Indragandhi, V. (2022). A Comprehensive Framework for Direct Lightning-Structure-Human Interaction Modeling in Heritage Monuments and Safety Assessment. *Energies*, 15(19), 7053. 10.3390/en15197053

Sun, W., He, J., Gao, Y., Zeng, R., Wu, W., & Su, Q. (2000). Optimal Design Analysis of Grounding Grids for Substations Built in Non-uniform Soil. *IEEE Transactions on Power Apparatus and Systems*, 1455–1460.

Sverak, J. G. (1984). Simplified analysis of electrical gradients above a ground grid. *IEEE Trans.*, PAS-103(1), 7–25.

Szakalos, P., & Seetharaman, S. (2012). Corrosion of copper canister, Technical Note 2012:17. Germany, www.stralsakerhetsmyndigheten.se

Talukder, S. (2011). *Mathematical Modeling and Applications of Particle Swarm Optimization, Master Degree*. Blekinge Institute of Technology.

Thabet, A. (2002), Grounding systems of electric substations in non-uniform earth structure with new analysis, M. Sc. Thesis, High Institute of Energy, Aswan.

Thabet, A. A. (2002). Grounding Systems of Electric Substations in Non-uniform Earth Structure with New Analysis, M. Sc. Thesis, High Institute of Energy, Aswan.

Thanasaksiri, T. (2004), Improving the Lightning Performance of Overhead Distribution Lines. TENCON 2004, IEEE Region 10 Conference, 21-24 Nov., 2004, Chiang Mai, Thailand.

Thapar, B., Gerez, V., Balakrishnan, K., & Blank, D. A. (1991). Simplified Equations for Mesh and Step Voltages in AC Substation IEEE Transaction on Power Delivery, 6(2), 601–607.

Tom, M. (2004). Lightning Protection Design Workstation. *Electric Power Research Institute, LPDW*, V5, 2004.

Towne, H. M. (1929). Impulse characteristics of driven grounds. *General Electric Review*, 1929, 605–609.

Tullmin, M., & Roberge, P. R. (1995). Tutorial corrosion of metallic. *IEEE Transactions on Materials Reliability*, 44, 271–278. 10.1109/24.387383

Unde, M. G., & Kushare, B. E. (2014), Effect of compression ratio on touch voltage in layered soil, 2014 6th IEEE Power India International Conference (PIICON).

Visacor, S. (2007). A Comprehensive Approach to the Grounding Response to Lightning Currents,||. *IEEE Transactions on Power Delivery*, 22(1), 381–386. 10.1109/TPWRD.2006.876707

Visacor, S. (2012), How the frequency dependence of soil parameters affect the lightning response of grounding electrodes, *International Conference on Lightning Protection*, 1-4.

Viscaro, S. (2005). *Lightning: an Engineering Approach*. ArtLiber EditSao Paulo.

Viscaro, S. (2007). Direct Strokes to Transmission Lines: Considerations on the mechanisms of Overvoltage Formation and their Influence on the Lightning Performance of Lines. *Journal of Lightning Research*, 1, 60–68.

Wang, J., Liew, A. C., & Darveniza, M. (2005). Extension of dynamic model of impulse behavior of concentrated grounds at high currents. *IEEE Transactions on Power Delivery*, 20(3), 2160–2165. 10.1109/TPWRD.2004.839645

Yassin, S. (2017), Performance of Grounding Grids at Normal and Abnormal Conditions, Master thesis, South Valley University, Qena, Egypt.

Yves, Jaues, Fortin, Lehoux, & Simard. (2005). Earth Potential Rise Influence near HV Substations in Rural Areas, CIRED (Turin).

Zalhaf, A. S., Mansour, D. E. A., Han, Y., Yang, P., & Darwish, M. M. F. (2022). Numerical and Experimental Analysis of the Transient Behavior of Wind Turbines When Two Blades are Simultaneously Struck by Lightning, IEEE Transactions on Instrumentation and Measurement.

Zheng, R., He, J., Hu, J., Lu, G., & Luo, B. (2002).The theory and implementation of corrosion diagnosis for grounding system, *Industry Applications Conference*, 2, 1120-1126. 10.1109/IAS.2002.1042700

Zhongdong, Yao, Wen, & Xu. (2005), The Optimum Design of Grounding Grid of Large Substation, Proceedings of the XIVth International Symposium on High Voltage Engineering, Tsinghua University.

About the Contributors

Ossama El-Sayed Gouda is the professor of electrical Power engineering and high voltage in the Dept. of electrical power and machine, Faculty of Engineering, Cairo University since 1993.He teaches several courses in Power system, High voltage, Electrical machine, Electrical measurements, Protection of electrical power system &Electrical installation. He is a consultant of several Egyptian firms. He conducted more than 140 papers in the field of Electrical power system and High voltage engineering. He supervised about 55M.SC. & Ph.D. thesis. He conducted more than 150 short courses about the Electrical Power, Machine & High voltage subjects for the field of Electrical Engineers in Egypt & abroad.

* * *

Abhaya Gupta is a dedicated M.Sc. Forensic Science student at Mody University of Science and Technology, Lakshmangarh, Rajasthan embodies a passion for unraveling mysteries through scientific investigation. She is poised to make significant contributions to the field, armed with both theoretical expertise and hands-on experience gained through rigorous academic training.

Sameer Saharan is a dedicated researcher in Forensic Science, specializing in fingerprint analysis, cyber forensics, and DNA fingerprinting. He holds a Ph.D. in Forensic Science from Amity University, Haryana, and has authored several research papers and book chapters. With a passion for continuous learning, Dr. Saharan actively contributes to the advancement of forensic science through his expertise and knowledge. He is committed to nurturing the next generation of forensic scientists and making a positive impact in the field.

Index

C

D

F

G

I

M

O

P

R

S

V

www.ingramcontent.com/pod-product-compliance
Lightning Source LLC
LaVergne TN
LVHW080559300926
849162LV00005B/998